68th Annual Safety in Ammonia Plants and Related Facilities Symposium 2024

AIChE Technical Manual 2024 Volume 64

San Diego, California, USA
9 - 12 September 2024

ISBN: 979-8-3313-0264-1

Printed from e-media with permission by:

Curran Associates, Inc.
57 Morehouse Lane
Red Hook, NY 12571

Some format issues inherent in the e-media version may also appear in this print version.

Printed with permission by Curran Associates, Inc. (2025)

AIChE
120 Wall Street, FL 23
New York, NY 10005-4020

Phone: (800) 242-4363
Fax: (203) 775-5177

www.aiche.org

Additional copies of this publication are available from:

Curran Associates, Inc.
57 Morehouse Lane
Red Hook, NY 12571 USA
Phone: 845-758-0400
Fax: 845-758-2633
Email: curran@proceedings.com
Web: www.proceedings.com

TABLE OF CONTENTS

Lessons Learned from Investigations of Ammonia Incidents

The purpose of this paper is to identify the most important lessons from investigations of major incidents involving ammonia production, storage, transportation, and utilization. These lessons are presented in Safety Investigation One-Pagers (SIOP) linked to a general framework for preventing major accidents, including developing safety barriers within the phases of design, construction, operations, and maintenance to build knowledge across and towards partners and contractors. A team of in-house safety professionals reviewed around 150 open-source reports of ammonia incidents. They selected 17 representative events based on criteria such as the likelihood of major hazards for the environment, severe harm to individuals and the quality of the available information.

V. Alipour Tabrizy, Berit Moltu, Arild Førland and, Sandra H. Nilsen
Equinor ASA, Norway

Introduction

Equinor has set a clear climate ambition for absolute reduction of net carbon intensity from scope 1, 2, and 3 emissions by ~ 40% in 2035 compared to 2023, and near zero emission in 2050 via profitable growth in renewables, decreasing carbon emission from oil and gas production activities, developing low carbon solutions including building markets for carbon capture and storage (CCS) and low carbon intensity hydrogen and ammonia products [1].

Hydrogen and ammonia are known to be key enablers for both reducing emissions from hard-to abate sectors (to generate heat and electricity) [2] and they are seen as useful as an energy storage medium for transportation of renewable electricity [3]. Many studies have also pointed to green and blue ammonia (NH_3) as a potential fuel for shipping since ammonia is a carbon-free molecule and therefore burning it in an internal combustion engine or using it in fuel cells has no CO_2 or CH_4 emission [4-10]. Ammonia is also a sulphur-free fuel, and it does not require any SO_x removal system to avoid sulphur emission [10]. Furthermore, any NO_x and N_2O generated from ammonia combustion and non-combusted ammonia can be removed from exhaust gases with selective catalytic reduction (SCR) and Ammonia Slip Catalyst (ASC) technologies [4].

Ammonia is also seen as a convenient medium for transportation and storage of hydrogen due to high hydrogen density and the possibility for storage and transportation at ambient temperature and pressure below 10 bara (145 psi), or at ambient pressure and temperature ~ -33 °C (27.4 °F) (compared to the boiling point of H_2, which is -253 °C (-423.4°F)) [10]. The transportation and storage costs for ammonia are estimated to be only one third based on the cost pr GJ of transported energy compared to that for hydrogen [11]. Ammonia has approximately 45% higher volumetric energy density (12690 mJ/m^3 or 340 BTU/cf) compared to liquid hydrogen (8500 or 227 BTU/cf mJ/m^3), based on low heating value (LHV) [11]. Ammonia could be used directly as a fuel in high temperature fuel cells (e.g., Solid Oxide Fuel cells (SOFC)) or as a fuel/ fuel blend in combustion engines and gas turbines [5-7]. Ammonia could be also cracked into hydrogen if the end consumer requires hydrogen as a final fuel [11].

The use of ammonia as a fuel presents distinct safety risks when compared to conventional fuels. Conventional fuels have a higher potential risk for fire and explosion than ammonia, while ammonia is a toxic fluid and a corrosive chemical and poses different health and environmental hazards, unlike conventional fuels. Despite the extensive experience of the fertilizer industry in producing and transporting ammonia and urea over many decades, the energy sector lacks the same level of experience in terms of safety risks across the entire value chain, including production, transportation, storage, and end use of ammonia. As a result, the safety risks associated with designing and implementing ammonia as a fuel are still in the early stages, and international safety regulations have not yet been established.

This paper aims to highlight some critical takeaways from investigating significant accidents involving ammonia. The objective is to align these findings with a framework for preventing major accidents, which involves the establishment of safety barriers during the design, construction, operations, and maintenance phases. This knowledge can be used collaboratively with partners and contractors to enhance safety measures across all stages. In addition to developing knowledge, it is equally crucial to ensure that the information is user-friendly and easily accessible. To achieve this, an interactive tool named Safety Investigation One-Pagers (SIOP) has been created in a one-page format specifically for ammonia incidents. Among 17 SIOP's, five major accidents are discussed in more details in this paper. Going into new value chains, handling new materials with new properties and new risks, we need to build knowledge as an organization across disciplines, departments, and projects to be able to build the correct barriers and mitigate the correct risks. We need to build organizational learning on ammonia safety. Building organizational learning across is also a way of building a framework of safety culture to prevent major accidents.

Ammonia safety, health, and environmental risks

The main risk of ammonia for life and health is toxicity. Ammonia is classified as a toxic chemical in international classification rules. Some relevant gas concentration level for ammonia gas release in air are summarized as follows [12, 13]:

- Time Weighted Average (TWA) in any 8-hour work shift and 40-hour week: 25 ppm
- STEL (Short Term (15 min) Exposure Limit): 35 ppm
- IDLH (Immediately Dangerous to Life or Health): 300 ppm

Ammonia health hazards can be classified into two main categories:

1. Leaks resulting in large gas clouds causing toxic exposure (respiratory failure) outside immediate vicinity. Such leaks have major hazard potential.

2. Chemical burns or frostbite of skin, eyes, or respiratory failure due to leaks exposing personnel in immediate vicinity of a leak. Most of these injuries are caused by minor leaks leading to inhalation, burns, or even death.

It should be mentioned that exposure of ammonia at concentrations that do not represent an acute risk might feel very uncomfortable and might lead to panic for persons exposed (pungent odor). This is probably why the IDLH value is a relatively low concentration compared to fatal concentrations (> 2500 ppm) and this may cause irrational behavior, e.g. when 3rd parties are exposed, through incident. IDLH values are established to: (1) ensure that the worker can escape from a given contaminated environment in the event of failure of the respiratory protection equipment and (2) to indicate a maximum level above which only a highly reliable breathing apparatus, providing maximum worker protection, is permitted by CDC NIOSH guidelines [14]. Ammonia is also labelled as a very toxic substance to aquatic life with long lasting effect

[15,16]. When liquid ammonia is spilled directly into water, part of the ammonia will dissolve into the water forming a balance of mostly ammonium hydroxide and ammonia. The remaining ammonia will evaporate above the water level resulting in a potentially toxic gas cloud [15]. The actual effects of ammonia spillage on the water are difficult to predict due to several factors [16]. First, the course of the reaction is dependent on the pH of the water and its temperature. The higher the pH of the water and its temperature, the greater the concentration of ammonia in relation to the ammonium ion [16] thus; ammonia is more toxic to aquatic life at higher temperature and pH values [16]. Apart from the concentration of ammonia in water, its negative impact on living organisms depends mainly on the time of exposure. The concentration of ammonia should not exceed 2.9-5 mg/l for short-term exposure (within few hours to few days) and 0.26-3 mg/l for long-term exposure (for a month or longer), depending on temperature, pH and fish species based on the recommendation by United States Environmental Protection Agency (US EPA) [17]. It is also assumed that a concentration lower than 0.02 mg/l for un-ionized ammonia is safe for fish reproduction [17].

Auto ignition temperature and flammability limits for ammonia are compared to auto ignition temperature and flammability limits for hydrogen and hydrocarbon fuels as shown in Figure 1. The lower flammability limit of ammonia is significantly higher than for conventional fuels. This means that hazardous areas (ex-zones) in general will be smaller than for hydrogen and hydrocarbons. The ignition energy is also higher and the flame speed lower for ammonia compared to hydrocarbons. In other words, ammonia has lower fire and explosion risk than hydrogen and hydrocarbons. It is worth mentioning that even though ammonia is not considered as flammable as hydrocarbons or hydrogen, explosions have occurred in enclosed areas [18,19].

Ammonia Stress Corrosion Cracking (SCC) has been considered as the main cause of ammonia leakage from storage tanks and pipelines, when corrosive agent (i.e. ammonia), residual stress, oxygen and susceptible materials are present simultaneously 20]. The main mechanism of NH_3 SCC is believed to be due to straining cracks of the iron oxide film [20]. Brittle fracturing is another risk for vessels containing refrigerated ammonia or pipes transporting refrigerated ammonia.

Data incident sources

The following data sources have been considered to identify relevant accidents:

- Accident Search Results Page | Occupational Safety and Health Administration [21]
- Ammonia Incident Summaries | Minnesota Department of Agriculture [22]
- EUROPA - eMARS Accidents Search - European Commission - including Major Accidents (under Seveso) [23]
- CSB Chemical Incident Database [24]
- Case Study: Ammonia Release Incidents (2007 – 2017) | Technical Safety BC - refrigeration systems [25]
- S&P Global (previously IHS database) for marine accidents [26]
- ARIA - La référence du retour d'expérience sur accidents technologique [27]

Incident review

Based on the data source basis, approximately 150 investigation reports with ammonia have been reviewed and a selection of 17 representative events have been made, shown in Table 1. Safety Investigation One-Pagers (SIOP) have been created in a one-page format specifically for ammonia incidents to summarize main causes, consequences, safety elements/barriers, causal factors (underlying reason and root cause(s)) as well as the main lesson learned from the incident. Among these 17 incidents, five incidents are reviewed in this paper with more details, covering the production, storage, transportation, and end use (in refrigeration systems) value chains.

Safety Investigation One-Pagers

SIOPs are concise one-page summaries created specifically for ammonia incidents. Each SIOP summarizes the incident, including actual and potential consequences, a key lesson, and specific lessons for each element in the major accident prevention framework described above, as well as for each relevant barrier in the generic ammonia accident bowtie. This covers underlying root causes for the incidents. Each SIOP is plotted in an ammonia value chain drawing, from production to storage, transportation, and end-user (Figure 2). Figure 3 shows an example SIOP for a peanut oil plant ammonia incident in Senegal in 1992, which resulted in 129 fatalities and 1500 injuries. This allows users to read the SIOPs interactively, depending on their interest or role. It is observed that most ammonia incidents tend to occur at the end-user stage. A value chain perspective on ammonia safety is important. SIOPs are intended to make lessons from accidents easily available when designing and operating new ammonia value chains. The five selected incidents from SIOPs are as follows:

Hydrogen explosion in ammonia production plant, Norway (1985) [28, 29]

In the summer of 1985, a severe hydrogen-air explosion occurred in an ammonia production plant in Norway with two fatalities [28, 29]. The event started when a gasket in a water pump was blown out. The water pump was situated inside a 100 m long, 10 m wide, and 7 m high building. The pump was feeding water to a vessel containing hydrogen gas at a pressure of 30 bars. This pressure caused a back flow of water to flow through the pump and out through the failed gasket. The hydrogen reached the leakage point after about 3 minutes. The hydrogen gas cloud was probably ignited due to a hot bearing. 3.5 to 7 kg/s hydrogen leaked. Hydrogen was discharged into the building between 20 and 30 seconds before the explosion occurred, with a total mass of hydrogen estimated to 10 to 20 kg. As a consequence of this incident, window glass was broken up to 700 m from the center of the explosion. Concrete blocks, weighing 1.2 metric tons were thrown up to 16 meters. Explosion followed by a large horizontal jet fire lasting about 30 seconds. The main lesson from this incident is to ensure ventilation of areas with potential hydrogen leaks. The main causal factors for this incident are as follows:

- Insufficient ventilation in buildings allows H_2 gas build-up.
- The risk of misdirected H_2 flow into pump room seems not to be fully addressed. The emergency response was not dimensioned for an accident of this magnitude.
- Potential hydrogen leak sources should not be situated within confined areas, location of adjacent buildings, including offices and windows design. A design was considered without consideration of adequate safety distances to e.g., control room.
- The building was originally designed without walls to ensure ventilation of eventual hydrogen gas leaks. Later walls were added. There was lack of risk understanding when this change was made.
- Lack of maintenance (function test) of isolation valve at pump outlet. Lack of maintenance on the check valve.

Rupture of cryogenic ammonia storage tank, Lithuania (1989) [30]

A cryogenic NH_3 tank weighing 10,000 tons and filled to 70% capacity suddenly experienced a pressure rise and burst at its base [30]. A few hours prior to the accident, one of the two liquefaction turbochargers was shut down for some lengthy repair work. One hour before, the second turbocharger was stopped for a short repair job. Operators were not easily able to activate the backup piston compressor and rerouted the NH_3 flow to a pressurized storage area. Fourteen tons of hot NH_3 (+ 10°C) were nonetheless introduced into the lower part of the cryogenic tank, whose gaseous atmosphere rose quickly in pressure. Despite the presence of relief valves, the tank bottom deformed and then burst. Reference [31] investigated the rollover mechanism versus thermal overload mechanism for this incident. As a consequence of this incident, A 70-cm high pool of liquid NH_3 spread over the site and took 12 hours to fully evaporate. The toxic cloud contaminated an area extending over 400 km². This incident had 7 dead and 57 injured, 32,000 people were displaced. A flare stack ignited the vapours emitted. The main lesson from this incident is to know that mixing warm and cold ammonia may cause rapid pressure increase. The main causal factors for this incident are as follows:

- It took 25 minutes before external emergency authorities were alerted. This response time was not optimal.
- Use of water curtains to contain gas clouds and evacuation of private citizens seems well handled.
- The risks of a NH_3 plant might not have been properly understood since the accident caused 7 fatalities and 57 wounded amongst operating personnel. NH_3 leakages can form large toxic gas clouds and these types of chemical plants require escape masks and Personal Protection Equipment (PPE).
- Lack of understanding risk during maintenance work when introducing warm NH_3, due to cooling circuit start-up failure.
- Bund/protection walls were modified during construction to reduce materials and labor costs which seems to be in non-conformance with the specification.
- The flare stack location seems to be a potential ignition source.

Rupture of ammonia pipe in a refrigeration plant, USA (2010) [32]

Hydraulic shock (hammer effect) caused a roof-mounted 12-inch ammonia suction pipe to fracture [32]. The hydraulic shock was enabled during the restart of the plant's ammonia refrigeration system following a power outage. As a consequence of this incident, 15 tons (32,100 Ibs) of ammonia were released. One employee sustained injuries after briefly losing consciousness from ammonia inhalation. Nine ship crew members and 143 of the offsite contractors downwind reported exposure to ammonia. Of the exposed victims, 32 required hospitalizations, and 4 were placed in intensive care. The main lesson from this incident is to know that the severity of dynamic effects in pipes can be difficult to predict. Knowledge, design guidelines and good simulation tools are important. The main causal factors for this incident are as follows:

- Piping/equipment did not withstand the pressure surge and rupture occurred. Investigation of transient consequences of pressure surge could answer if inherent design of piping and equipment can be achieved to avoid detrimental effects or if other mitigating measures were necessary.
- No outdoor NH_3 detection was installed. Outdoor detection might have triggered an alarm earlier (the leak lasted for 4 hours), enabling central control room (CCR) operator to intervene and reduce gas cloud size and alarm offsite personnel.
- No ESD was triggered during the event. Power outage/blackout should automatically initiate trigger an emergency shutdown safe state (e.g. ESD and depressurization).
- No water curtains installed. Water curtains might have confined/restricted the gas cloud.
- Dispersion analysis could be performed to establish toxic hazard zones and investigate

both internal and domino effects. This might have indicated that the decontamination site was located within a NH_3 toxic hazardous zone.

Leak from coupling between ship manifold and loading arm during ammonia loading, Australia (2018) [33]

During the hot-gas purging step of the import process, abnormal flow of hot ammonia vapor and the inadvertent closing of the import valve resulted in the Emergency Release Coupling (ERC) ball valve closing, and then hammering in the loading arm and the downstream pipeline [33]. When the ERC ball valve closed, pressure in the quick connect/disconnect coupler elevated to the purging ammonia vapor supply pressure of 300-500 kPa (43.5-72.5 psi) and the connection to the loading arm decoupled, resulting in the ammonia release from the ship's manifold. The release was estimated to be 1200 kg (2645.5 lbs) over 24 seconds. Five personnel were present at the jetty and were taken to the hospital for medical examination. All were released from hospital the same day. The main lesson from this incident is to consider that rapid closing of valves in ammonia transfer systems may cause pipeline hammer effect which causes rupturing of the weak connections. The main root causes for this incident are as follows:

- A manual hand jack used to control root station valve for hot gas purging was operated using an improper sequence. Abnormal hot gas purging rate occurred due to manual operation and poor communication.
- Ensure clear step-by-step procedures, including checks throughout the process to make sure each procedure has been done correctly. Close communication between operators and contractors is key.
- Improper clamping of quick connect/disconnect coupling (QCDC) was the main cause. This was due to design faults of the control panel ('QCDC clamped' lamp on the control panel was not derived from actual close condition (lack of visual indicator of 'closed' valve position); inadequate procedure that did not require the 'QCDC locked' position indicator to be checked, together with incomplete preventative maintenance on QCDC springs resulted to this incident.
- Critical equipment involved in high-risk tasks should have hard controls and measurements (e.g. instrumented flow rate measurement of hot purge gas, proximity switches for clamps) to prevent assumptions or mistakes, in addition to preventing the system from progressing in an uncontrolled state (e.g., Emergency Shut Down system with trip logic).

Leak from refrigeration system from fishing vessel, Norway (2021) [34]

An ammonia leak incident was reported 14th December 2021 on a vessel in the Barents Sea, Norway [34]. Ammonia was used as a refrigerant in this vessel. No one in the crew was seriously injured, but several had to go to hospital for checkups and follow-up. The ship had to go to the shore for repair after a major ammonia leak had occurred on board for about 5 hours. The leak started after maintenance on the refrigeration system. Most likely due to corrosion of a plug in a water collector under an ammonia tank or a not fully closed valve. Isolation of the leak was difficult due to poor visibility and ice forming on the gloves. The incident happened either because a brass plug had corroded and loosened, or someone had opened the valve by mistake in connection with the oil boil-off or the valve had leaked a little and deteriorated. The main lesson learned is to ensure thorough training of personnel in handling ammonia and prevent galvanic corrosion in plugs. The main causal factors for this incident are as follows:

- Ammonia should be stored locked away in a tightly closed container with ventilation and protection from heat sources.

- Install gas detector(s) that ensure early warning of leaks.

Ammonia accident bowtie

A bowtie diagram, shown in Figure 4, can be used to illustrate how different types of safety barriers can be used to reduce the risk of a leak or hazardous accident. The left-hand side shows preventive barriers which are used to reduce the probability of an accident and on the right-hand side, mitigating barriers which are used to reduce the consequence of an accident are shown. Although the barrier diagram was originally focused on the hazards of flammable fluids, many of the barriers are relevant for hazardous fluids in general, like the containment barrier, leak detection, emergency shutdown, layout considerations etc. Some of the barriers are focused on the hazards of flammable fluids, like fire detection, firefighting, and control of ignition sources. Barriers connected directly to the toxicity of ammonia might be areas/buildings/shelters where operation personnel can escape to and stay for the needed amount of time. PPE is also a barrier connected to the toxicity and corrosive properties of ammonia, to ensure that personnel can safely escape to a safe area. It is considered that the barriers shown in this diagram indicated by orange color might need extra focus due to the properties and hazards of ammonia, and where technology qualification might be needed. Requirements to ignition source control, fire and explosion protection might be less strict for ammonia than for hydrocarbons. However, in ammonia production plants, there will also be large volumes, containing flammable hydrocarbons gases, and hydrogen (blue and green). In these areas, requirements to ignition control cannot be relaxed.

Major accident prevention framework

Preventing significant accidents entails more than just technical considerations; it also involves people, organizational processes, and how the organization operates. Therefore, preventing major accidents requires a system perspective on safety. See Figure 5 for an illustration of the major accident prevention framework. SIOP's are also available for each element in the major accident framework and each barrier in the Ammonia Bowtie. An example of SIOP for an element in the major accident prevention framework is shown in Figure 6. Equinor has established a framework for major accident prevention based on three main pillars:

A) Leadership, culture, and organizational framework conditions

Leadership refers to the ability of management to set the right culture, policies, processes, capacities, competence, assurance, and learning systems in place to ensure safety is a top priority in the organization. Culture in this context refers to the values, beliefs, attitudes, and behaviors that promote safety within the organization. As an example, in the incident of ammonia leakage from coupling between ship manifold and the loading arm (SIOP 13), the organizational framework did not ensure clear communication between operator and contractors, via establishing robust step-by-step procedures for ammonia loading.

B) Safe practices and design

Safe practices refer to risk management, procedures operational limits, technical documentation, maintenance processes and emergency preparedness that are put in place to promote safety in the workplace. Design, on the other hand, refers to the way that infrastructure is designed and constructed. Design factors such as materials, equipment, and management of change can impact the likelihood and impact of accidents. As an example, in hydrogen explosion incident in ammonia production plant (SIOP 1), the risk of misdirected H_2 flow into a pump room and sufficient ventilation seems not to be fully addressed in the design phase. In the event of rupture of a cryogenic ammonia storage tank (SIOP 2), the bund/protection walls were modified not according to the conformance safety practices.

C) Safety barriers

Safety barriers are designed to prevent hazardous situations from developing or to mitigate the consequences of the accident. Bowtie diagrams can be used to illustrate how different barriers are functioning. As an example, in the incident of ammonia pipe in a refrigeration plant (SIOP 6), implementing of several safety barriers such as outdoor NH_3 detection, ESD, water curtain and gas dispersion analysis could have reduced the likelihood and consequence of this accident.

Organizational learning

Organizational learning is a multidisciplinary field and is the process where members of an organization are engaged in sharing of learnings. Organizational learning is more than adding up individual learning which is necessary but not a sufficient condition. Two different concepts of knowledge are information processing [35] and communities of practice [36]. Information processing, as per Huber's approach [35], suggests that generic knowledge is possible and can be transferred like information. Knowledge is viewed as "hard" and real, and language is seen as a vehicle for conveying knowledge.

On the other hand, communities of practices see knowledge as soft, personal, and embedded in individuals' experiences, making it more difficult to transfer [36]. This approach emphasizes shared learning experiences within a community, resulting in the creation of new knowledge that is unique to that community. Moreover, organizational learning, according to communities of practice, is not just about collecting, interpreting, and storing information but rather involves changing common practices and perspectives that are shared by the community. In this work, we tried to build communities of practices across different disciplines from ammonia incidents to create a culture of continuous learning and improvement.

Conclusions

Unlike conventional fuels, toxicity is the main risk of ammonia as a fuel, whereas the fire and explosion risks for ammonia are lower compared to conventional fuels. The energy sector is comparatively less experienced in managing the safety risks linked with ammonia at every stage of its value chain. The period before full capacity and competence are implemented in an organization for a new value chain (such as ammonia) might be vulnerable and requires risk assessment and extra measurements.

The major risk during ammonia production is linked to the safe handling of H_2 gas with adequate ventilation system as a major safety barrier. Several incidents were also reported for Air Separation Units (ASU) due to accumulation of hydrocarbons from neighboring petrochemical plants in liquid oxygen in the main condenser unit [37]. The primary reasons for incidents that occur during the transportation and storage of ammonia are over pressurization and stress corrosion cracking. Ammonia has not been yet utilized as a fuel in combustion engines or fuel cells (in a commercial scale); however proper training of workers dealing with ammonia and understanding the main lessons learned from previous ammonia incidents, among other necessary safety barriers, are essential to prevent major accidents. Preventing major accidents has three main pillars: i) Organizational framework conditions ii) Safe practices and design, and iii) Safety barriers.

Acknowledgement

The valuable contribution of P. E. Tvedt, T. Ryberg, I. Fosse, G. S. Østby, O. Skalmeraas, L.

Z. E. Yeoh, T. Abrahamsen, L. Tronstad, D. Osnes, G. Hopland, and E. K. Elde for reviewing of incidents from data sources are highly appreciated. Equinor AS is appreciated for financial support and granting permission to publish this paper.

References

[1] Equinor 2024 Capital Markets Update https://www.equinor.com/investors/capital-markets-updates

[2] Salmon, N., & Bañares-Alcántara, R. (2021). Green ammonia as a spatial energy vector: a review. Sustainable Energy & Fuels, 5(11), 2814-2839.

[3] Assessment of the Norwegian Full-Scale CCS project's benefits (2020) https://www.regjeringen.no/contentassets/971e2b1859054d0d87df9593acb660b8/the-role-of-ccs-in-a-carbon-neutral-europe.pdf.

[4] Maritime Decarbonization Strategy (2022), Mærsk McKinney Møller Center for Zero Carbon Shipping.

[5] Machaj, K., Kupecki, J., Malecha, Z., Morawski, A. W., Skrzypkiewicz, M., Stanclik, M., & Chorowski, M. (2022). Ammonia as a Potential Marine Ammonia: A Review. Energy Strategy Reviews, 44, 100926.

[6] Hammer, L. S., Leisner, M., Eide, M., Sverud, T., Mjos, N. (2020). Ammonia as a Marine Fuel Safety Handbook.

[7] Alfa Laval, Hafnia, Haldor Topsøe, Vestas, Siemens Gamesa, Ammon fuel: an Industrial View of Ammonia as a Marine Fuel (2020).

[8] Guidelines for Ships using Ammonia as Fuels (2021) Korean Register (KR).

[9] Potential of Ammonia as Fuel in Shipping (2022). https://www.emsa.europa.eu/publications/item/4833-potential-of-ammonia-as-fuel-in-shipping.html.

[10] Mengyao, Y., Clara N. K. L., and Ming L., Ammonia as a Marine fuel– Bunkering, Safety and Release Simulations (2022). Nanyang Technological University.

[11] Zamfirescu, C., & Dincer, I. (2008). Using Ammonia as a Sustainable Fuel. Journal of Power Sources, 185(1), 459-465.

[12] CDC - Immediately Dangerous to Life or Health Concentrations (IDLH): Ammonia - NIOSH Publications and Products

[13] ACGIH (American Conference of Governmental Industrial Hygienists) https://www.acgih.org/about/.

[14] National Institute for Occupational Safety and Health (NIOSH) https://www.cdc.gov/niosh/idlh/default.html

[15] Tsang, S. C. E., Ayvali, T., & Van Vrijaldenhoven, T. (2021). The Position of Ammonia in Decarbonizing Maritime Industry: An Overview and Perspectives: Part II: Costs, safety and environmental performance and the future prospects for ammonia in shipping. Johnson Matthey technology review, 65(2).

[16] Levit, S. M., (2010). A Literature Review of Effects of Ammonia on Fish, Center for Science in Public Participation, Bozeman, Montana.

[17] USEPA, Aquatic Life Ambient Water Quality Criteria for Ammonia - Freshwater 2013, vol. 13, United States Environ Prot Agency, 2013, pp. 1–255.

[18] Anhydrous Ammonia Explosion in an Ice Cream Plant, Max H. McRae, Houston Fire Department, 410 Bagby Street, Houston, TX 77002.

[19] Elk Grove Village, IL Yogurt Plant Ammonia Engine Room Explosion-YouTube.

[20] Lunde I. and Nyborg R. (1989) Prevention of stress corrosion cracking of carbon steels in ammonia, Safety in Ammonia Plants and Related Facilities Symposium, AIChE, Paper No. Aiche-30-007.

[21] Occupational Safety and Health Administration https://www.osha.gov/ords/imis/AccidentSearch.search?acc_keyword=

[22] Ammonia Incident Summaries | Minnesota De-partment of Agriculture https://www.mda.state.mn.us/ammonia-incident-summaries

[23] EUROPA - eMARS Accidents Search - Euro-pean Commission - including Major Accidents (under Seveso) https://emars.jrc.ec.europa.eu/en/emars/accident/search

[24] CSB Chemical Incident Database https://www.muckrock.com/foi/united-states-of-america-10/csb-chemical-incident-database-5658/

[25] Ammonia Release Incidents (2007 - 2015)https://www.ipe.org/docs/default-source/refrigeration/bc/sos_2016_casestudy_ammonia_final.pdf?sfvrsn=fc5315d1_0

[26] S&P Global (previously IHS database) for ma-rine accident https://www.spglobal.com/marketintelligence/en/mi/products/ship-and-port-data.html

[27] ARIA - La référence du retour d'expérience sur accidents technologique https://www.aria.developpement-durable.gouv.fr/the-barpi/the-aria-database/?lang=en

[28] Mjaavatten, A., & Bjerketvedt, D. (2005). A Hydrogen-Air Explosion in a Process Plant: A Case History.

[29] Samdal, U. N., Grainger, D., Hamborg, E. S., Nilsen, S., and Sommersel, O. K. (2021). Hydrogen Safety Strategies and Risk Management in Equinor.

[30] https://www.aria.developpement-durable.gouv.fr/fiche_detaillee/717_en/?lang=en.

[31] Dharmavaram, S., Duisters, H., & Daugherty, T. L. (2023) A Fictional Expectation of a Rollover in an Ammonia Storage Tank.

[32] https://www.csb.gov/assets/1/20/final_csb_casestudy_millard_0114_0543pm.pdf

[33] Patel N. (2019) Ammonia Release During Ammonia Import Activity CSBP limited.

[34] GSP pilot ammoniakk-drevet Tråler https://grontskipsfartsprogram.no/wp-content/uploads/2023/06/Ammoniakkdrevet-traler-Leroy.pdf

[35] Huber, G. P. (1991). Organizational learning: The contributing processes and the literatures. Organization science, 2(1), 88-115.
[36] Brown, J. S., & Duguid, P. (1991). Organizational learning and communities-of-practice: Toward a unified view of working, learning, and innovation. Organization science, 2(1), 40-57.
[37] Air Separation Units – Preventing explosions in liquid oxygen https://new.abb.com/products/measurement-products/measurement-analytics/industrial-gas/preventing-air-separation-unit-explosions-n.

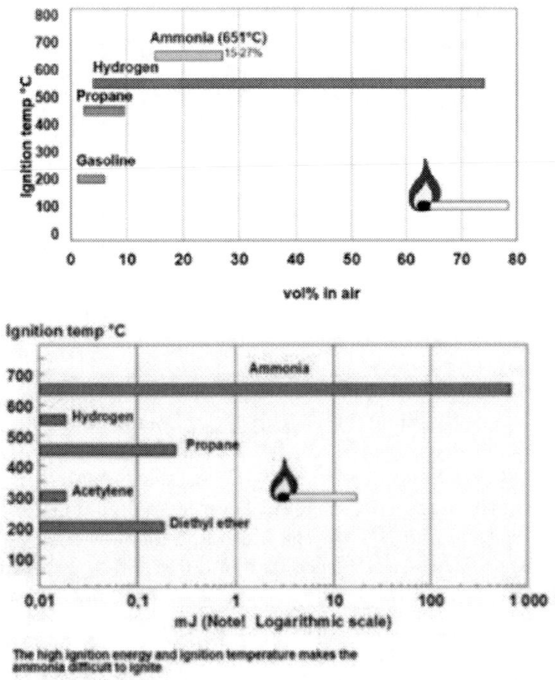

Figure 1. Autothermal ignition temperature and flammability limit of ammonia compared to other fuels.

Table 1. Selected investigation of ammonia incidents

#	Date	Unit/system	Incident	Key lesson	Hazard
1	01.01.1985	Ammonia production, Norway	Hydrogen explosion	Ensure ventilation of areas with potential hydrogen leaks	Fire/explosion, two fatalities
2	01.01.1989	Storage tank, Lithuania	Rupture of cryogenic ammonia tank	Mixing of warm and cold ammonia may cause rapid pressure increase	Overpressure - ammonia inhalation
3	24.03.1992	Peanut oil plant, Senegal	Overpressure in ammonia tanker truck	Ensure safe distances to populated areas	Overpressure - ammonia inhalation
4	18.01.2002	Freight train, USA	Derailment of ammonia tank cars	Ensure robust railway program and practices for inspection and maintenance for railway transportation	Technical integrity - ammonia inhalation
5	10.06.2008	Heat exchanger, USA	Rupture of ammonia heat exchanger during maintenance	Heating of ammonia may cause large pressure increase	Overpressure, impact, inhalation
6	23.08.2010	Refrigeration plant, USA	Rupture of ammonia pipe	The severity of dynamic effects in pipes can be difficult to predict. Knowledge, design guidelines and good simulation tools are important.	Overpressure, - ammonia inhalation
7	30.05.2013	Electrical power plant, Norway	Leak during maintenance on ammonia storage tank	Work on ammonia systems to be managed as work on normally pressurised systems	Ammonia leak inhalation
8	30.09.2013	Food refrigiration plant, China	Leak from pipe end cap	Escape routes must be available, and workers trained	Ammonia leak inhalation
9	07.12.2015	Storage tank, Canada	Leak during maintenance of an ammonia storage tank	Wear PPE if there is any risk of ammonia sudden ammonia exposure	Ammonia leak inhalation
10	17.10.2016	Pipeline, USA	Pipeline rupture	Ensure regular inspections/tests of pipelines	Ammonia leak inhalation
11	07.07.2017	Refrigiration, USA	Ammonia leak and explosion	Establish strategies for handling of ammonia leaks including ignition prevention	Ammonia leak inhalation and explosion
13	21.07.2018	Harbour loading, Australia	Leak from coupling between ship manifold and loading arm	Fast closing of valves in ammonia transfer systems may cause pipeline hammer effects and rupture weak connections	Ammonia leak inhalation
14	29.08.2018	Refrigiration, Canada	Leak in heat exchanger	System changes must be managed	Ammonia leak inhalation
15	25.09.2020	Office building, Norway	Ammonia leak in cooling unit	Clear routines must be established for the event of leakage of ammonia	Ammonia leak inhalation
16	14.12.2021	Fishing vessel, Norway	Leak from refrigiration system	Be prepared for ammonia leak	Ammonia leak inhalation
17	26.05.2022	Ice distribution facility, Canada	Leak during removal of equipment	Only those with the necessary skills and knowledge should be conducting activities with hazardous work	Ammonia leak inhalation

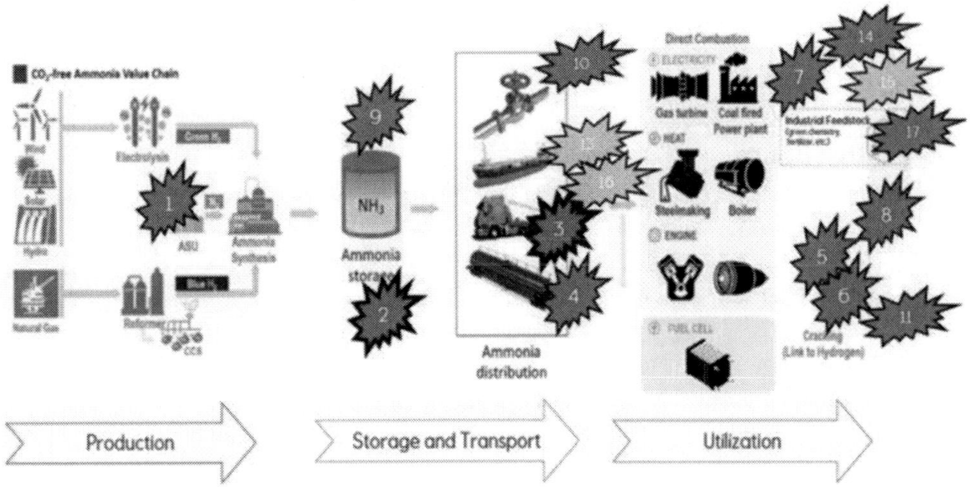

Figure 2. Ammonia value chains with investigated incidents.

Peanut oil plant, Senegal	Overpressure in ammonia tanker truck	24.03.1992

Consequence actual/potential

Uncontrolled discharge/emissions

22 m.t. of pressurized ammonia.

Injury

41 people died immediately. Ultimately, the incident caused 129 fatalities and 1,150 injuries.

Major accident

Anhydrous ammonia, was stored in a tank thruck. The tank suddenly burst open along the middle, with the two portions propelled in different directions. Anhydrous ammonia from the tank was released almost instantaneously. A two-phase flow of ammonia fluid (vapor plus liquid as a fine aerosol) formed a dense vapor cloud which spread well beyond the facility into the industrial and residential neighborhoods. The dense plume settled over the oil mill, nearby offices, and adjacent restaurants where people were present at lunchtime.

Ensure safe distances to populated areas

Safety-elements/Barriers	Causal factors
Apply procedures, respect operating limits and use stop-work authority	Failure to understand the gravity of an abnormal situation and poten tial consequences
Safety leadership and culture	Lack of safety concerns at senior leadership levels. Failure to be open/receptive, bad safety culture. No sense of vulnerability and failure to equip plants with required re sources
Organizational capacity and competence	Lack of training & competency development
Management system	Lack of process safety regulations, standards. Absence of toxics substance management policies & procedures. No policies, procedures, or guidance documents related to process safety.
Assurance, indicators and improvement	Poor implementation of safety audits & recommendations
Manage risk	Lack of risk assessment & management practices. Inadequate or no hazard reviews, consequences, and risk analysis
Demonstrate technical integrity from design to operations	Improper design and utilization of equipment and protection systems; inadequate design basis documentation. Improper testing and inspection of equipment & control systems
Prepared for emergency response	Poor emergency management and lack of coordination of community response. P oor emergency response planning & procedures.
PS01 Containment	Frequent overfilling of the tank, which was authorized to hold 17.7 m.t. The overfilling led to overpressure and crack formation that was detected in 1991 during a maintenance inspection. Maintenance workers welded the tank to repair the crack. After the repairs, the truck continued to be overfilled. The day before the incident, the tank was filled with 22 m.t. of liquid ammonia under pressure. Debris also damaged process equipment and resulted in an extended release from a hose failure.
PS14 Escape, evacuation and rescue	Most of the injuries and fatalities resulted from inhalation of ammonia at concentrations that caused respiratory lesions, edema (fluid buildup) in the lungs, and skin/eye irritation. Near the release location, many fatalities resulted from direct skin exposure, cold burns, and inhalation of high concentrations of ammonia. The cause of death for many victims was identified as the aftereffects of pneumopathy (pulmonary infection, bronchiectasis, and pulmonary fibrosis). The intensity of lesions and mortality was proportional to the quantity of inhaled ammonia per m3 of air.
PS15 Layout design principles	Adhoc siting of hazardous industrial operations. Lack of controlling land use and poor zoning of land use. An analysis of the onsequences of the ammonia release scenarios demonstrates that the estimated distances for potential fatalities (1 km) and injuries (4 to 5 km).

Figure 3. SIOP for peanut oil plant ammonia incident in Senegal in 1992.

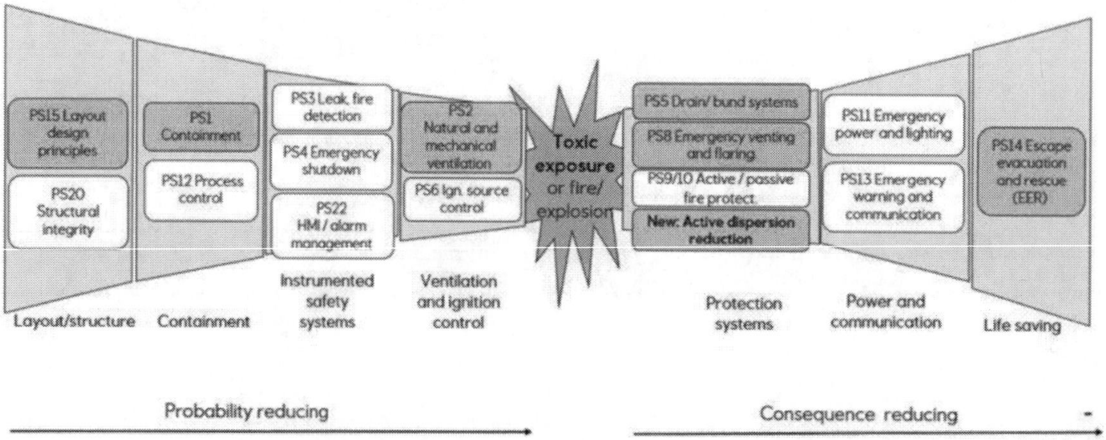

Figure 4. Ammonia accident bowtie diagram

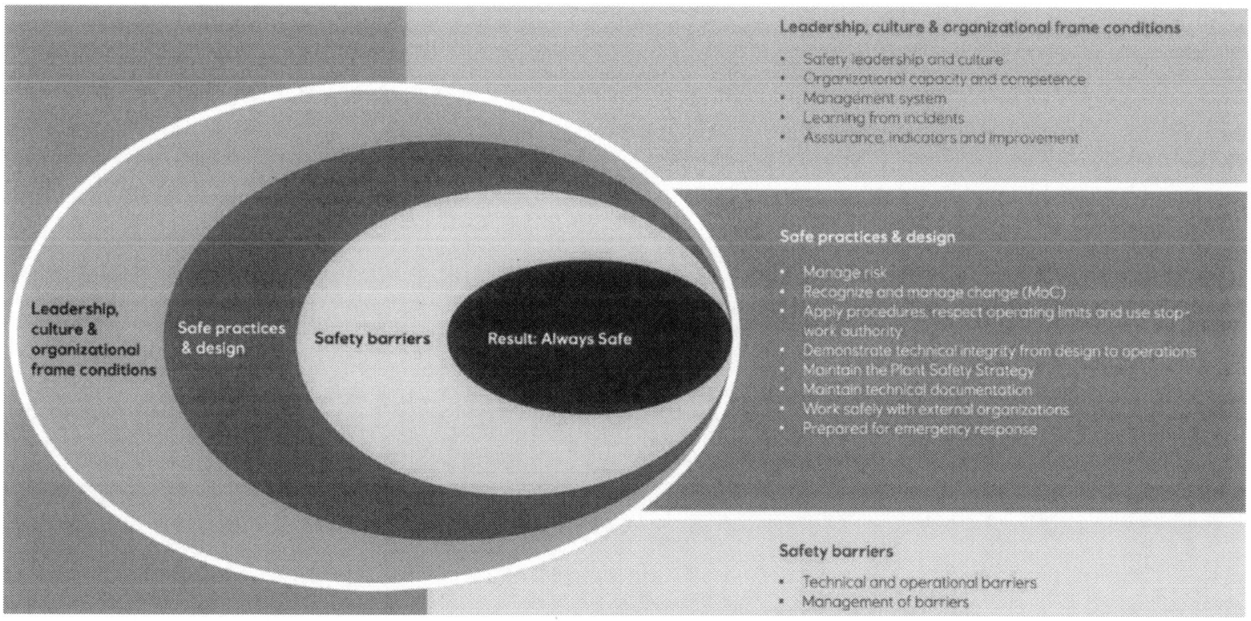

Figure 5. Equinor Framework for major accident prevention

Prepared for emergency response

Performance standard

Emergency planning and preparedness are essential for the protection of people, the environment, and values in case of incidents

Be prepared for emergencies

Causal factors from accidents	**SIOP#**
The emergency response was not dimensioned for an accident of this magnitude.	1
The risks of a NH3 plant might not have been properly understood since the accident caused 7 fatalities and 57 wounded amongst operating personnel. NH3 leakages can form large toxic gas clouds and these types of chemical plants require escape masks as a part of the PPE. It took 25 min. before external emergency authorities was alerted. This response time was not optimal. Use of water curtains to contain gas cloud and evacuation of private citizens seems well handled.	2
Poor emergency management and lack of coordination of community response. P oor emergency response planning & procedures.	3
Emergency responders must understand importance of SCBAs and PPE and have access to them Quick triggering of emergency response, including informing and working closely with local authorities Educate local residents about how to handle release of ammonia (when to stay indoors, when to evacuate).	4
•A malfunction in the badge tracking system delayed supervisors from immediately retrieving the list of personnel in their area. •Companies should conduct worker headcount drills according to the emergency response plans •Company procedures must account for breakdowns in automated headcount tracking systems to ensure that all workers inside a facility are quickly accounted. •Perform drills that simulate such a malfunctions should be conducted to verify that all lines of responsibility and alternate verification techniques will account for workers in a real situation	5
If an ammonia release cannot be promptly isolated, the emergency shutdown system should be activated. This will de-energize pumps and stop circulation of ammonia, thus limit the release from components running upstream of the failed equipment/piping.	6
A number of employees had not participated in emergency response exercises at the facility. • The emergency preparedness plan incorporates a general point on chemical discharges, but ammonia is not mentioned. •Two people ran through ammonia vapour without protective gear at an early stage in the incident while seeking to assist one of those injured. This does not accord with the description in the emergency preparedness plan. •Several employees had not participated in emergency response exercises at the facility. Some had taken part in such exercises a number of years ago.	7
Quick triggering of emergency response, including informing and working closely with local authorities. Local authorities where pipeline runs should establish a tactical response plan to release of ammonia, including how to alert, advise and evacuate residents (depending on concentration of ammonia in their area). Local rescue departments should also use appropriate PPE when entering areas of ammonia release. Ensure clear guidelines for handling leak detection and emergency responses	10
Clear routines should be established for entry of machinery spaces in the event of leakage of ammonia. Limit values for ammonia concentrations should be set for the various measures. Personal protective equipment must be available and used.	15
Establish emergency accident procedures and conduct drills accordingly. Have preparedness for handling minor injuries, equipment and routines for first aid in place.	16

Figure 6. Example of SIOP for an element in the major accident prevention framework.

Improvement of Mitigation Barriers in Fire Case: A Study Case from the Ammonia Converter

Ammonia Converter is the main equipment of ammonia synthesis which has been operating for 21 years without experience of catalyst change. The Ammonia converter leaked from its bolt and nut connection on the top cover in 2015 and created a continuous leak of flammable syngas during normal operation. A hazard identification study was carried out and concluded the condition as "High Risk". Tightening the connection during shutdown window did not solve the problem. Following the leakage, the fire occurred five times from some of the leakage points. Some mitigation barriers are in place to manage released gas safely.

Zaki S. Rahman, Gilang H. Pratomo, Dimas D. Ananda, Rahmatullah Q. Gobel, & Baskara A. Nugraha

PT. Kaltim Parna Industri, Bontang, East Kalimantan - Indonesia

Introduction

PT. Kaltim Parna Industri (KPI) based in Bontang, East Kalimantan – Indonesia, is an ammonia producer using Topsoe license for the process with the capacity of 1,500 MTPD and built by Mitsubishi Heavy Industries (MHI). The ammonia converters (R-0501 and R-0502) are radial-type converters with the gas flowing through the catalyst bed in a radial direction. R-0501 is an S-200 converter which contains two catalyst beds and an interbed heat exchanger (see Figure 1). R-0502 is an S-50 converter consisting of one catalyst bed. The S-50 converter is installed downstream of an S-200 converter with a steam superheater between the two converters.

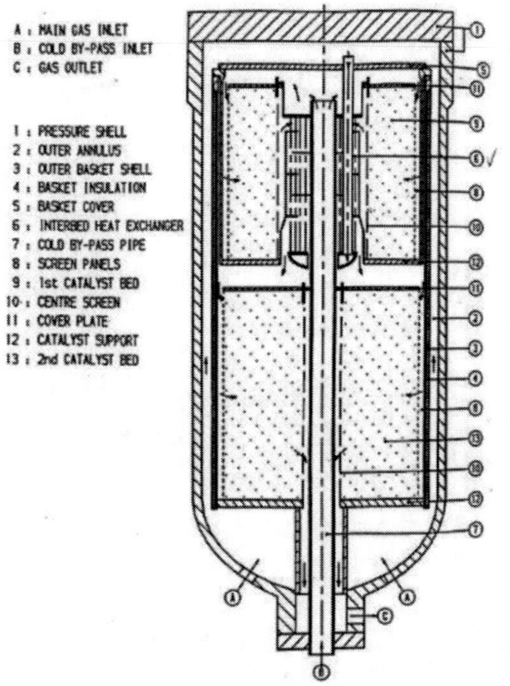

Figure 1. S-200 Ammonia Synthesis Converter [1]

The ammonia synthesis loop has been designed for a maximum pressure of 2204 psig (155 kg/cm^{2}g). The normal operating pressure will be 1191 psig (140 kg/cm^{2}g) in the 1st ammonia converter and 1963 psig (138 kg/cm^{2}g) in the 2nd ammonia converter, depending on load and catalyst activity. Normal operating temperatures will be in the range of 680-900^{0}F (360-482^{0}C) for the first bed and 697-822^{0}F (369-439^{0}C) for the second bed in R-0501. For R-0502, the range will be 375-481^{0}C.

History

The ammonia converter (R-0501) has been operating for over 20 years since this company operated commercially in 2002 without any catalyst changed. In December 2015, the first fire was found in the top cover of ammonia converter towards the primary reformer. Fire cases at the top cover ammonia converter can be shown in the table below.

Year	Source
December 2015	Thermowell 4 leakage
February 2018	Bolt No.8-13 and No.19-29
July 2019	Bolt No. 23-25
April 2021	Bolt No. 26

Table 1. 1st Ammonia Converter Fire Cases History

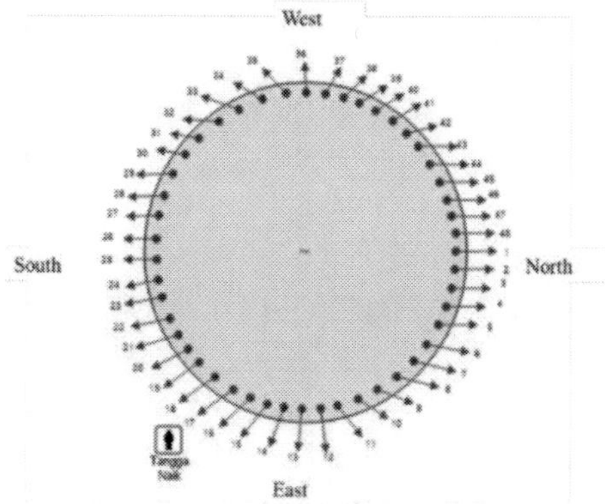

Figure 2. Sketch of Top Cover Ammonia Converter

The gas composition of flammable gases inlet 1st ammonia converter and physical properties of these gases can be shown in the table below.

Gas	Gas Composition (%Vol)	Lower Flammability Limit (%)	Upper Flammability Limit (%)	Auto-Ignitio (oF)
H2	55,0	4,0	74,2	1076

Table 2. Physical Properties of Flammable Gas of Inlet 1st Ammonia Converter

Since syngas contains flammable gases such as hydrogen and methane, the daily gas check is carried out to monitor and ensure the concentration of gas release does not exceed its flammability limit. Ammonia was also monitored, but the gas detector used can only detect maximum at 200 ppm or only 1,3% of its Low Flammability Limit so that the flammability of ammonia can't be identified.

Case Study

In December 2018, after the fire happened, the gas check was carried out and found the release gas around the leakage point exceeded the lower flammability limit of hydrogen (>4%) at 0 cm and 5 cm from the leakage point. The gas check result and leakage mapping can be shown in Table 3 and Figure 3.

Bolt	0cm			5 cm		
	H_2 (ppm)	NH_3 (ppm)	%LEL CH_4	H_2 (ppm)	NH_3 (ppm)	%LEL CH_4
Bolt 8	10000	200	70	0	0	0
Bolt 9	>40000	>200	100	>40000	>200	100
Bolt 10	>40000	>200	100	>40000	>200	100
Bolt 11	>40000	>200	100	>40000	>200	100
Bolt 12	>40000	>200	100	>40000	>200	100
Bolt 13	>40000	>200	100	0	0	0
Bolt 19	>40000	>200	100	>40000	>200	100
Bolt 20	>40000	>200	100	>40000	>200	100
Bolt 21	>40000	>200	100	>40000	>200	100
Bolt 22	>40000	>200	100	>40000	>200	100
Bolt 23	>40000	>200	100	>40000	>200	100
Bolt 24	>40000	>200	100	>40000	>200	100
Bolt 25	>40000	>200	100	>40000	>200	100
Bolt 26	>40000	>200	100	>40000	>200	100
Bolt 27	13000	200	70	0	0	0
[illegible]	7000	200	[illegible]	0	0	0

LEL : Lower Explosive Limit

Table 3. Gas Check Result on Bolt 8 to 29 After Fire Case in 2018

In July 2019, the concentration of hydrogen at the fire point (bolt 23 to 25) was above its flammability limit. The gas was detected around 5 cm from the bolt.

In April 2021, the fire happened for the third time. The fire was found around bolt 23 to 26 and caused severe damage to the elements of temperature indicator (TICA-0514).

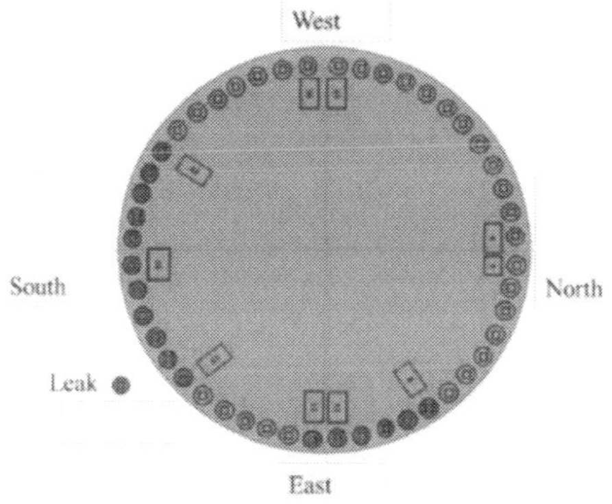

Figure 3. Leakage Mapping After Fire Case in 2018

Date	Bolt 24			Bolt 25			Pressu Synloc (psig
	H2 (ppm)	NH3 (ppm)	CH4 (%LFL)	H2 (ppm)	NH3 (ppm)	CH4 (%LFL)	

Table 4. Gas Check Result on Bolt 24 and 25 After Fire Case in July 2019

The indication became apparent when the alarm on DCS alerted the operator, informing them that the operating mode of controller TICA-0514 changed from AUTO to MANUAL. Field Operator also found the droplets of aluminum from the top cover 1st ammonia converter that indicated the fire catching the cover of insulation below the grating and caused the aluminum sheet to melt.

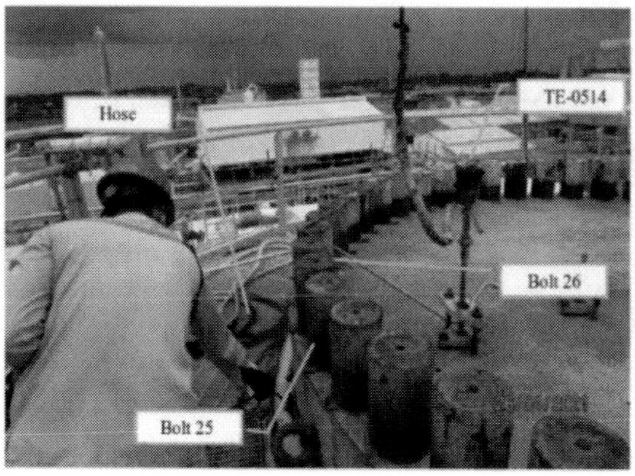

Figure 4. Field Check on Top 1st Ammonia Converter After Fire in April 2021

Risk

The risk analysis was done qualitatively through Hazard Identification (HAZID) Study. The worst credible scenario identified from this case was continuous release of syn-gas leading to fire and explosion with potential for fire escalation resulting in personnel fatality and asset loss, if ignited. This scenario was categorized as High Risk or Consequence in the Company Risk Matrix.

The scenario was further investigated in the Quantitative Risk Assessment (QRA) and was identified as one of the highest risk outcomes from the study. The study used a continuous leak of flammable gas from the ammonia converter top cover through a 10 mm hole size. The consequences analysis of this scenario shows that the jet fire will generate length of flame of only 4.75 meters and the radiation heat will no higher than 4 kW/m^2 which the level of lethality less than 3% of population. The impacted radius from the flash fire is about 2.8 meters from the leakage point.

The risk raised from this scenario is the flammable gas is released continuously and if a delayed flash fire happened is about 4.50 E-3/year and if followed by explosion, the risk is 3.00 E-3/year.

The scenario contributes higher Individual Risk per Annum (IRPA) for the ammonia field operator which works around the leakage point daily. The IRPA for the ammonia field operator is 7.01E-03/year, higher than the acceptable level for personnel at 1.00 E-03/year.[2] The overall risk contour from the Quantitative Risk Assessment study can be seen in Figure 5.

Both of those studies provided the recommendation to repair the leak on 1st ammonia converter to lower the risk. However, limited options for repairing the connection on top cover ammonia converter became an issue since the repair required removal of the top cover, unloading of the catalyst and replacement once contacted with air.

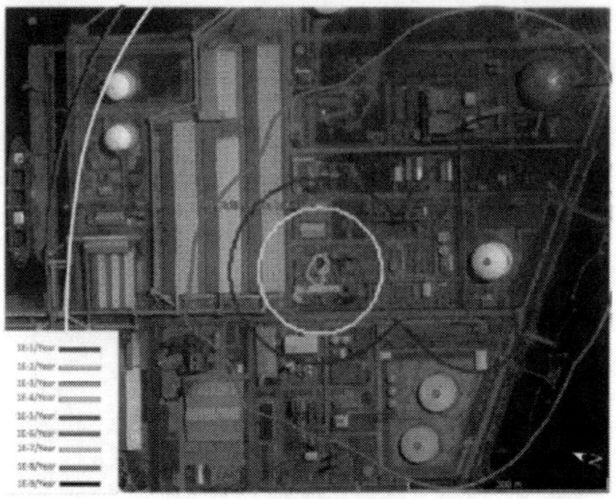

Figure 5. Overall Risk Contour of PT Kaltim Parna Industri [3]

The identification of ignition sources of the fire is excluded in this paper.

Post Release Mitigation

Preventive action was taken to eliminate or reduce the leakage. During the shutdown window in 2018, the physical check and cold bolt tightening at bolt was carried out. However, the result was insignificant, the flammable gases were still found at the bolts. The gas check result before and after bolt tightening are shown in Table 5.

Since the preventive action could not solve the leakage, post-release mitigation was required to reduce the probability of releasing gas that could ignite fire and minimize the impact of worst-case scenario. Center of Chemical Process Safety (CCPS) defined a countermeasures technique when the released material had formed a vapor cloud.[4] The technique includes containment or suppression, which involves limiting the amount of material that is released into the air and countermeasures, which are applied once the released material has formed a pool and/or a vapor cloud.

No	Monitoring Location	ppm H_2		ppm NH_3		%LEL CH_4	
		Before CB (*)	After CB (**)	Before CB(*)	After CB (**)	Before CB (*)	After CB (**
1	Bolt 1	0	20000	0	200	0	10
2	Bolt 2	0	18000	0	200	0	10
3	Bolt 3	0	3600	0	200	0	25
4	Bolt 8	10000	0	200	0	70	0
5	Bolt 9	40000	0	200	0	100	0
6	Bolt 10	40000	0	200	0	100	0
7	Bolt 11	40000	0	200	0	100	0
8	Bolt 12	40000	0	200	0	100	0
9	Bolt 13	40000	0	200	0	100	0
10	Bolt 19	40000	0	200	0	100	0
11	Bolt 20	40000	0	200	0	100	0
12	Bolt 21	40000	0	200	0	100	0
13	Bolt 22	40000	0	200	0	100	0
14	Bolt 23	40000	0	200	0	100	0
15	Bolt 24	40000	0	200	0	100	0
16	Bolt 25	40000	0	200	0	100	0
17	Bolt 26	40000	0	200	0	100	0
18	Bolt 27	13000	0	200	0	70	0
19	Bolt 28	7000	0	200	0	50	0
20	Bolt 29	2000	0	50	0	5	0

Table 5. Gas Check on Top Cover 1st Ammonia Converter Before and After Cold Tightening

From those techniques, there are several additional mitigation barriers and actions determined to safely manage this condition.

Syn-loop Pressure Reduction

The main purpose of reducing the syn-loop pressure is to reduce the leakage rate. The mass flow rate of gas charged through a hole or leakage point is determined by several factors. CCPS has defined the equation to estimate the mass flow rate in Figure 6. [5] From the equation in Figure 6, it can be known that the mass flow rate (m) is directly related to upstream or internal pressure (P_1)

$$\dot{m} = C_D A P_1 \sqrt{\frac{2 g_c M}{R_g T_1} \frac{k}{k-1} \left[\left(\frac{P_2}{P_1}\right)^{2/k} - \left(\frac{P_2}{P_1}\right)^{(k-1)/k} \right]}$$

where

- $\dot{m}$ is mass flow rate of gas through the hole (mass/time)
- C_D is the discharge coefficient (dimensionless)
- A is the area of the hole (length^2)
- P_1 is the pressure upstream of the hole (force/area)
- g_c is the gravitational constant (force/mass-acceleration)
- M is the molecular weight of the gas (mass/mole)
- k is the heat capacity ratio, C_p/C_v (unitless)
- R_g is the ideal gas constant (pressure-volume/mole-deg)

Figure 6. Gas Mass Flow Rate Through A Hole Equation [5]

The mass flow rate will decrease when the pressure of the syn-loop is decreased. Since the hole size of leakage is unknown, with this concept, reducing the syn-loop pressure can reduce the gas release. The mass flowrate can be reduced up to 7% when depressurizing the syn-loop from 1992 psig (140 kg/cm^2.g) to 1850 psig (130 kg/cm^2.g).

When the fire occurs, the operator will decrease the pressure of the syn-loop to 1850 psig by opening pressure control valve (PV-0213) to release some syn-loop feed through a cold vent stack. However, during a fire case in 2021, the operator needed to reduce pressure syn-loop from 1905 psig (133,9 kg/cm^2.g) to 1895 psig (133,9 kg/cm^2.g) to help extinguish the fire.

Since the mass rate of gas release reduces, the concentration of flammable gas exits the leakage point can be reduced to below its flammability limit after mixed with the steam dilution.

Steam Dilution System

As the gas check result found the concentration at leakage point above flammability limit, dilution system using non-flammable substance is required. The dilution system is intended to reduce the released flammable syn-gas concentration to below the lower explosive limit before contact with the air, so the fire can be prevented. Several non-flammable gases such as nitrogen, carbon dioxide, and steam can be used to dilute the leakage gases. [6]

After a fire case in December 2015, an ejector was installed around leakage point and several non-flammable gases were used as dilution system.

Fluid	Pressure (psig)	H2 Concentration (% vol)	
		Inlet Ejector	Outlet Ejecto
Nitrogen Low Pressure Combined With Medium Low Pressure Steam	58.8	6.73	1.78
Nitrogen High Pressure	117.6	6.24	2.34

*With dry basis

Table 6. Non-Flammable Gas Dilution

In 2018, an additional steam ring was provided to dilute the flammable gases through Bolt 7 to 30. The used steam was low pressure steam (3 kgf/cm^2). This steam ring successfully reduced concentration of flammable gases to below their flammability limit as shown in table below.

However, after the fire case in 2019, steam was switched to medium pressure steam and replacing the hose that used to deliver the low pressure steam to permanent piping.

Sampling Date	Bolt 23			Bolt 24			Bolt 25		
	5 cm			5 cm			5 cm		
	H_2 (ppm)	NH_3 (ppm)	%LEL CH_4	H_2 (ppm)	NH_3 (ppm)	%LEL CH_4	H_2 (ppm)	NH_3 (ppm)	%LEL CH_4
4-Mar-18	8000	50	4	6000	40	2	6000	40	2
5-Mar-18	8000	50	4	6000	40	2	6000	40	2
6-Mar-18	8000	50	4	5000	38	2	5000	37	2
7-Mar-18	8000	50	4	6000	40	2	6000	40	2
8-Mar-18	8000	50	4	6000	40	3	6000	40	2
9-Mar-18	8000	50	4	6000	40	3	6000	40	2
10-Mar-18	7990	50	2	6000	38	2	5950	40	2
11-Mar-18	8000	50	2	6000	38	2	6000	40	2
12-Mar-18	8000	48	2	6000	38	2	6000	40	2
13-Mar-18	8000	48	2	6000	40	2	6000	40	2
14-Mar-18	8000	50	2	6000	39	2	6000	40	2

Table 7. Flammable Gas Concentration After Diluted with Low Pressure Steam

Since the leakages are not found only in one spot, a cover that was made by aluminum and isolation was provided around lip seal of 1st ammonia converter to relocate the release gas. Two specific holes in the cover were made as the exit way for the diluted gas. The cover was connected to a permanent steam ring to dilute the release gas in the cover.

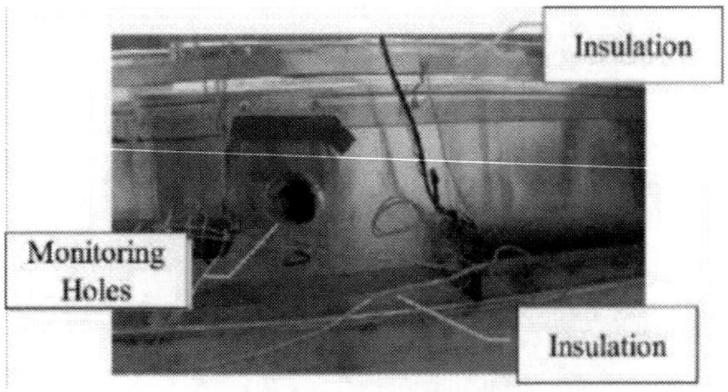

Figure 7. Cover Lip Seal 1st Ammonia Converter with Cover (above) and Steam Box (below)

The steam box was fabricated to replace the cover to enhance the integrity of the steam dilution system since the cover was made by aluminum sheet that has melting point around 1220^{0}F (660^{0}C) and exceeds the flame temperature of hydrogen that up to 4712^{0}F (2600^{0}C).

After the fire case in 2021, the monitoring area on the top cover of 1st ammonia converter was divided into 4 points as follows.

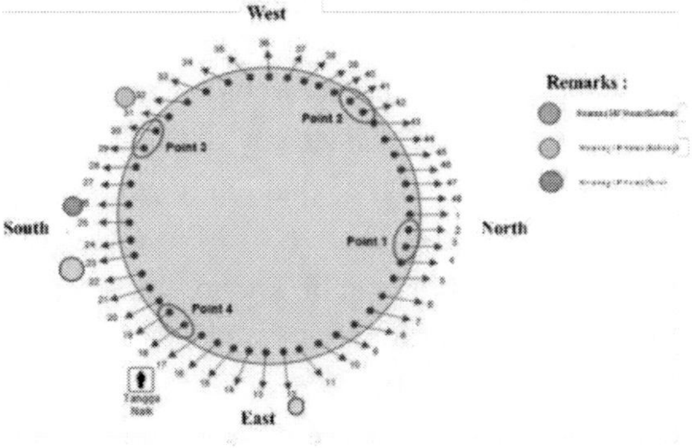

Figure 8. Flammable Gas Monitoring Point After Fire Case In 2021

To pull out the diluted gas in the cover, an ejector is also installed at the outlet of hole. A low medium pressure steam is used as the motive fluid. This facility is also intended to control fires and help the operator and the emergency response team to focus on specific locations to extinguish the fire.

Figure 9. Ejector From Steam Box Outlet

During the fire, the amount of steam dilution can be added into the steam box to help extinguish the fire.

Temporary Fire Monitor and Sprinkler

One of countermeasures technique is plume dilution. The aim of this technique is to minimize the impact area of hazardous vapor release. The dilution can be done by using air and steam curtains, water sprays and curtains, etc. Water sprays are simple, reliable, and inexpensive and can be either fixed or mobile. [4]

The water sprays and curtains also have a role as a fire suppression required to limit the fuel or oxygen supply or reduce the flame zone's temperature. The water will extinguish the flame by these mechanisms by heat extraction (cooling), oxygen displacement by water vapor, and blocking of radiant heat. [7]

The top cover of the ammonia converter is about 100 feet from the ground. This condition gives several challenges to the emergency response team and operations team to extinguish the fire as below.

a) KPI did not have a dedicated fire water pump and relied on fire water supply from another party. The distance between the fire water pump and KPI plant is about 1 km. On a standby basis, there is a jockey pump run to maintain the fire water supply pressure at 73 psig (5 barg). Based on fire water test, it needs 8-10 bar pressure for the fire water at KPI plant to achieve that elevation from a fire monitor at the ground level. The supplier shall switch from the jockey pump to the main fire pump to achieve that pressure and this operation takes more than 5 minutes.
b) The team needs more than five minutes to climb the monkey ladders to get to the location. The maximum time for the firefighting team to arrive at the fire location is 5 minutes and 20 seconds. [8]

Figure 10. Access Ladder to Top Cover 1st Ammonia Converter

Facing these challenges, the temporary fire monitor and sprinkler facility were installed as water sprays and curtains during the fire on the top of the ammonia converter. Fire monitor nozzles and hose lines can be used for vapor mitigation.[4]

The temporary fire monitor consists of two fire nozzles that are installed toward bolt 20-23 (east side) and bolt 41-42 (west side) where the specific holes from the cover are provided. The nozzles are connected to the nearest fire hydrant and valves are provided at the hydrant connection. The operator or emergency team should only open those valves to activate the nozzles. The nozzles are in spray position. Since the first fire occurred in 2015, the water curtain has already been installed. The illustration of water curtain activation can hbe seen in **Figure 11.**

Figure 11. Water Curtains Illustration on Top Cover 1st Ammonia Converter

Since the water curtain using fire hose is activated manually, a water sprinkler is provided as a water suppression that can be activated automatically in accordance with its thermal sensitivity. The specification of the water sprinkler is as follows.

Type	**Pendent**	
Size	Inch	1/2
Max Ceiling Temperature	F (C)	225 (107)
Thermal Sensitivity	F (C)	286 (141)
Finishing		Chrome

Table 8. Specification of Water Sprinkler

Thermal CCTV and Flame Detector

Since the ammonia plant is not equipped with a flame and fire detector, a real time monitoring is required to detect the fire on the top cover of ammonia converter as fast as possible. From 2015 to 2019, the fire was only detected if the operator went to routine check every two hours and might see fire occurred on the top cover of ammonia converter. Since the gas composition mostly contains hydrogen, the flame is invisible to the naked eye during daylight.

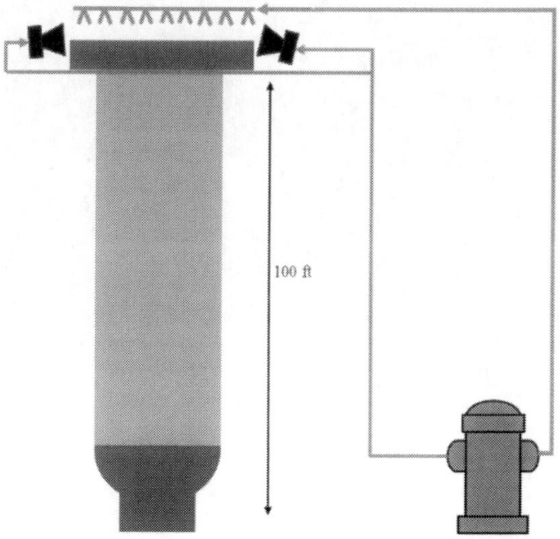

Figure 12. The Illustration of Water Sprinkler and Temporary Fire Monitor on The Top Cover Ammonia Converter

A thermal CCTV is installed to monitor the leaking converter and may found a fire on the top cover ammonia. The thermal CCTV shows temperature around the top cover. When the fire occurs, the temperature will increase until the set point at 572^0F (300^0C) and the alarm will alert the operator in the Control room. The thermal CCTV will show a red mark on the fire point.

Figure 13. Thermal CCTV and Normal CCTV With Top Cover 1st Ammonia Converter View

Figure 14. Flame Detector on The Top Cover 1st Ammonia Converter

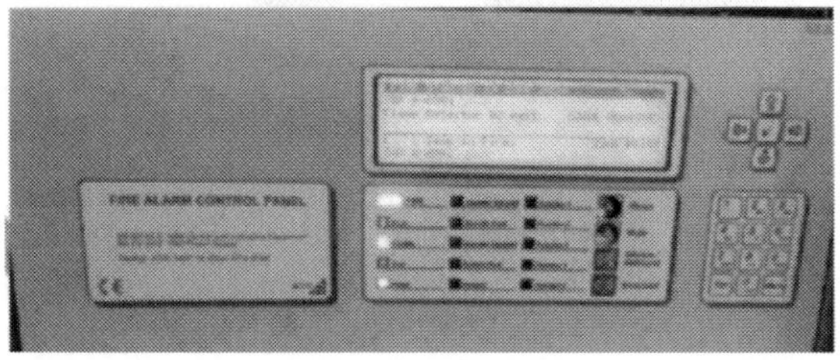

Figure 15. Existing Fire Alarm Panel When Detect Fire on The Top Cover of 1st Ammonia Converter

The specifications of the flame detector and design for flame detector installation are below.

Material	Stainless Steel
Detects	Hydrogen and Hydrocarbon Fires
Detection Range	Fuel Hydrogen – max. 131 ft (40 m)
Minimum Fire Dimension	0,75 High, 0,25 wide plume fire
Coverage area	58400 ft^2 (5426 m^2)
Standard Response Time	Typically, <2 secat (131 ft)

Table 9. Specifications of Flame Detector

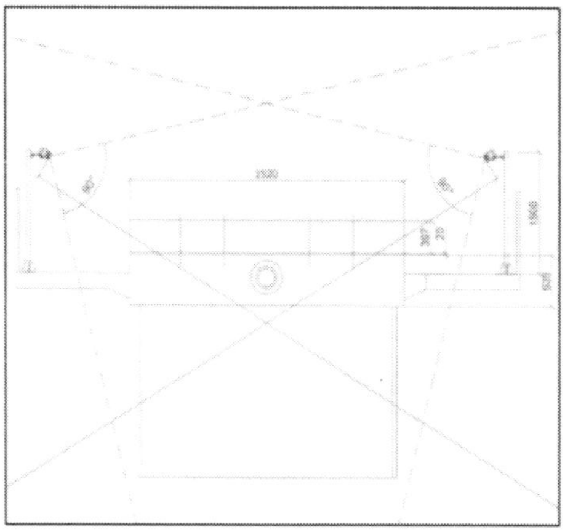

Figure 16. Flame Detector Installation Design on The Top Cover 1st Ammonia Converter

Evaluation

The post release mitigation is evaluated to know how effective the implementation.

a) Company experiences show that reducing syn-loop pressure during fire could help to reduce duration of fire. However, plant shutdown is still required if the fire escalates.
b) The steam dilution system can effectively dilute the flammable gases to below their lower flammable limit. The cover with two specific exit holes also gives opportunity help on fire detection and firefighting strategy.
c) The temporary fire monitor can be activated immediately by the operator and emergency response personnel and the fire water can be supplied in less than five minutes. The water sprinkler is also successfully activated and being an indication of fire on the top cover 1st ammonia converter.
d) The thermal CCTV and flame detector can improve the response time of operator in case of fire at the top of ammonia converter. Before the equipment installation, the fire could only be detected if any field operator sees the fire during field surveillance every two hours or if he found the

water sprinkler was active. When the CCTV thermal and flame detector alert the operator in control room then the mitigation tasks will be carried out immediately, starting with reducing the synloop pressure and activating the water curtains from the ground level.

Conclusions

The post release mitigation is successfully implemented to improve the mitigation barriers in case of fire on the top cover of ammonia converter. Notes should be taken that the mitigation barrier is active after the event of fire. A fire can still occur since the root causes of fire can't be identified.

References

[1] Operation Manual of 1500MTPD Ammonia Plant for KPI Project. 2000. Mitsubishi Heavy Industries, LTD.

[2] Reducing Risks Protecting People. 2001. HSE Executive UK.

[3] PT Kaltim Parna Industri Quantitative Risk Assessment (QRA) Report. 2023. PT Keselamatan Proses Indonesia

[4] Guidelines for Post Release Technology In The Chemical Process Industry. 1997. CCPS of AICHE.

[5] Guidelines for Consequence Analysis of Chemical Releases. 1995. CCPS of AICHE.
[6] Biennal Report, "Hydrogen safety barrier and measures". 2006. Hysafe.

[7] Alageel, K.S. 1999. Mitigation of Compartment Jet Fire Using Water Sprays.

[8] NFPA 1710. 2020. Standard for the Organization and Deployment of Fire Suppression Operations, Emergency Medical Operations, and Special Operations to the Public by Career Fire Departments.

101-B Primary Reformer Riser Transitional Assembly Failure

A loss of containment occurred on the #3 Riser Transitional Assembly (RTA) of the 101-B Primary Reformer. This paper will discuss the events preceding the failure, the inspection findings, and subsequent repairs. The operational risk management strategies implemented to bring the plant back online until the next turnaround and the findings from the failure analysis will also be discussed.

Carissa Ramdin
Nutrien
Point Lisas, Trinidad and Tobago

Samuel Okulaja
Nutrien
Calgary, Alberta, Canada

Introduction

In September 2023, a routine pyrometric scan of the Nutrien Trinidad 04 Ammonia Plant Primary Reformer, 101-B, identified an abnormal temperature profile along the No. 3 Riser, which prompted burner adjustment. The frequency of temperature profile monitoring was increased to observe the Riser tube. Shortly after, a fire was seen emanating from the Primary Reformer penthouse. The necessary steps were taken to shut down the plant. This paper includes the inspection findings, the failure analysis, the risk management strategies employed, and the long-term recommendations for the No. 3 Riser Transitional Assembly (RTA).

Background

The Nutrien Trinidad 04 Ammonia Plant was commissioned in 1998, utilizing the Kellogg Advanced Ammonia Process design with a capacity of 1850 MTPD (2039 STPD). Optimization has been made since 1998 to increase the plant's capacity and it currently operates at 2230 MTPD (2458 STPD).

In the Primary Reformer, hot mixed feed gas from the Mixed Feed Preheat Convection Coil, 101-BCX, enters at 621°C (1150°F) and is distributed among the catalyst tubes which are suspended within the radiant section of the top fired furnace. The feed gas is fed over a bed of nickel catalyst and reacts to form H_2, CO and CO_2. The reformed gas exits via the internalized outlet manifold where it continues to pick up heat in the riser tubes while exiting the radiant section at approximately 820°C (1508°F). There are five riser tubes that are connected to the Effluent Chamber Transfer line (107-D) via the Riser Transition Assemblies (RTAs). A design temperature and pressure of 941°C (1725 °F) and 4 MPa (580 psi) are specified for the Riser tubes. See Figure 1 for a cross-sectional view of the radiant section of the Primary Reformer.

The 107-D and the RTAs, apart from its transition cone section, have been in service since the commissioning of the plant in 1998. The transition cone section of the RTAs and the radiant harps were replaced in 2009.

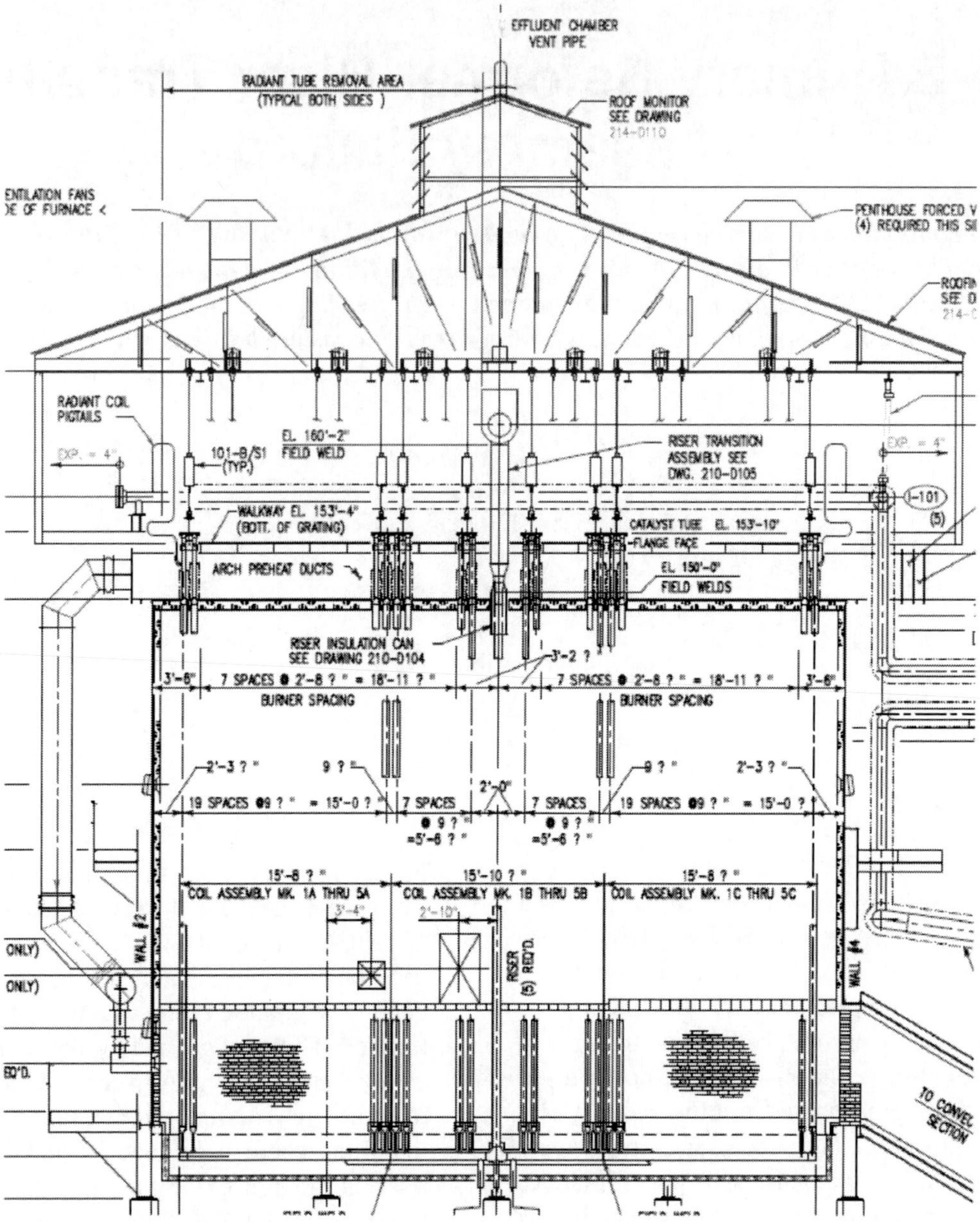

Figure 1: Showing cross sectional view of Primary Reformer radiant section.

The RTAs are of dissimilar metal configuration and consist of a centrifugally cast G4859 Micro (20Cr-33Ni-Nb) transition cone welded to an ASME SA 312 Gr. TP 310 stainless steel transition pup piece connecting to ASME SA 106 Gr. B carbon steel. The internal of the RTAs are lined with alumina castable refractory and have an inner alloy 800 HT liner with a slip joint design. During the 2009 transition cone upgrade, the alumina castable refractory was recessed within the stainless steel transition pup piece to allow for field fit-up of the transition cone. This recessed area created a void that was filled with compressed ceramic fiber during installation.

Incident Description

In August 2023, an upward trend in temperature was noted on the No. 3 Riser. The pyrometric scans revealed temperatures as high as 932°C (1709°F). Burner flame adjustments near this Riser were conducted to reduce the firing and bring the temperature closer to the calculated Tube Metal Temperature (TMT) of 885°C (1625 °F). In October 2023, a visual inspection revealed a brighter glow on the No. 3 Riser, isolated to one side of the tube with evidence of bowing. This later resulted in a failure of the No. 3 Riser Transition Assembly.

Inspection Findings

Inspections were initiated to investigate the extent of damage and effect the necessary repairs to return the plant to production. The post-incident inspection activities included inspection of the penthouse and furnace of the Primary Reformer.

Penthouse Inspection

Visual inspection observed a breach on the No. 3 Riser Transition Assembly. The four accessible welds (Welds 1-4) of the RTA were subjected to Liquid Penetrant Inspection (LPI); see Figure 2 and Table 1 for details. Access to Weld No. 5 is limited because the Riser Insulation encloses its can.

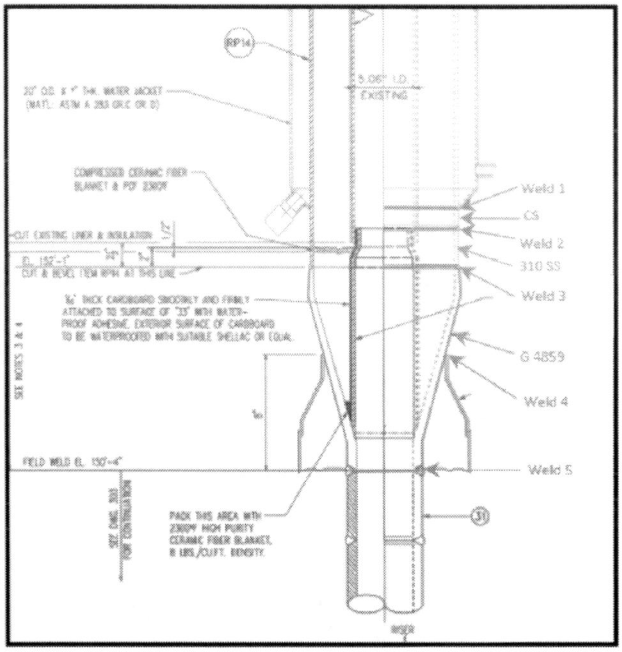

Figure 2: Showing cross-sectional view of RTA assembly.

Weld #	Base Materials	Base Materials
1	Water Jacket (ASTM A 283 Gr C)	Upper Riser Transition Pipe (ASME SA 106 Gr. B)
2	Upper Riser Transition Pipe (ASME SA 106 Gr. B)	Lower Transition Pup Piece (ASTM A312 GR. TP 310)
3	Lower Transition Pup Piece (ASTM A312 GR. TP 310)	Transition Cone G4859 Micro (20Cr-33Ni-Nb)
4	Transition Cone G4859 Micro (20Cr-33Ni-Nb)	Riser Insulation Can (Alloy 800H)
5	Transition Cone G4859 Micro (20Cr-33Ni-Nb)	G4852 Micro Cast Tube (25Cr 35Ni Nb Micro)

Table 1: Showing cross-sectional view of RTA assembly.

LPI revealed two cracks on Weld No. 3, as shown in Figure 3. The first crack was observed at the toe of the weld and extended about eighty per cent of the weld's circumference, while the second was a centerline crack measuring approximately 6 to 8 inches in length. LPI further revealed a crack also on Weld No. 2, which was on the weld toe of the Upper Transition Pipe side (ASME SA 106 Gr. B). The findings prompted further inspection of the remaining four RTA welds and the radiant harps.

Figure 3: Showing crack found via LPI on Weld No.3 of the RTA assembly.

Inspection scope comprised of the following:

- LPI on all RTA welds.
- Phased Array Ultrasonic Testing (PAUT) was done only on Weld No. 2 of all RTAs due to the absence of a specialized calibration block for inspecting high nickel alloy welds.
- Ultrasonic B-scan on the base metal of the upper and lower riser transition pipe.
- Hardness testing
- LEO-Scans of the radiant harps
- Replica Metallography on various catalyst tubes
- Visual inspection of spring hangers.

Inspection did not detect any discontinuities on Riser Nos. 1 and 4. On Riser No. 2, LPI revealed a linear indication at the toe of the No. 2 Weld. The indication measured approximately 5 inches long and was located on the side of the Upper Riser Transition Pipe.

An indication was found on the base material of the Transition Cone, approximately 0.5 inches away from Weld No. 3 on Riser No. 5. The indication was linear and measured approximately 1 inch. The fillet Weld No. 4 also exhibited linear indications at the toe of the Riser Insulation Can.

Furnace Inspection

Upon visual investigation within the furnace, nine catalyst tubes observed randomly across rows 1, 2, and 3 were found to be bowed. The No. 3 Riser tube also exhibited bowing to a lesser extent. Several tubes exhibited scaling, primarily visible in the lower half of the tube.

Replica testing performed on the bowed tubes showed evidence of metallurgical degradation due to significant carbide coarsening, but no OD (Outside Diameter) connected indications.

LEO-Scans, a technique comprising of eddy current and outside diametrical measurement, were conducted on the radiant harps. Eddy current detected two minor indications on Risers No. 1 and 3. The indications could not be classified within the calibration criteria due to their small size, and as such, were deemed as not a cause for immediate concern. The indications were not present during the previous LEO Scans in 2018.

Outside diametrical measurements recorded most of the tube expansion in Row No. 3 which accounted for sixty percent of the measurements taken. The average increase in expansion from the 2018 inspection was found to be 0.3%, with three tubes in Row No. 1 experiencing a 0.5% increase. Row No. 1 experienced the second highest increase in tube expansions, with 20% of the tubes exhibiting growth compared to the 2018 inspection.

Failure Analysis

A multidisciplinary team was established to determine the root cause of the failure while commissioning a metallurgical failure analysis on the failed No. 3 Riser Transition Assembly.

Visual and Stereoscopic Examination
The internal refractory exhibited spalling, and there was no evidence of the compressed ceramic fiber installed in 2009. The inner alloy 800 HT liner was found intact with no evidence of deformation. Shallow pitting was observed on the ID of the transition cone.
The visual assessment observed that Weld No. 2 exhibited excess root penetration for almost the entire circumference. Weld No. 3 displayed insufficient root penetration throughout the circumference for up to fifty per cent of the joint thickness. A cross-section of Weld No. 3 was etched for further analysis, and it showed evidence of multiple welding defects and poor joint preparation, including lack of side wall fusion, linear mismatch, and poor bevel preparation, see Figure 4.

Microstructural Analysis
Weld No. 2, the joint between the Upper Riser Transition Pipe (ASME SA 106 Gr. B) and the Lower Transition Pup Piece (ASTM A312 GR. TP 310) exhibited high hardness (282-300 HB) in stark contrast to the base materials. The carbon steel section recorded a maximum hardness of 162 HB, and the stainless steel section recorded a maximum of 206 HB. No discernible degradation of the weld and adjacent base materials microstructure was observed in the weld. Positive material identification revealed the chemistry of the electrode utilized was within the composition of Inconel 617 (ENiCrCoMo-1).

Weld No. 3, the joint between the Lower Transition Pup Piece (ASTM A312 GR. TP 310) and the Riser Transition cone (G4859 Micro: 20Cr-33Ni-Nb), exhibited multiple fatigue cracks initiating from the weld root area. The microstructure of the stainless base material adjacent to Weld No.3 showed a widespread sigma phase dispersed throughout the microstructure coupled with secondary OD intergranular cracks. In contrast, the same material adjacent to Weld No. 2 did not exhibit any significant microstructural changes.

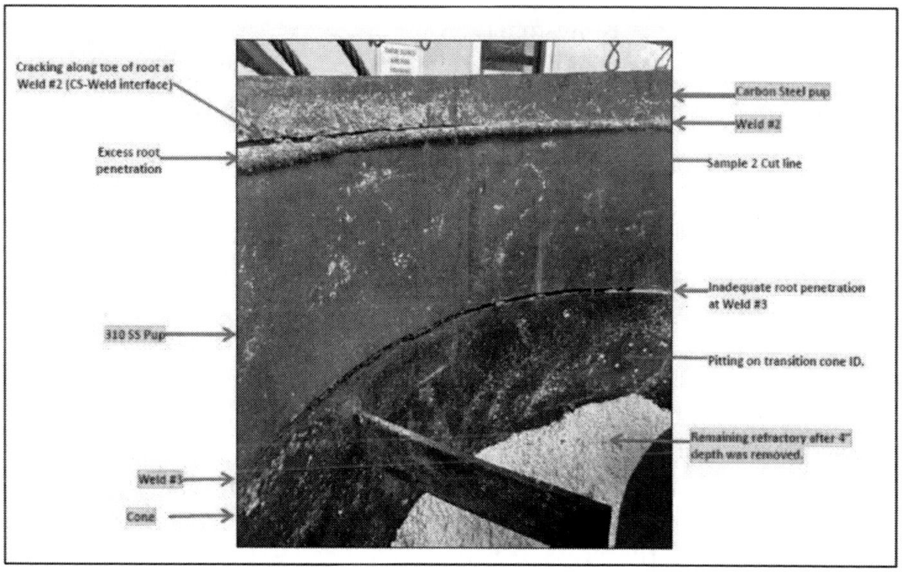

Figure 4: Showing internal RTA assembly

The fractured surface of Weld No. 3 was analyzed using Scanning Electron Microscopy (SEM). According to the SEM images, the cracking, which started on the ID and moved to the OD, was consistent with fatigue cracking. It also revealed that sigma phase precipitation was significantly higher within 0.25" of the weld on the stainless base material and reduced as it moved towards Weld No. 2.

The hardness of the adjacent stainless base material measured higher (257 HB) than the typical expected hardness for materials in similar service (190-210 HB).

The transition cone was analyzed where the shallow pits (1mm deep) were found, and a uniform layer of carburization was observed.

Discussions

The following summarizes the suggested chain of events that resulted in the failure:

- The 2009 turnaround (TAR) saw the installation of an improved RTA transition cone design (elimination of additional weld on the conical section); with this installation, new Weld Nos. 3, 4, and 5 were made. Evidence of volumetric inspection conducted on Weld No. 3 was not found for this installation. It is evident from the failure analysis, that the many weld discontinuities created during installation compromised the integrity of the weld.

- The 2009 modification left a gap in the refractory that was filled with compressed ceramic fiber, composed of 53% silica and 45% alumina. To prevent insulation deterioration, industry practice advises that insulation used in hydrogen (H_2) and carbon monoxide (CO) services have a low silicon concentration. The inner Inconel 800HT liner inherent slip joint design places it directly across from Weld No. 3 and, where the location of the ceramic fiber was installed. It is suspected that process gas, H_2 and CO bypassed through the slip joint that compromised the ceramic fiber insulation and exposed Weld No. 3 and the transition cone to the process. The suspected bypass of the process gas exposed the transition cone to carburization temperatures above 480 °C (900 °F), which manifested in shallow pitting. Carburization is a damage mechanism that occurs to metals in a high temperature, carbon rich environment, resulting in reduced ductility and increased hardness, leading to metal loss over long periods. The gas bypass also contributed to sigma phase formation in the exposed section of the stainless-steel metal adjacent to Weld No. 3. Sigma phase embrittlement is a damage mechanism affecting susceptible alloys, such as 300 series stainless steel, operating at temperatures between 540 °C to 925 °C (1000 °F to 1700 °F) that results in reduced ductility and toughness and subsequent cracking.

- The investigation findings noted that there were gaps in the Inspection Test Plan (ITP) for the RTA welds. As a result, the weld discontinuities went undetected since installation.

- The Riser No. 3's elevated temperatures in September 2023 and the Riser's subsequent bowing also added more stresses on the already compromised RTA Weld No. 3. The thermal stresses induced by the dissimilar metal configuration of Weld No. 3 further contributed to the thermally induced fatigue cracking and subsequent loss of containment.

It was determined that insufficient quality assurance and control (QA/QC) during the installation of the 2009 transition cone were the primary causes of the failure.

Repairs

Temporary repairs were conducted to return the plant to service during the 2023 outage with the objective of replacing all RTAs during the next scheduled TAR. The No. 3 RTA was replaced by cutting from weld No 5 and directly below weld No. 1 to accommodate the installation of the new in-kind Riser Transition Cone and the Lower Transition Pup Piece (ASTM A312 GR. TP310). Weld repairs were also conducted on the remaining RTA welds where weld discontinuities were detected.

Mitigation Strategies

An Operational Risk Management (ORM) review was conducted to evaluate the risks related to continued operations. Developing an Abnormal Equipment Procedure (AEP) was recommended to capture the interim safeguards and effectively manage risks. Highlights from the AEP are as follows:

- Limit the number of personnel conducting tasks within the Primary Reformer penthouse.
- Limit and prioritize activities within the penthouse, such as burner adjustment, unless the task warrants immediate attention.
- Limit the firing of burners near Riser tubes to prevent flame impingement and bowing of tubes.
- Increase the frequency of pyrometric scans on the radiant harps to a weekly basis as opposed to the previous monthly inspection.
- Install thermocouples near the RTA welds to monitor the skin temperature and establish alarm setpoints.
- Ensure all forced draft fans are operational to allow for effective air circulation within the penthouse and remove any insulation blankets near the RTA.
- Install infrared cameras within the penthouse to monitor any loss of containment.
- Every reasonable attempt should be taken to minimize the frequency of start-ups and shutdowns to reduce stresses associated with thermal cycling.

Conclusions & Recommendations

The radiant harps have been in service since 2009. The Effluent Chamber Transfer line (107-D) and the Riser Transition Assemblies (apart from the transition cone section that was replaced in 2009) have been in service since commissioning in 1998, and as such, it was recommended to replace the radiant harps, the Effluent Chamber, and the RTAs at the next TAR opportunity.

Improvements in RTA design, including eliminating the Inconel 800HT liner slip joint and dissimilar metal joints when designing future replacements, will prevent such failures in future.

The failure accentuated the importance of providing adequate and detailed diligence in QA/QC during fabrication and field installation. Assuring the integrity of equipment during its lifetime requires the development of comprehensive repair plans and ITPs including all relevant Non-Destructive Examinations (NDE).

Unusual Fire Beneath a Steam Methane Reforming Furnace

This paper describes an unexpected fire which occurred two weeks after start-up of a hydrogen plant following a Major Plant Turnaround. The fire occurred beneath the furnace box as a result of loss of containment from a radiant section process gas tube. In this paper we will share the sequence of events leading up to the fire, damage caused by the fire, and most importantly the incident investigation results and how they have been applied to prevent recurrence.

Bill Ross (retired), Ken Wohlgeschaffen
Chevron Products Company (a division of Chevron U.S.A., Inc.)

Introduction

A large fire developed suddenly and unexpectedly during normal operation of a hydrogen plant following a major turnaround. This paper will detail the sequence of events leading up to the fire, damage caused by the fire, the incident investigation findings, and the key lessons learned.

Background

This incident occurred in one of two 1960s vintage hydrogen plants in one of our refineries (Figure 1). There are 112 radiant section tubes in each of the three furnace cells (F-305A, B, and C in the PFD in the Appendix) for a total of 336 tubes per furnace.

In March 2005, one plant was shut down for a Major Turnaround in which the reforming catalyst in the furnace radiant section tubes was replaced.

Figure 1.Hydrogen Plants, furnace in red box

Five reforming catalyst tubes were also replaced in B cell. During catalyst loading, the catalyst support cones at the bottom of seven other furnace tubes in B cell were found to be damaged. The bottom flanges of these tubes were opened and the support cones replaced prior to loading with fresh reforming catalyst. The gaskets on all twelve tubes were replaced with standard 3 inch ASME B16.5 flange sized corrugated metal core graphite layered ("CMG") gaskets that were not self-centering in the custom dimensioned catalyst tube flanges. The CMG gasket design seals at much lower bolt loads than the standard spiral wound gasket. Our Standard Instruction recommended self-centering CMG gaskets as an option to spiral wound gaskets for critical flange connections or flanges with a history of leaking.

Operations performed a leak check as normal operating practice using plant nitrogen at 150 psig (10.3 barg), as a last check to verify that all connections were leak tight. This step was part of the normal pre-startup procedure. No leaks were found.

Incident Summary

The plant started up normally. During stable operation two weeks later, at 1 pm on April 8, 2005, several furnace tube process gas exit temperature indicators failed simultaneously. This sounded alarms in the control room. The Outside Operator was sent to the furnace to investigate. The Outside Operator saw fire at the bottom of B cell of the reforming furnace and immediately called the Board Operator to shutdown process feed. The Board Operator quickly shutdown process feed at 1:15 pm, isolating the source of fuel to the fire, and started an emergency shutdown of the plant. The main fire stopped under the furnace box when the natural gas feed was removed. The refinery fire department quickly arrived and extinguished secondary ground fires in the area using water from fire hoses.

Damage was limited to instrumentation at the bottom of the furnace, including the actuator on the waste heat boiler temperature control (internal bypass) valve, refractory, and the furnace structure (Figures 2 and 3). Fortunately, no one was injured as no one was in the area when the fire occurred, and actions were immediately taken to safely shut down the plant and extinguish the fire.

Figure 2. Damage to the furnace structure - buckling of I-beams

Figure 3. Damage to the furnace structure - bulging of the furnace floor

Investigation Findings

Subsequent investigation revealed the fire originated from a leaking flange on the bottom of one of the reforming furnace radiant section tubes. The initial leak was from tube #144 (Figure 4) which was one of the twelve that had been worked on during the Turnaround. This leak was caused by lack of gasket seating stress which allowed 1500ºF (816ºC) syngas to leak from the bottom of the tube.

Figure 4. Bottom flanges of tubes #143/#144

The leaking, hot syngas then auto ignited. The heat from the fire elongated the stud bolts and allowed the gap between the flanges to open up from 1/16 in to 1/4 in (1.6 mm to 6.4 mm). Fire impinging on the flanges of adjacent tube #143 elongated the stud bolts and allowed the gap between the flanges to open up from 1/16 in to 1/8 in (1.6 mm to 3.2 mm). The gasket in tube 143 also failed, but given the small amount of metal remaining (Figure 5), this failure likely happened after the fire started.

Thirty other tube flanges in B cell, fifteen East and fifteen West of tubes #143/4, were found loose from the heat of the fire. None contributed to the fire due to the expediency of pulling feed.

Figure 5. Blown out gasket

The flange gasket sealing surfaces looked to be in good condition (Figure 6). These two CMG gasket cores failed from high gas velocity which resulted in increasing the quantity of released gas. The remaining graphite not blown off by high gas velocity was burned off by the heat of the fire.

Figure 6. Flange gasket sealing surfaces

However, the flange gaskets were found to be off center (Figure 7). The gasket stress was higher in the areas of minimal gasket contact with the flange raised face and the gasket initially sealed because of this reduced amount of surface area in contact with the flange. The gasket failed at the corner where the higher contact area provided the lower amount of gasket stress.

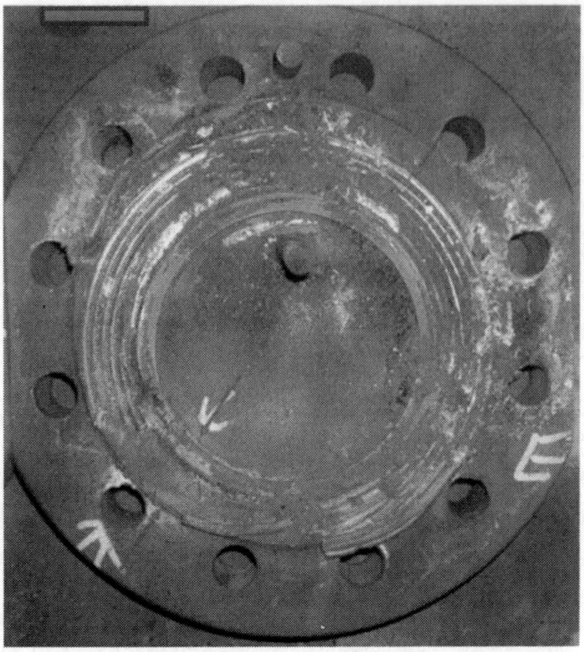

Figure 7. Gasket not centered correctly

The studs were not properly tightened (Figure 8). Some of them were hand tightened only, with the expectation that another maintenance crew following behind the first crew would fully tighten them with a torque wrench. See the stud in the center of Figure 7 which was a newly installed stud. It was determined that those studs that were hand tightened only had about 20 foot-pounds (21.7 Nm) of torque whereas 90 foot-pounds (122.4 Nm) is required. Additionally, many of the nuts were installed upside down which reduced the final stud loads. See the stud on the right hand side of Figure 8 (red arrow). The "washer" built into the nut must be against the flange to reduce friction, which clearly was not the case here. It was also noted that anti-seize was not used and many old studs were reinstalled. Both of these actions can cause lower stud loads.

Tubes #143 and #144 were opened at the bottom and their catalyst support cones replaced during the Turnaround. It was found that six of the twelve bottom flanges worked on during the shutdown were essentially loose. A leak could have occurred from any of these six tubes but thankfully was confined to tubes #143 and #144.

Figure 8. Loose stud (center of photo), upside down nut (right side of photo)

The contractor's policy was to tighten the bolts when the flange is put on before leaving the job, then hang an aluminum tag with the initials of the fitter performing this work. The foreman would then check the craftsmanship and tightness of the bolted joint and initial the tag. A Quality Assurance/Quality Control ("QA/QC") Inspector would check the tag, check the joint, check the tightness, sign and remove the tag, and put initials and date on the flange. Four bolts were found loose on tube #144 after the QA/QC was completed and while catalyst was being loaded into the tube. Several witnesses did not recall tags being attached to the flanges. No QA/QC marks, initials, or any other identification marks were observed on the flanges.

CMG gaskets were used on the tube bottom flanges instead of the usual spiral wound gaskets. CMG gaskets have excellent sealing capability at very low bolting loads. This explains why no tubes were found leaking during the nitrogen pressure test. Furthermore, with the CMG gaskets the core blowout resulted in a much bigger leak than if a spiral wound gasket had been used. Consequently, we have changed our standard to specify CMG gaskets are not to be used in raised face piping flanges.

A Management of Change ("MOC") had been processed the previous year to allow standard 3 inch ASME flange spiral wound gaskets on the top tube flanges of A-train if custom sized spiral wound gaskets would delay the startup. This was incorrectly translated into a recommendation to use all standard 3 inch ASME flange gaskets on both top and bottom tube flanges during the Turnaround without an adequate engineering review.

A generic Engineering Work Order ("EWO") had been issued to replace furnace tubes. This EWO did not include information for gaskets, bolts, or torque requirements for the uniquely designed tube end closure blind flanges.

Causes

1. The bottom flanges were not tightened adequately to ensure tight closure while in operation.
   - The evidence indicated the contractor assigned to tighten the bottom flanges did not execute their own standard work practice.
   - The contractor did not follow a strict policy to never leave a flange joint once they start hand tightening until final tightening is completed.
   - There were workmanship issues on flange closure details including nuts that were installed upside down, old stud and nuts re-used, lack of thread lubricant, use of short studs, improper use of the impact gun, and lack of thread engagement into the nut.
   - The Investigation Team also found two completely loose studs on one flange and one completely loose stud on three other flanges that were worked on during the Turnaround and were all outside the fire zone.
2. The Quality Control processes used were inadequate to identify the compromised enclosure tightening.
   - The shutdown contractor appeared to have not executed their own QA process to audit that their enclosure practice was utilized or that the bolts were insufficiently tightened.
   - Our QA process relied on the contractors audit practice with only a spot audit of the contractor's records.
3. Engineering failed to recognize corrugated metal core graphite gaskets have excellent sealing capability at very low bolting loads and this design of gasket experiences higher (20%) relaxation than a spiral wound gasket when process temperatures exceed 200°F (93°C) due to softening of the graphite covering layers. This style of gasket can mask a potential leak from an untightened flange using the standard low pressure nitrogen check for leaks by operators prior to startup.

Corrective Actions

1. The contracting company developed a certified training program to familiarize all levels of workers on contractors' procedures.
2. The contracting company developed a bolting training program and competency training assessment for its workers.
3. We enhanced our in-house Quality Assurance process with defined roles and responsibilities.
4. We communicated to Design Engineers that work direction not covered by standards or policies must be included in EWOs.
5. We developed a specific EWO for bolting top and bottom tube flanges in the hydrogen plants.
6. We developed an engineering guideline on expectations of what is required for performing a design review in an MOC.
7. We modified our standard to include a statement that all piping gaskets must be self-centering.
8. We updated our Global Refining standard to not allow corrugated metal core graphite gaskets to be used in raised face flanges.

Key Lessons Learned

- It is important to have proper training in place for critical maintenance procedures, in this case the bolting of radiant tube flanges, and to have specific EWOs for this work if it is not already covered by company standards or policies.
- Critical maintenance work by contractors needs to have an independent audit by company representatives to ensure the quality of their work meets company standards.
- Written guidelines should be in place and followed for clear expectations of engineering reviews of MOCs.
- Written standards for gaskets on hydrogen plant radiant tubes should include the need for self-centering gaskets, and specify that corrugated metal core gaskets shall not be used in raised face flanges.

There have been no incidents of this nature in our hydrogen plants since this fire incident.

Appendix – Furnace PFD

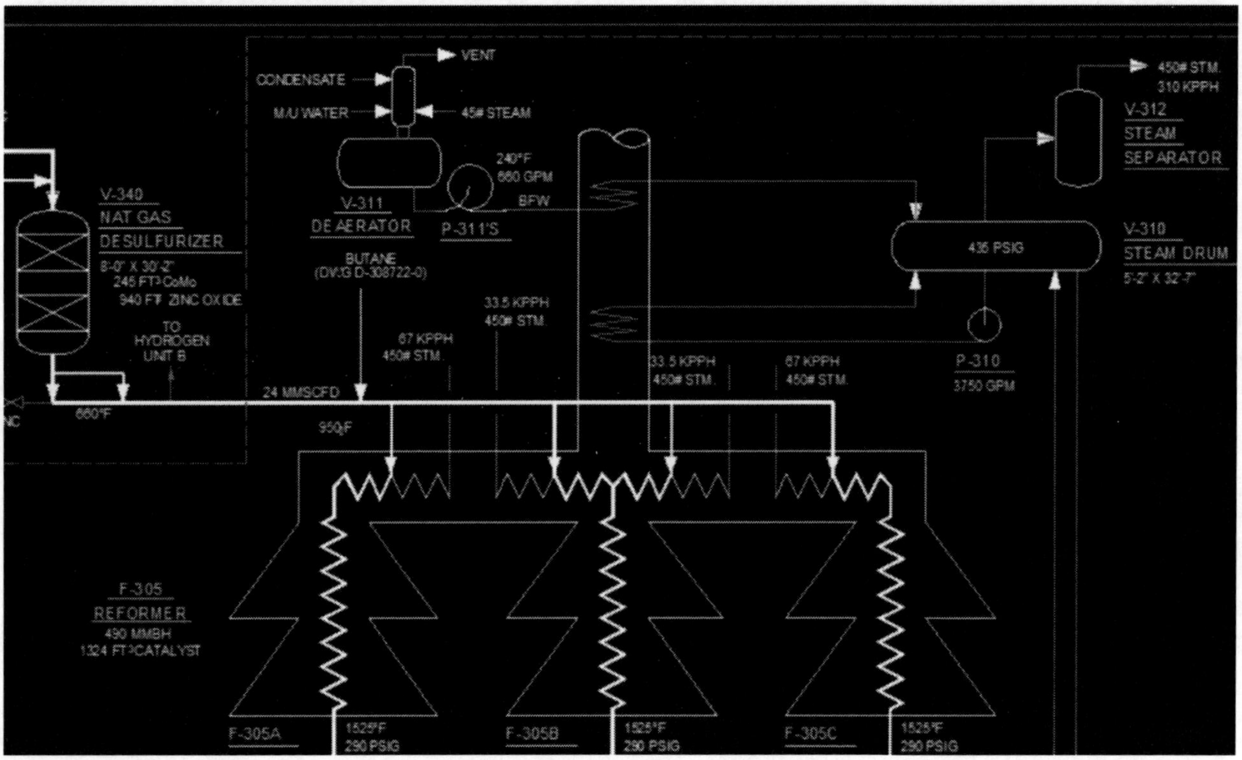

Reliability of Outlet Systems of Primary Reformer for the Safe Operation of a Syngas Plant

The safe and reliable operation of primary reformers is a key issue for syngas plants. Considerable effort has been expended on development of inspection and life assessment methodologies for the catalyst tubes. The hot outlet collection system, typically consisting of pigtails, manifolds, and bull-tees, is equally important to ensuring overall plant safety and reliability.

Pigtails are universally fabricated using alloy 800H or 800HT, and tube design is typically based on API 530, accounting for creep resistance to internal pressure at high operating temperatures. However, other damage mechanisms not accounted for in design may lead to unanticipated failures, including thermal loading/bending, stress relaxation cracking, and nitridation. At least in part, these damage mechanisms can be controlled by fabrication practice and purchasing specification. Furthermore, the role of alloy 800H as the predominant material of choice for pigtails can be challenged.

Similarly, the material of construction for outlet manifolds and bull-tees is dominated by a single material, CT15C. This alloy is provided by a range of manufacturers using proprietary designations. There are no standardized properties for this alloy, and design is based on properties provided by the manufacturer. As in the case of pigtails, the design basis rarely accounts for the predominant damage mechanisms that lead to failure. The absence of detailed materials data has made fitness-for-service assessment of such manifolds difficult.

This paper describes the damage mechanisms and material limitations that impact the safety and reliability of primary reformer outlet systems and the improvements that may be implemented in a syngas plant for safe and reliable operation.

Keywords: hot reformer outlet system, pigtails, manifolds, damage mechanisms, fitness-for-service

Charles Thomas, Olivia Chung
Quest Integrity

Pavan Ravulaparthy
Quest Integrity

Introduction

The primary reformer in petrochemical plants is a critical asset that enables the conversion of hydrocarbons (e.g. natural gas or methane) into hydrogen-rich synthetic gas (syngas). The main chemical reactions that occur to generate syngas occur within a reformer furnace containing spun cast HP50 tubes filled with a catalyst (referred to as catalyst tubes). The main damage mechanism for catalyst tubes is creep from internal pressure. Literature describes the metallurgical changes that occur for HP50 catalyst tubes, the effect this has on HP50 creep properties, and the life assessment methodology based on this understanding [1][2].

Downstream of the reformer furnace is the hot collection system. This outlet system typically consists of the following equipment:

- Outlet pigtails. small-diameter tubing, known as pigtails due to their often convoluted geometry, is used to accommodate for thermal expansion between the large catalyst tubes and manifolds during start-up and shut-down.
- Manifold
- Bull-tee

A schematic showing an example hot collector system and materials of construction is shown in Fig. 1.

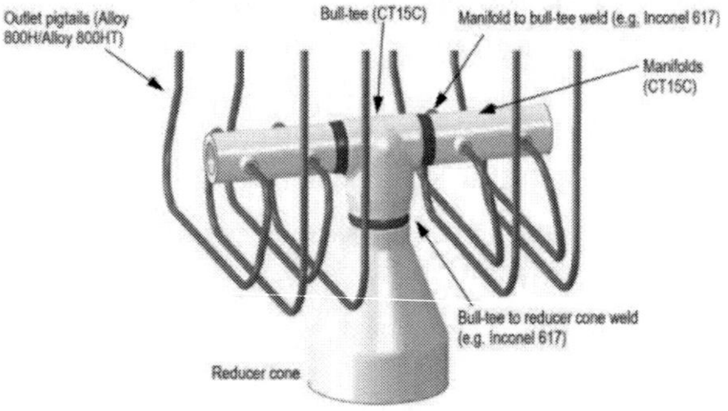

Fig. 1: Example of a hot collector system and representative materials of construction.

The hot collector system is equally important, as the equipment operates at a sufficiently elevated temperature for creep to occur, and complex stresses are generated during operation. Methods used to assess fitness-for-service (FFS) and remaining life of high-temperature components are well established and embodied in codes such as API 579-1/ASME FFS-1 Fitness-for-Service (API 579) [3]. However, metallurgical changes to the materials used for the hot collector equipment and other damage mechanisms not accounted for in the design of the hot collector system present additional aspects to incorporate into the assessment. There is a need to understand the useful remaining life of these components to ensure continued safe operation and reliability, especially for aging plants.

This paper discusses other damage mechanisms and metallurgical factors not accounted for in the design of reformer hot collector systems and the material limitations that affect an FFS assessment.

Outlet Pigtails

Outlet pigtails are predominantly fabricated from Alloy 800H, Alloy 800HT, or their proprietary variants. The nominal material properties for Alloy 800, Alloy 800H, and Alloy 800HT, as specified in ASTM B407, are summarized in Table 1 [4]. The outlet pigtails may also be dual specified as Alloy 800H/800HT due to the similarities in the nominal specification.

Alloy 800 was developed from high value but scrap Nimonic (nickel-chromium) alloys with additions of aluminum and/or titanium and diluted with iron [5]. The Nimonic alloy contained significant levels of aluminum and titanium which effectively became contaminants in the resulting Alloy 800. Nonetheless, these elements are reported to generate benefits [6]. The following 800H and 800HT variants are modifications that offer improved material properties over Alloy 800 [6]:

- 800H: a more restricted carbon concentration and grain size than Alloy 800. Introduced as a variant after it was recognized that the creep and rupture properties of higher-carbon Alloy 800 were superior to those of low-carbon material.

- 800HT: A subsequent variant of with further restrictions on carbon, aluminum, and titanium concentrations was also developed to offer further improvement creep and rupture properties due to the higher carbon content and higher aluminum and titanium content to enable higher levels of precipitation in the relevant temperature range. An ASTM grain size specification of size 5 or coarser is also specified for Alloy 800HT.

Alloy 800 and its variants are solid solution alloys, and the microstructure of these alloys consists of wrought equiaxed austenitic grains with annealing twins (see Fig. 2). Grain boundary and cuboidal orange/pink intragranular precipitates (e.g. Ti[C,N]) are also present in the microstructure.

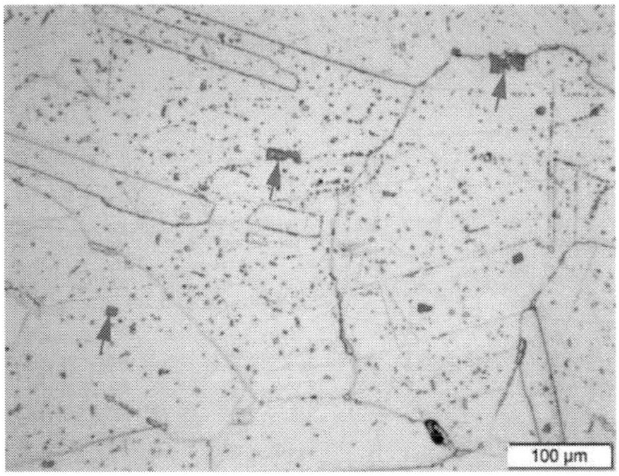

Fig. 2: Representative microstructure. Examples of Ti(C,N) precipitates are shown by the red arrows.

Damage that occurs in outlet pigtails can be grouped as creep, creep fatigue, environmental attack, and relaxation cracking [7]. The main damage mechanism for pigtails covered in design codes and standards such as API 530 Calculation of Heater-tube Thickness in Petroleum Refineries [8] is creep. Classical creep damage accumulates in a pigtail as a change in diameter from internal pressure (i.e. hoop stress). A minimum wall thickness can be determined using standardized material properties, conservative design temperatures, and pressures to obtain a nominal design life to 100,000 hours. In-service creep damage from hoop stress can be monitored by measurement of diametral expansion of the pigtail tube at nominally the hottest part of the outlet tube (see Fig. 3).

An example of diametral expansion of pigtails due to creep was presented by M. J. Smillie et al [9]. In that case study, diametral expansions of up to 13% were measured on the pup-pieces after 7 years of service whereas, the pigtail tubes had significantly lower diametral expansions (e.g. 1.5% compared to 13 % for the pup-piece expansion). The difference in diametral growth was attributed to a variation in grain size between the different pigtail tube sections examined, despite both sections conforming to the specification.

Fig. 3: Observed diametral expansion of a pigtail.

Creep damage in the circumferential direction can also occur in pigtail tubes, where axial stresses are greater than hoop stress. Axial stresses can be from transient thermal stresses experienced during start-up/shutdown, bending loads, and/or other system loads. Additional factors from fabrication (e.g. ovality) may also contribute to locally higher axial stresses. Creep damage from axial stresses can be difficult to monitor between plant shutdowns and will likely involve on-line monitoring methods.

Table 1. Specified properties of ASTM B407 Alloys 800, 800H, and 800HT Error! Reference source not found..

<table>
<tr><th colspan="2">Material property</th><th>UNS N08800 Alloy 800</th><th>UNS N08810 Alloy 800H</th><th colspan="2">UNS N08811 Alloy 800HT</th></tr>
<tr><td rowspan="11">Chemical composition (weight %)</td><td>C</td><td>0.10 max.</td><td>0.05 - 0.10</td><td colspan="2">0.06 - 0.10</td></tr>
<tr><td>Al</td><td>0.15 - 0.60</td><td>0.15 - 0.60</td><td>0.15 - 0.60</td><td rowspan="2">Al + Ti = 0.85 - 1.20</td></tr>
<tr><td>Ti</td><td>0.15 - 0.60</td><td>0.15 - 0.60</td><td>0.15 - 0.60</td></tr>
<tr><td>Si</td><td>1.0</td><td>1.0</td><td colspan="2">1.0</td></tr>
<tr><td>S</td><td>0.015 max.</td><td>0.015 max</td><td colspan="2">0.015 max</td></tr>
<tr><td>Cr</td><td>19.0 -23.0</td><td>19.0 - 23.0</td><td colspan="2">19.0 - 23.0</td></tr>
<tr><td>Mn</td><td>1.50 max.</td><td>1.50 max</td><td colspan="2">1.50 max</td></tr>
<tr><td>Fe</td><td>Balance</td><td>Balance</td><td colspan="2">Balance</td></tr>
<tr><td>Ni</td><td>30.0 - 35.0</td><td>30.0 - 35.0</td><td colspan="2">30.0 - 35.0</td></tr>
<tr><td>Cu</td><td>0.75 max.</td><td>0.75 max</td><td colspan="2">0.75 max</td></tr>
<tr><td></td><td></td><td></td><td colspan="2"></td></tr>
<tr><td colspan="2">Average grain size</td><td>-</td><td>ASTM 5 or coarser</td><td colspan="2">ASTM 5 or coarser</td></tr>
<tr><td colspan="2">Tensile strength</td><td>520 MPa (cold-worked annealed) 450 MPa (hot-finished annealed)</td><td>450 MPa (all conditions)</td><td colspan="2">450 MPa (all conditions)</td></tr>
<tr><td colspan="2">Yield strength</td><td>205 MPa (cold-worked annealed) 170 MPa (hot-finished annealed)</td><td>170 MPa (all conditions)</td><td colspan="2">170 MPa (all conditions)</td></tr>
<tr><td colspan="2">Elongation</td><td colspan="4">30% (all conditions)</td></tr>
</table>

The extent of creep damage and material ductility is affected by the grain size distribution within the pigtail tube's microstructure. For Alloy 800H and Alloy 800HT, an average grain size of ASTM #5 or coarser is routinely specified. The material's creep performance is dependent on the grain size distribution to achieve this specification and the relative uniformity of the grains. Coarse-grained regions have better creep strength but lower creep ductility (i.e. "creep brittle" behavior) and are unable to deform without voiding and/or cracking. Fine-grained regions have better creep ductility and can deform with less voiding. An example of a mixed grain size distribution of an Alloy 800H/800HT pigtail cross section that meets the ASTM average grain size specification, and the observed circumferential creep damage is shown in Fig. 4. This wide variation in grain size frequently leads to cracking either due to the poor ductility of coarse grains or the concentration of stain into fine-grained zones. In addition to the requirement to maintain grain size coarser than ASTM grain size #5, it is recommended to limit grain size variability.

The variation in grain size is best managed during fabrication. Grain growth response during heat treatment is strongly influenced by the extent of prior cold work. Cold bending will automatically generate variable degrees of cold work, hence variable response to subsequent solution annealing. Fabrication methodologies should be developed to minimize grain size variation.

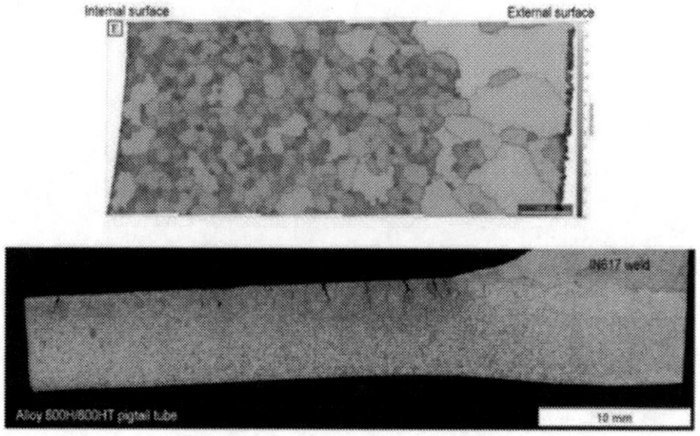

Fig. 4: Example of a mixed grain size distribution and the corresponding circumferential creep damage observed. Average ASTM grain size = 2.99. Evidence of bending is also shown in the pigtail cross section.

The presence of creep voids and cracking can also provide a path for nitridation to occur in Alloy 800H and Alloy 800HT pigtails. Nitridation is the formation of nitrogen-rich precipitates. The following two main forms of nitrides are formed in Alloy 800H and Alloy 800HT materials (see Fig. 5):

1. Titanium carbo-nitrides (Ti[C,N]), formed during fabrication and generally observed as blocky orange/pink precipitates. These are stable precipitates.

2. Aluminum nitrides (AlN) typically formed in-service and generally fine, acicular needle-like precipitates near the external surface. The primary source of nitrogen for in-service pigtail tubes is the external atmosphere.

The formation of nitride-rich acicular precipitates in service have been observed around creep cracks and have contributed to pigtail failures [7][9][10]. Nitridation can reduce ductility and cause embrittlement, although it is acknowledged that published literature validating the effect of in-service nitridation on mechanical and creep properties of Alloy 800H and 800HT is very limited. The rate of formation is a function of the exposure of the pigtail material to air at elevated temperatures, and an initial model of the kinetics that enable nitridation to occur in-service for Alloy 800H has been provided by Young et al [11].

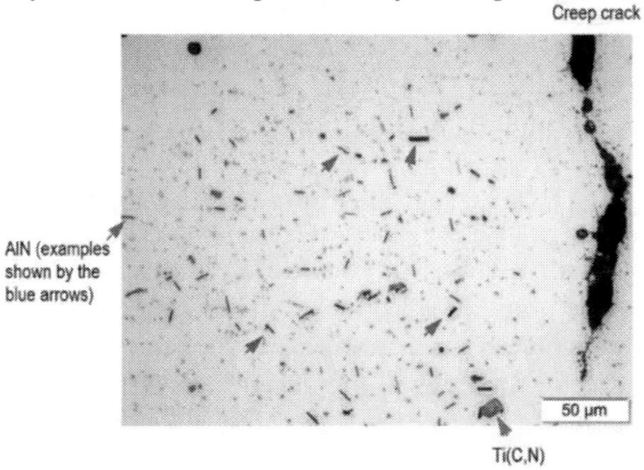

Fig. 5: Representative microstructure showing nitride-rich precipitates. A creep crack is also shown.

Stress relaxation cracking can occur in Alloy 800H and Alloy 800HT at temperatures between 500°C and 750°C due to grain boundary strain [7]. This is generally associated with thick-walled components with an intergranular crack morphology with a Fe-Ni-rich layer sandwiched between chromium/oxide layers (see Fig. 6). Susceptibility to stress relaxation cracking is also increased for outlet pigtail tubes due to the use of Inconel 617 weld filler material; this filler material has relatively improved creep properties than Alloy 800H/Alloy 800HT and, therefore, result in a mismatch in material performance as observed by a band of creep voids near the weld fusion line in the parent material (see Fig. 7). The risk of stress relaxation cracking can be reduced by using a matching filler material in addition to consideration of procurement of material with a thermal stabilization heat treatment and stress relieving/post-weld heat treatment of welds and cold-worked sections. For Alloy 800H, the risk of in-service cracking can be further reduced using a base metal and weld filler material with a combined aluminum and titanium content less than 0.7%.

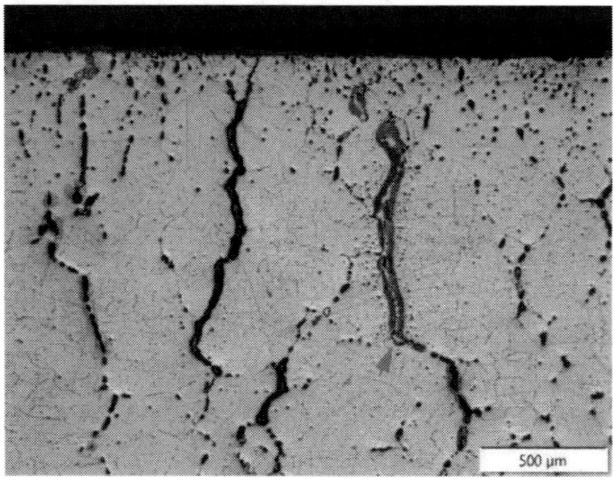

Fig. 6: Example of stress relaxation cracking in an Alloy 800H/Alloy 800HT pigtail, highlighted by the red arrow. Creep voiding and cracking is also shown.

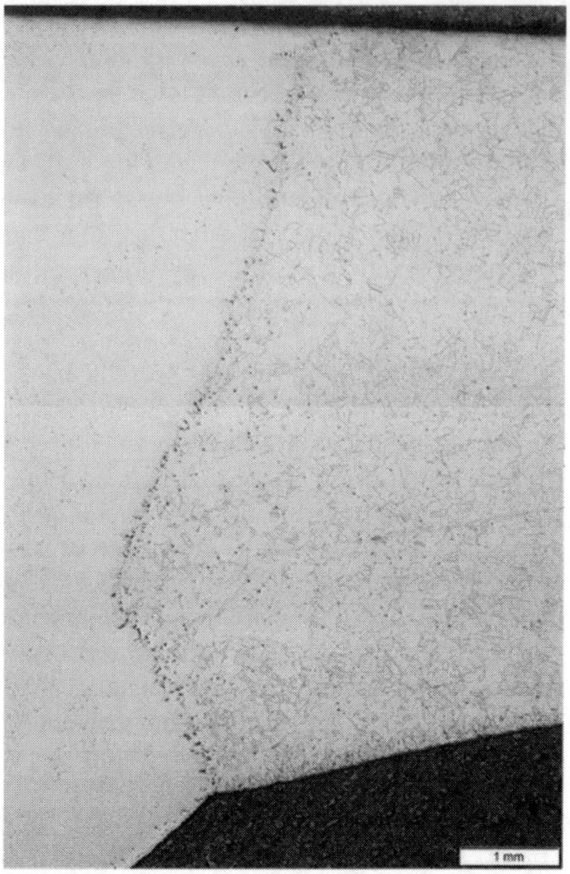

Fig. 7: Creep voiding observed along the fusion line of a pigtail weld joint.

Manifolds and Bull-Tees

By far the most common material of construction of steam reformer outlet manifolds and bull-tees is CT15C, as described in ASME SA-351/SA-351M [12]. (SA-351/351M is identical to ASTM specification A351/A351M). It is typically marketed by a range of suppliers using their proprietary trade designations. This alloy was developed as a cast version of Alloy 800H/Alloy 800HT and followed catalyst tube developments in being further strengthened by the addition of niobium. Niobium adds to strength by the formation of a fine distribution of niobium carbides.

The material is often considered to be a cast version of Alloy 800 and its variants based on its similarity in chemical composition. The adoption of a cast variant saw the use of similar levels of the prime alloying elements, chromium, and nickel. Titanium and aluminum were not used in the cast variant. However, in parallel with the development of spun cast reformer tube alloys involving alloying with small amounts of niobium, CT15C similarly was alloyed with approximately 1% of niobium to improve creep properties by encouraging the formation of a fine distribution of niobium carbide particles. The similarities in the compositions of CT15C and Alloy 800HT, as listed in Table 2, are obvious.

Table 2. Nominal chemical compositions of CT15C and Alloy 800HT.

Element (weight %)	Material Specification	
	UNS N08151 (CT15C) **Error! Reference source not found.**	UNS N08811 (Alloy 800HT) **Error! Reference source not found.**
C	0.05-0.15	0.10 max.
Cr	19.0 - 21.0	19.0 -23.0
Ni	31.0 - 34.0	30.0 -35.0
Si	0.50 - 1.50	1.0
Mn	0.15 -1.50	1.50 max.
S	0.03 max.	0.015 max.
P	0.03 max.	-
Nb	0.50 - 1.50	-
Ti	-	0.15 -0.60
Al	-	0.15 -0.60
Fe	Bal.	Bal.

As for pigtails, the manifold design is typically based on API 530, which allows the calculation of required material thickness based on material creep properties at the design pressure and temperature. However, there are no standardized creep properties for alloy CT15C, and design is based on creep data provided by the relevant manufacturer. Fig. 8 shows a comparison of published stress rupture data for the cast material (based on a typical manufacturer's data) and the wrought Alloy 800HT (based on WRC/API 579 data [3]). According to these data, the expected life of the cast material is an order of magnitude higher than that of the wrought material at typical design stress levels around 10 MPa.

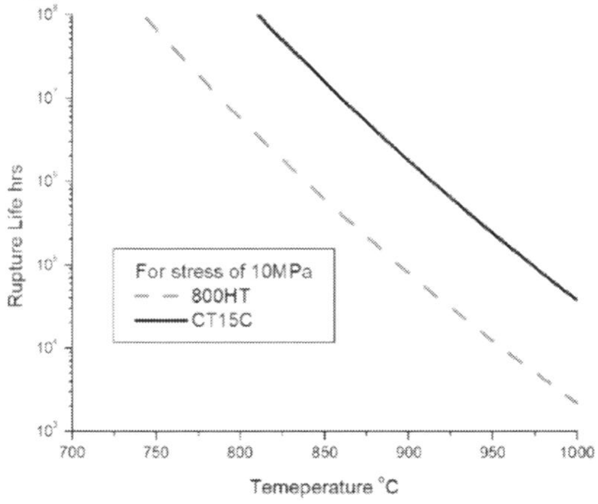

Fig. 8: Comparison of rupture life for Alloy 800HT And CT15C.

Given this improvement in strength and the lower cost of manufacture of the cast material, the benefits of CT15C are clear. This alloy, however, is not without problems. Even after relatively short periods in service, routine inspection at scheduled plant outages frequently leads to the discovery of cracking at the weldments connecting the main manifold arms to the bull-tee (see Fig. 9).

Fig. 9: A typical manifold bull-tee weldment, cracking revealed by dye penetrant and an extreme example of that cracking in cross-section [13][14].

Fig. 9 shows a typical bull-tee and the cracking that may occur. Such cracking has been attributed to bending of the manifold arms relative to the bull-tee [13][15]. Under such circumstances, the weldments act as a fulcrum point, concentrating bending loads at the bull-tee to manifold arm welds. The external bending loads leading to this damage are due to thermal movement of the system caused by both start up and shut down cycles and day-to-day temperature variations.

This cracking has led to a need to determine FFS or a need for repair. Both options are potentially problematic. FFS relies on standards such as API 579, which requires knowledge of material properties such as creep strength, creep crack growth rate, and fracture toughness. These data are not readily available for alloy CT15C. Furthermore, experience has shown that ex-service material properties of CT15C may vary significantly from that provided to enable initial design.

Fig. 10 shows creep and toughness data obtained following laboratory testing of ex-service manifold material from an ammonia plant that had operated at 760°C. Testing revealed that the creep properties had degraded significantly, and the material toughness was extremely low [14]. Subsequent experience has shown that the material properties shown in Fig. 10 were at the extreme low end of the range, highlighting the uncertainty associated with selecting data for use in FFS assessments.

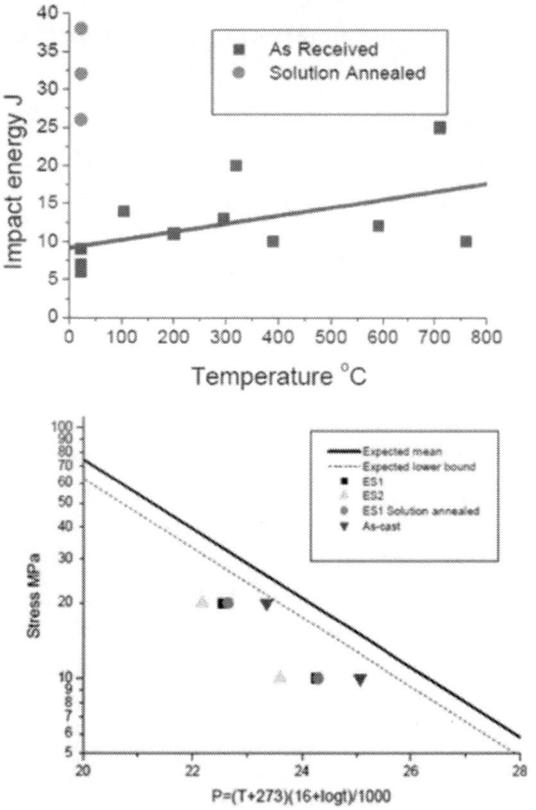

Fig. 10: Charpy impact and creep data for ex-service CT15C before and after solution anneal heat treatments (ES is Ex-service) [14].

Cracking repair has shown itself to be equally problematic. Information regarding manifold repair is available in API 942-A [16]. The difficulty lies in the low temperature toughness experienced in ex-service CT15C (Fig. 10) with Charpy impact energies below 10 Joules. This has led to serious welding problems. Repair welds typically result in further cracking. The data in Fig. 10 shows that some toughness can be restored by solution annealing, which demonstrates that solution annealing can restore weldability. This is not trivial, requiring temperatures greater than 1100°C.

The loss of ductility is due to the in-service formation of G-phase [15]. Analogous to the more familiar sigma phase in stainless steel, G-phase is an intermetallic formed between niobium and silicon. In wrought niobium containing stainless steels, it has been shown that G phase forms most rapidly at approximately 750°C to 800°C [17], which is very close to the outlet temperature of many steam reformers.

Given the significant impact of G-phase in modifying toughness, and possibly creep properties, by absorbing niobium and the resulting uncertainty around the resulting material properties, it is suggested that alloying with niobium could be avoided. A niobium-free alloy would, in principle at least, not embrittle in service due to G-phase. Alloying with niobium has become standard due to the improvement it offers in creep strength as measured on as-cast material. Manifolds are not restricted to a particular thickness and adequate strength while using a niobium free alloy could be obtained by design. It is significant that failure due to pressure generated creep damage is not unknown but is very rare compared to cracking at the bull-tee to manifold welds. Creep data for a niobium-free material has not been found, nor has information relating to embrittlement processes in such an alloy. Nonetheless, the improvement in embrittlement resistance and the variability of ex-service properties may offset the reduction of initial creep strength.

The choice of weld consumable may also affect weld performance. API Technical Report 942-A [16] describes the weld consumables used in manifold welds. Inconel 82 has been used extensively in the past. This weld metal has relatively low creep strength and creep failures within the weld metal has resulted a migration to the high-strength alloys such as 617. Inconel 617 has significantly higher creep strength than the parent CT15C material. The consequence of this appears to be relaxation of weld residual stress concentrated into a narrow zone of parent material adjacent to the weld fusion line. Inconel 617 can lead to a form of stress relaxation cracking that will exacerbate any bending loads caused by thermal movement of the manifold. This damage is also noted for outlet pigtail joints, as previously described. Although using a high-strength weld metal appears to have merit, it is considered more beneficial to use a weld consumable of matching strength to the parent material, thereby more evenly distributing welding residual stress.

Discussion

The choice of material of fabrication for both pigtails and manifolds is often customary. In the case of pigtails, Alloy 800H and/or 800HT and in the case of manifolds and bull-tees CT15C and its proprietary variants are dominant material choices. Both alloys have performed well in their respective roles, but each has unique problems.

In the case of Alloy 800HT and its application in outlet pigtails, the aluminum and titanium content has led to long-term nitridation damage, ductility drop issues, and, consequently, an increased risk stress relaxation cracking. Given the spurious origin of these alloying elements in Alloy 800 and its subsequent variants, alternative alloys such as Alloy 330, are worthy of consideration.

Alloy 330 is effectively Alloy 800 without the aluminum and titanium additions and, hence, may avoid some of the deleterious aging characteristics. It is normally used for high temperature corrosion resistance. However, its creep properties without the benefit of grain size control are similar to those of the basic Alloy 800 (Fig. 11) [5] and if similarly processed to generate a coarse grain size, similar or superior creep properties are likely to be obtained.

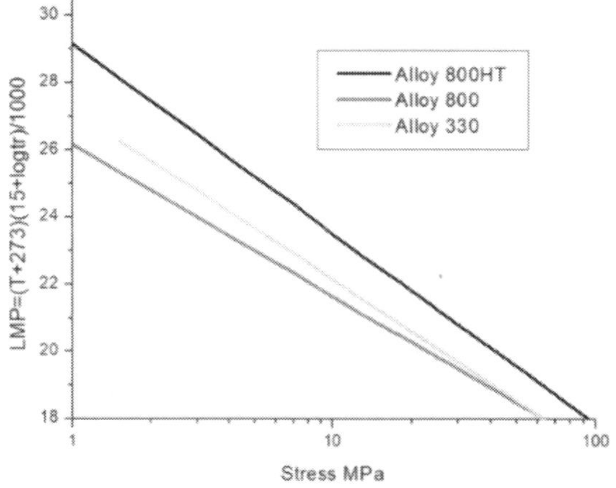

Fig. 11: Comparison of Alloy 800HT, Alloy 800, and Alloy 330 creep properties [5].

Controlling grain size in Alloy 800H and 800HT is of prime importance to obtain the expected properties. This is normally managed by specifying grain size to ASTM grain size #5 or coarser. However, this limitation been found to be oversimplified given that grain size variation within this range can lead to low ductility cracking issues. Grain size variability is mainly caused by fabrication methods, as grain growth during heat treatment is strongly affected by the degree of cold work prior to the solution anneal heat treatment. Fabrication approaches may need to be modified to control this effect; for example, hot bending pigtails in addition to conducting quality checks of the product during fabrication.

In the case of manifold materials, the use of CT15C as the material of choice has merit due to its good creep performance. However, its propensity to aging that leads to variable material properties and embrittlement has caused problems. The prime reason for this aging is believed to be the formation of G-phase in service. It absorbs niobium, removing it from its prime role of improving creep strength, and leads to low temperature embrittlement and weld repair difficulties. Eliminating the source of G-phase (i.e. niobium) would likely require compensation by design to increase wall thickness. This approach could reduce in-service degradation of material properties.

Choice of weld metal is also important. Inconel 617 has found favor when it comes to counteracting creep failures encountered in weaker weld metal. It is suggested, however, that stronger is not necessarily better, and the increased strength results in high residual stress and consequential relaxation cracking in the adjacent parent metal. A matching weld metal is more likely to provide long-term, reliable performance.

The material issues highlighted are rooted in attempts to improve high-temperature performance by alloying (i.e. aluminum and titanium in Alloy 800[H/HT] and niobium in alloy CT15C). These alloying additions, unfortunately, result in long-term material property instability. When selecting alloys for stability, it appears worthwhile to keep the material as simple as possible unless absolutely necessary to maximize strength.

Summary

An overview of damage mechanisms and metallurgical factors not accounted for in the design of reformer outlet systems has been presented. Although the equipment operates at sufficiently elevated temperature for internal pressure-driven creep to occur, other damage mechanisms and material limitations include creep from axial loading, stress relaxation cracking and nitridation for outlet pigtails, and degradation in creep and toughness properties attributed to the formation of G-phase for manifolds and bull-tees. A mismatched choice of weld consumable to join outlet pigtails, manifolds, and bull-tees could lead to stress relaxation cracking.

Limitations on the available information to support an FFS assessment have also been described to improve understanding of challenges associated with determining remaining life for reformer outlet systems. In addition, considerations for minimizing and managing in-service damage, including possible material substitution, have been provided to help ensure continued safe operation and reliability of reformer outlet systems.

References

[1] C. W. Thomas, "Factors Influencing Life of Reformer Furnace Tubes". Materials at High Temperatures, September 2007, Vol. 24 (3), p 233 – 240.

[2] C. Thomas, T. Hill, and T. Anderson, "Reformer Life Assessment - Development of a Materials Database". Presented at 53rd Annual Safety in Ammonia Plants and Related Facilities Symposium, San Antonio, September 2008.

[3] American Petroleum Institute and the American Society of Mechanical Engineers, API 579-1/ASME FFS-1 "Fitness-For-Service". December 2021.

[4] ASTM International, ASTM B407-08a (2019) "Standard Specification for Nickel-Iron-Chromium Alloy Seamless Pipe and Tube". 2019.

[5] S.F. Pugh, Book Review: Alloy 800. Proceedings of the Petten International Conference: Edited by W. Betteridge et al., North-Holland Publishing, Amsterdam (1978). Annals of Nuclear Energy, 1979, 6 (9-10)

[6] Anonymous, SMC-047 "INCOLOY alloy 800H & 800HT", 2004, Special Metals Corporation.

[7] C.W. Thomas and M.J. Smillie, "Failures of pigtails: Why they happen and how to avoid them". Ammonia Technical Manual, 2011, Vol. 52, p 101 – 111, AIChE.

[8] American Petroleum Institute, API 530 "Calculation of Heater Tube Thickness in Petroleum Refineries". December 2021.

[9] M.J. Smillie, O. Olatte, C. Thomas and T. Levi, "Investigation of in-service damage accumulation of Alloy 800HT methonol reformer outlet tubes". Materials at High Temperatures, 2007, Vol. 24, p 241 – 248.

[10] J.J. Hoffman and G.Y. Lai, "Metallurgical Evaluation of Alloy 800HT Pigtails". Paper NACE-05402, presented at Corrosion 2005, Houston, April 2005.
[11] A.M. Young, M.V. Kral, C.M. Bishop, "Time-Temperature-Precipitation Relations for Nitrides and Evaluation of Internal Oxidation Theory for Nitridation of Austenitic Stainless Steel". Metallurgical and Materials Transactions A, 2020, 51(9), p 4456 – 4470.
[12] American Society of Mechanical Engineers, SA-351/SA-351M "Specification for Castings, Austenitic, for Pressure-Containing Parts". 2023.
[13] C.W. Thomas, D.M. Knowles, D.J. Keen and Q.Z. Chen, "In Service Embrittlement of Cast 20Cr32Ni1Nb Components Used in Steam Reformer Applications". Pressure Vessels and Piping, 2004, Vol. 81, p 499 – 506.
[14] C.W. Thomas, D.M. Knowles, D.J. Keen, "Embrittlement of Cast Reformer Outlet Manifold Components". Ammonia Plant Safety, 2003, Vol. 44, p 136-143, AIChE.
[15] J.J. Hoffman, G.E. Gapinsky "Properties and Microstructures of Outlet Manifold Components". Ammonia Plant Safety, 2021, Vol. 42, p 10 – 21, AIChE.
[16] American Petroleum Institute, API Technical Report 942-A "Materials, Fabrication, and Repair Considerations for Hydrogen Reformer Furnace Outlet Pigtails and Manifolds". June 2014 (First Edition).
[17] T Sourmail, "Precipitation in Creep Resistant Austenitic Stainless Steels". Materials Science and Technology, January 2001, Vol.17, p 1 – 14.

Hydrogen Assisted Cracking (HAC) in the ammonia synthesis loop

Hydrogen Assisted Cracking (HAC) is a known form of hydrogen embrittlement that can occur in carbon or low ally steel when operating at high hydrogen partial pressure and metal temperature above 480 °F (250 °C). Welded areas and specifically heat affected zones are particularly susceptible to this failure mode. At high temperature and high hydrogen partial pressure, molecular hydrogen adsorbed at the surface may dissociate and diffuse into the metal structure resulting in local reduction of ductility and possibility for cracking. Temper embrittlement as well as nitriding can also play a role in promoting the phenomenon if the operating temperature is high enough. This degradation mechanism is extremely critical as cracks can propagate with high growth rate resulting in a leak. In this paper some recent case of HAC will be presented by analyzing the main contributing factors, possible mitigations actions during the design as well the most effective inspection techniques that can be applied in service. The methodology used for establishing the Minimum Pressurizing Temperature (MPT) curves will be also described with reference to API RP 934-F.

Stefano Cavalleri, Giulio Paci
Yara International

Richard Bosselaar, Cornelis Hamelink
Yara Sluiskil

Introduction

Hydrogen Assisted Cracking (HAC) is a known degradation that may occur in low alloy steel welded components in hydrogen service. HAC is normally seen in the Heat-Affected Zones (HAZ) of welds of equipment operating at high hydrogen partial pressure, i.e. >7 MPa (1015 psi) and operating wall metal temperatures >250 °C (482 F). This phenomenon can be considered a type of hydrogen embrittlement at high temperature resulting in formation of cracks. Fracture appears in most cases transverse to the welds and may be both transgranular and intergranular. Crack initiation and crack propagation rates depend on the hardness of the welds, material composition and fluid type. Presence of ammonia can be a detrimental factor for the crack nucleation. The failure mechanism is caused by the combined influence of hydrogen, high hardness nearby welded area and possibly promoted by nitriding as well as internal (thermal) stresses. Normally cracks start at or just below the surface and they mainly grow during start/stop cycles. The damage evolution can be summarized as follows: at high temperatures and pressures, such as in the synthesis section of ammonia plants, molecular hydrogen adsorbed on the steel surface may dissociate into atomic hydrogen which can diffuse through the lattice structure and become trapped. Macroscopically the presence of hydrogen in metals generally leads to reduction in fracture energy and ductility promoting brittle failure when the metal temperature is low enough e.g., during plant start-up and/or shutdown. Critical factors that influence HAC are:

1. Hydrogen partial pressure
2. Temperature
3. Hardness in the Heat Affected Zone (HAZ) of the welded joint
4. Effectiveness of Post Weld Heat Treatment (PWHT)
5. Presence and extent of nitriding
6. Yield strength (YS): experience indicates that high YS steels are more vulnerable to HAC
7. Material composition specifically when vanadium alloyed materials are used
8. Temper embrittlement (TE)
9. High stress level
10. Number of startup/shutdown cycles
11. Severity (e.g. pressure/temperature ratio) of the startup/shutdown cycles

PWHT will be discussed in the following section of this paper as it has been identified as one of the main causes of experienced failures. In general, its effectiveness may depend on whether the treatment is performed in a furnace or by local heating and how the metal temperature is controlled in the different position of the piece. Effectiveness of PWHT can influence the level of the hardness through the weld and in the circumference of the weld, hence the susceptibility to HAC. Nitriding could have a great impact on the nucleation of cracks that then propagate for HAC. When ammonia is present e.g., in the synthesis loop where the operating temperature could be 300 °C (572 F) or higher, nitriding can occur at the internal surface resulting in a brittle layer that can crack during thermal transients. Investigations have shown that the weld metal is more vulnerable to nitriding than the base material, so after crack initiation the crack growth rate in weld metal is expected to be higher compared to the crack growth rate in the base material.

By mitigating the critical factors, the HAC susceptibility can be reduced although not completely avoided. Typically, low strength materials are applied for the construction, high focus on PWHT is applied and Watanabe-J-factor and Bruscato-X-factor are controlled to prevent temper embrittlement.
A proper inspection plan shall be established to perform high confidence NDT on the equipment exposed to this failure mode. Normally internal accessibility on the equipment in the synthesis loop is not possible and, for that reason, external NDT are mainly applied. As welded joints are the areas where defects can occur, advanced ultrasonic testing (UT) technique should be considered. A combination of UT-Time of Flight Diffraction (UT-TOFD) and Phased Array Ultrasonic Testing (PAUT) are recommended. When internal inspection is possible i.e., during catalyst change on ammonia converters, Wet Fluorescent Magnetic Particle (WFMP) inspection is high effective for crack detection.

Review of reference standards and best practices

API RP 571 [1] can be used for information about Hydrogen Embrittlement (HE) as well as for Nitriding. API RP 934 [2]-[8] extensively reports best practices for construction of equipment in low alloy steels (1¼Cr- Mo, 2¼Cr-Mo, 3Cr- Mo and the V-modified variants). In the API RP 934-A [2] to prevent the Temper Embrittlement, the base material composition for 2¼Cr- Mo, 3Cr-Mo and the V-modified variants is regulated by the *J*-factor, while *X*-factor is used for welding consumables. For the same scope, the transition temperature curve shall be established via the Step Cooling Test. API RP 934-B [3] complements API RP 934-A and specifies additional fabrication considerations that should be observed when constructing a new heavy wall pressure vessel using vanadium-modified Cr-Mo materials intended for hydrogen service at elevated temperature and pressure. The vanadium-modified alloys should have higher strength especially at high temperature and lower susceptibility to

TE and HE, on the other hand they tend to exhibit characteristics such as higher hardenability which results in higher hardness in all areas around welds (weld, base metal, and heat affected zone), and the requirement for higher PWHT temperatures. The vanadium-modified alloys are also known for reduced notch toughness at lower temperatures in the as-welded condition. Inadequate PWHT holding times can also leave welds with unacceptable low toughness areas. Some critical aspects than are present when manufacturing equipment with these materials and extra care by a quality-focused and informed fabricator is needed during construction. To prevent re-heat cracking during fabrication, which can be somehow related to TE in operation, a *K*-factor to limit the content of Pb and Bi in the filler metal has been introduced. Although the PWHT temperatures recommended for vanadium-modified materials is slightly higher than the PWHT temperature for conventional Cr-Mo, the experience shows that the final hardness remains about 20% higher. API RP 934-C [4] covers the usage of 1¼Cr-Mo steels similarly to API RP 934-A. *X*-factor in this case is prescribed for both base material and welding consumables as it has been established that a maximum level of P of 0.007 % would be an effective limit for the control of temper embrittlement. For the same reason the step cooling is not required for these materials.

API TR 934-F [7]-[8] contains a large study on the effect of hydrogen on heavy wall reactors with the scope to provide recommendations for defining the Minimum Pressurization Temperature (MPT) curves during startup and shutdowns to mitigate HAC with specific reference to 2¼Cr-Mo steels. Fracture Mechanics approach for establishing the H concentration and temperature dependency if the threshold stress intensity factor (K_{IH}) is exceeded has been used. The analysis of experimental data and the theoretical model bult to predict the occurrence of HAC suggest the existence of a critical temperature (T_{crit}), above which subcritical crack growth due to trapped hydrogen does not occur. A correlation between concentration of hydrogen, Hydrogen-trap binding energy, hydrostatic stress and operating temperature is proposed and it is able to replicate the beneficial effect of increased temperature in reducing HAC. Considering all the above, a theoretical model called Al-Rumaih-Gangloff (AG) Model from its author's names is proposed in the document to correlate K_{IH} with the total hydrogen concentration dissolved. The model is built in form of master curves (refer to Figure 1) where the different scenarios given by the various curves are governed by the ratio between the diffusible hydrogen at the crack tip vs. the total bulk hydrogen concentration. Different master curves diagrams were developed for moderate and modern high purity 2¼Cr-Mo steels.

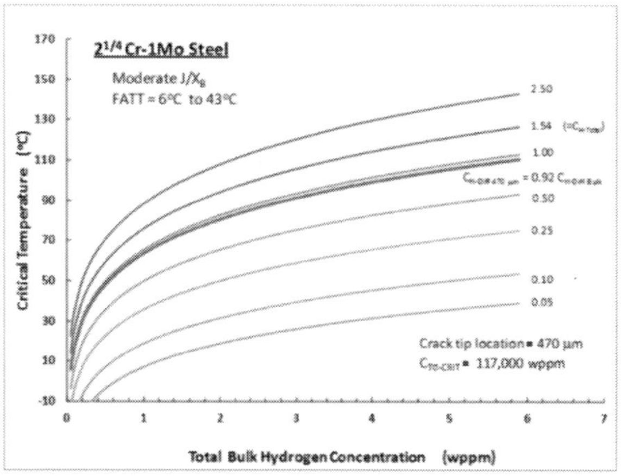

Figure 1. Effect of crack tip diffusible H concentration on the critical temperature for HAC elimination (Master curves from API RP 934-F)

A simplified model to correlate the bulk hydrogen concentration to the threshold stress intensity factor was also proposed by Anderson and Brown (AB) that resulted in a master-K_{IH} curve and temperature threohld (T_{IH}) predictions similar to the forms shown in Figure 1. For this approach, materials were grouped into three categories based on the thermally exposed fracture appearance transition temperature ($FATT_{Thermal}$) which is heavily correlated to the control on the material impurities. The results of the study are then summarized in a proposed architecture for a Fitness For Service (FFS) approach and Minimum Pressurizing Temperature (MPT) curve definition. A Level 1 procedure was proposed by considering following steps:

- Calculation of equilibrium concentration of H at the internal surface of the Cr-Mo base metal or weld metal.
- Correction due to the effect on the reduction of H concentration given by the weld overlay if present.
- Calculation of the threshold temperature (T_{IH}) with the assumption of steady-state H concentration as shown in Figure 2.

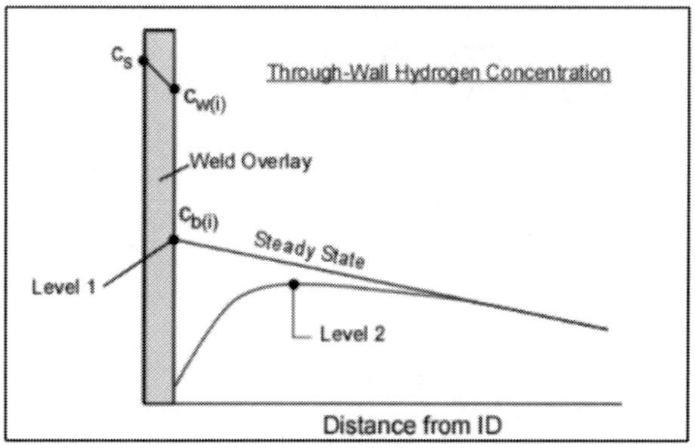

Figure 2. Through-wall hydrogen concentration profile in case of overlayed thick wall reactor (Ref. API RP 934-F).

More complex Level 2 and Level 3 assessments involving transient heat transfer Finite Element (FE) analysis and hydrogen diffusion simulations are also described. For the Level 3 analysis a full 2D or 3D FE model containing the surface crack is required. An example of ideal MPT diagram that could result from the FFS analysis is shown in Figure 3. Two criteria are defined:

- Critical temperature for avoiding subcritical crack growth (recommended to be lower than 20% of Maximum Allowable Working Pressure – MAWP)
- Pressure-Temperature curve for unstable crack growth

WRC 562 [9] bulletin also deal with HAC failure mode as it contains recommendations for establishing the minimum pressurization temperature (MPT) for pressurized equipment based on a fracture mechanics approach consistent with API 579-1/ASME FFS-1 [10] and based on the definition of Fracture Toughness Master Curve.

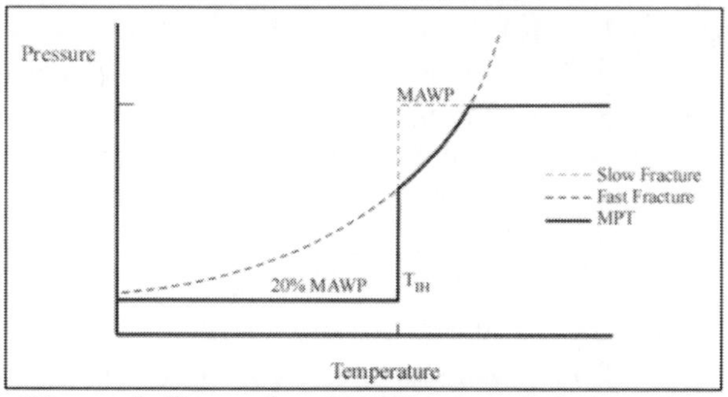

Figure 3. Example of MPT curve definition including subcritical and unstable HAC limits (Ref. API RP 934-F).

The fracture toughness estimation includes temper embrittlement, hydrogen effects on fracture toughness, and considers both fast and slow fracture in hydrogen environments so that MPT curves can be generated for high-pressure hydrogen processing vessels constructed from carbon steel, 1¼Cr-½Mo, 2¼Cr-1Mo, 2¼Cr-1Mo-V, 3Cr-1Mo and 3Cr-1Mo-V with and without austenitic stainless steel cladding. As the fracture mechanics approach is used, the resistance of the material tor cracking is function of the fracture toughness (K_{mat}) which can be defined based on Charpy-V impact test value and ductile-to-brittle temperature transition. The effect of aging, temper embrittlement and hydrogen embrittlement is accounted by properly shifting the transition temperature using estimated Fracture Appearance Transition Temperature (*FATT*) as shown in Figure 4 where the shift on a 2¼Cr-1Mo due to temper embrittlement is shown. Similarly to API 941-F, the slow fracture growth is calculated using the hydrogen availability at inner surface. Different approaches with progressively higher complexity are proposed to estimate the hydrogen concertation from a conservative assumed constant value to a complex FE analysis including crack simulation and hydrogen diffusion in transient conditions. Furthermore, the document also provides a methodology for the MPT curve modifications for flaws detected by in-service inspections.

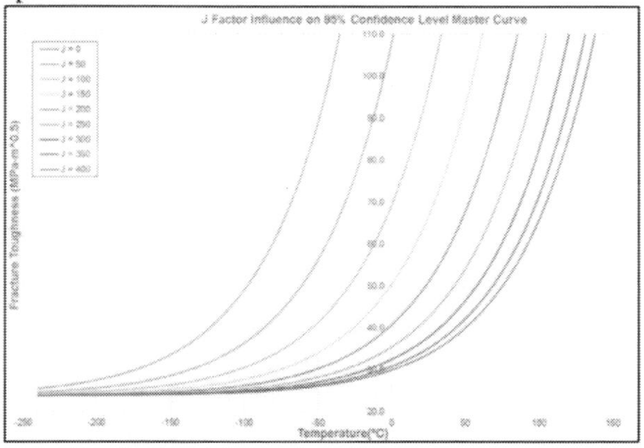

Figure 4. Example of the shift in The Fracture Toughness Master Curve (95%) Of 2.25Cr-1Mo Due to Temper Embrittlement Predicted by J (Ref. WRC 562).

Recent findings related to HAC

This section summarizes the recent HAC related failures occurred in the last years at Yara Sluiskil (NL).

R306: Hot-walled ammonia converter

In 2019 during a scheduled turnaround of one of the ammonia plants, the second synthesis loop converter was inspected for internal corrosion, nitriding and hydrogen assisted cracking (HAC) according to the Risk Based Inspection (RBI) plan [11]. Several cracks were found mainly at the connection between the manway and the top head. Cracks were uniformly distributed along the weld seam with a depth of about 40% of the total thickness. As no repair option was possible, reactor was taken out of service and bypassed in the process. The converter was a hot walled type, and the operating conditions were in the range of susceptibility for HAC (ref. to Table 1). The case was presented during the AIChE ammonia symposium of 2021 [12]. Time Of Flight Diffraction (TOFD) and Phased Array Ultrasonic Testing (PAUT) were applied as inspection techniques to all the reactor welds resulting in several indications as shown in Figure 5. A Root Cause Analysis was performed to identify the main factors contributing to the failure assisted by laboratory analyses performed on samples machined from the top manway where the deepest cracks were found.

Installation year	1996
Shell material	SA-336 Gr. F3V (3Cr-1Mo-0.25V)
Medium	Ammonia synthesis gas: H_2:63%, N_2:19%, NH_3:14%
Operating Pressure	170 barg (2466 psig)
Operating Temperature	380 °C (716 °F)

Table 1. R306 ammonia converter operating data.

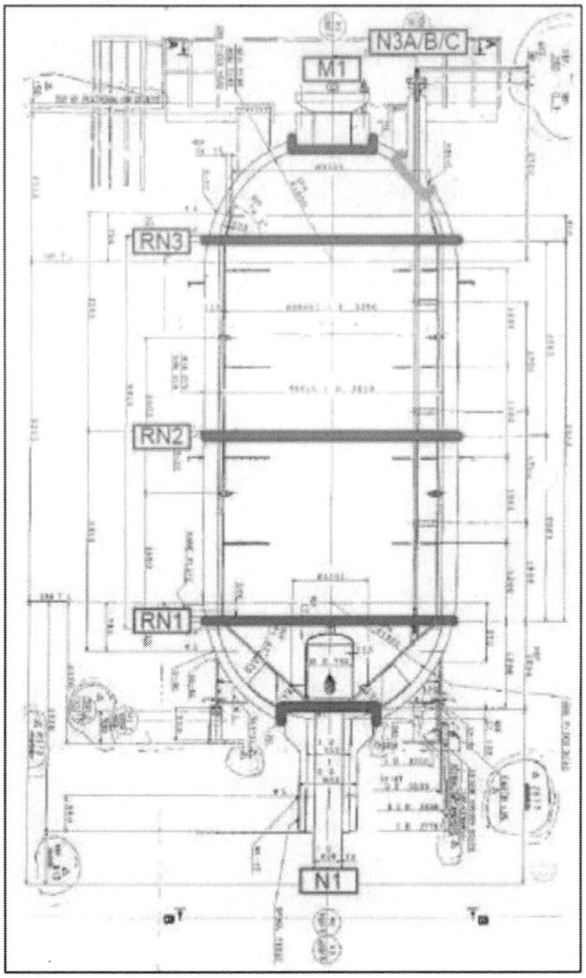

Figure 5. R306: Locations of cracks found during the 2019 turnaround.

A review of operating history as well as a FFS analysis was performed. The results of the investigation are summarized below:

- **Visual examination:** cracks were in radial direction (perpendicular to the weld) with a constant distance of about 100 mm (3.9 in) as shown in Figure 6 that highlights the cracks present in the sample taken from the top manway.

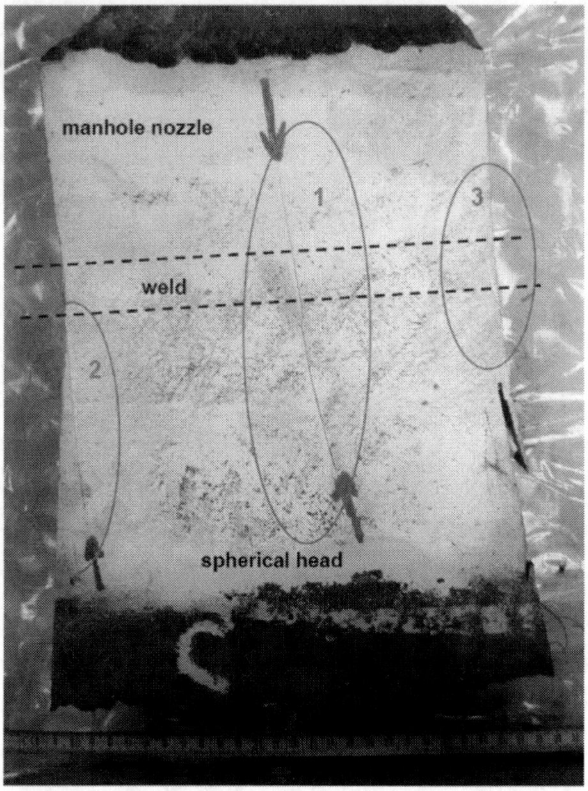

Figure 6. R306: portion of the ammonia converter with indication of the cracks.

- **Chemical analysis:** high content of nitrogen found at crack surface indicated, as expected, that nitriding reaction with the process fluid occurred at the inner surface. This was confirmed by the hardness profile.
- **Impact tests:** the comparison between the ductile-to-brittle transition curves of the "as-built" condition, taken from the material certificates and step-cooling tests, and "aged" material highlighted an evident temperature shift of about 40°C (72°F) and 60°C (108°F) respectively for the average Charpy V-Notch value (about 100 J) and for the upper-shelf CVN at about 150 J as shown in Figure 7.

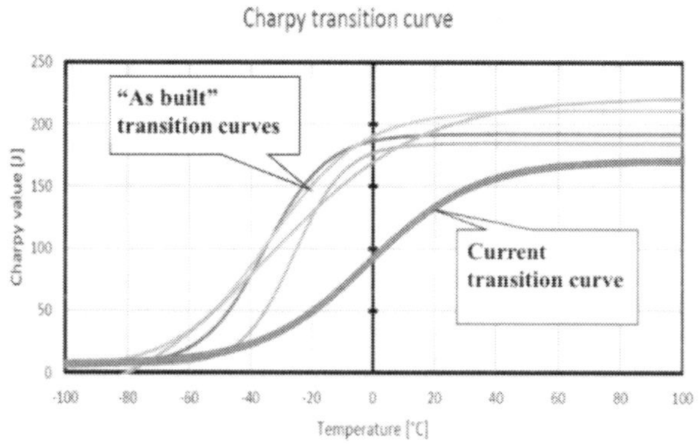

Figure 7. R306: plot of Charpy Transition Curves for Material in "As Built" and "Aged" Conditions.

- **FFS analysis:** a transient thermal-structural FE stress analysis was performed by considering a typical start-up/shutdown cycle. Level 2 assessment for crack like flaw according to API-579 was applied and the Failure Assessment Diagram (FAD) were developed. The criterion for unstable crack growth was met, however slow stable crack growth check according to WRC 562 pointed out that potential situations of operating conditions at start-up/shutdown were above the threshold temperature vs. pressure curve occurred more than incidentally.

Based on all the analysis performed, an interaction between nitriding and HE were considered as the failure modes for the cracking. Brittle nitrided layer that cracked during thermal cycles was the initiation for an hydrogen embrittlement crack propagation at the start-up/shutdown conditions were a combination of low temperature and high hydrogen partial pressure was present. V-modified material and particularly its susceptibility to notch effect and crack initiation due to embrittlement has been considered as a major influencing factor. To mitigate the risk for the new converter and the pieces of equipment in the same loop, a Minimum Pressurizing Temperature (MPT) curve was developed by applying considerations from WRC 562 as shown in Figure 8.

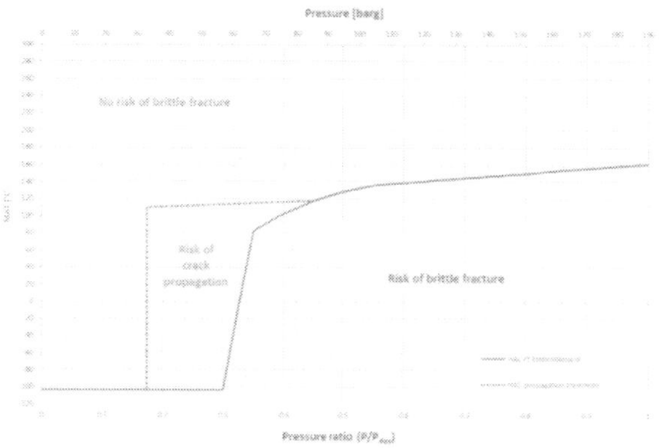

Figure 8. R306: MPT curve developed to mitigate HAC risk.

E416: Synthesis waste heat boiler

In 2020 during a scheduled turnaround of the second ammonia unit, an un-expected crack was found during the inspection of weld between the ammonia converter (R406) and the waste recovery boiler (E416). This could have potentially led to a leakage of syngas with safety risks. The configuration of the waste heat boiler object of the failure with the indication of the cracked weld is shown in Figure 9. As the site have multiple boilers having the same design with one spare unit, a preventive replacement and maintenance strategy is in place. The boiler object of the failure was removed – by cutting the welded connections – in 2012 from the third ammonia unit and installed in 2014 in the second unit after some repair activities at workshop.

The operating conditions are the following:

- Nozzle size: 830 mm (21 in)
- Thickness: 100 mm (3.94 in)
- Inlet Temp tube side: 470°C (878 °F)
- Outlet temp tube side: 321°C (610 °F)
- Shell saturated steam: 106 barg (1537 psi)
- Gas inlet: 380 °C (716 °F)
- Top temperature (R406): 390 °C (734 °F)

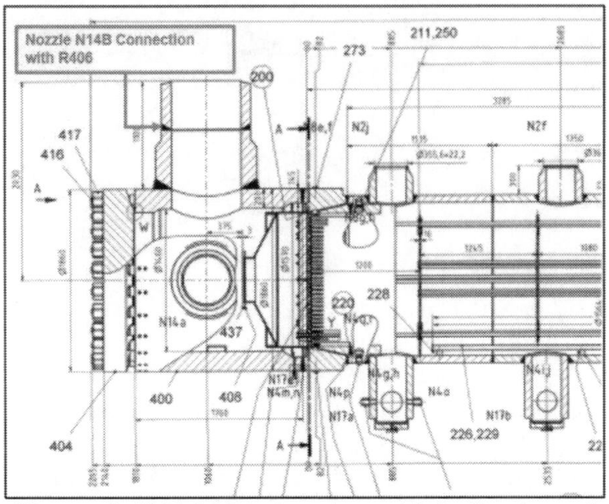

Figure 9. E416: GA drawing with indication of welded connection to the ammonia converter.

The crack found during the inspection was having a depth of about 20 mm (0.79 in) for a total length of 1340 mm (52.75 in) localized in correspondence of a weld repair. The repair and maintenance history indicated that, during the installation of 2014, the connection weld was rejected twice because of weld defects and a repair at the inlet surface involving excavation and rewelding for a depth of about 25 mm (1 in) was performed as shown in Figure 10. Local PWHT according to API standard was carried out. Failure investigation was performed on a sample of the cracked area and the analysis concluded that, from the appearance of the fracture (brittle with secondary cracks), as well as from the results of the metallographic examination, the EDS analyses and the hardness measurements, the nozzle weld has most likely suffered from Hydrogen Assisted Cracking (HAC). The cracking was restricted to the length of the repair weld, which suggests that the repair activity has resulted in unfavorable properties of the weld metal and HAZ that has not been sufficiently improved during the PWHT. There are no indications/signs of any mistakes in the PWHT procedure itself that was used in 2014. However, as the metal temperature was not measured at the internal surface during the treatment, it cannot be confirmed that adequate heating of the material took place and that can explain the high hardness as high as 296 HV that have been measured and which exceeds the maximum allowable hardness for welded low-alloy steel in hydrogen service, which normally assumed as 240 HV (225 HB).

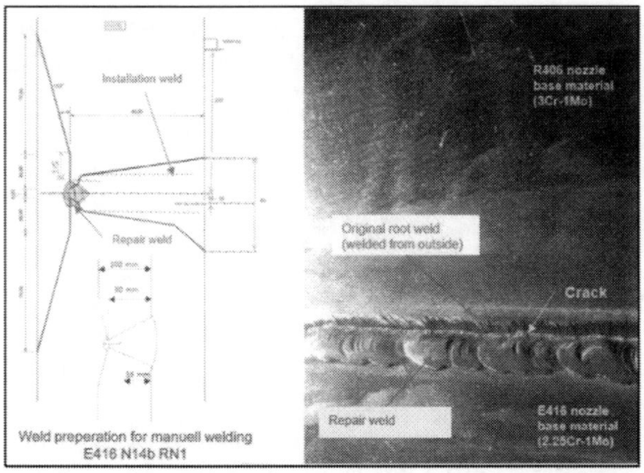

Figure 10. E416: Schematic drawing of the weld repair performed with indication of the welding defect.

As lesson learned and mitigation action for future projects as well as on the field welded connections and repair, the internal Yara PWHT technical standard was updated to include specific requirements for the installation of thermocouples at the inner surface of thick components during the treatment.

E516: Synthesis waste heat boiler

In 2023, during normal plant operation, a hot spot in the insulation of the outlet nozzle of the E516 waste heat boiler connected to the ammonia converter of the third unit (R507) was noticed by an operator indicating a possible leak of syngas. Plant was shut down and the inspection confirmed the presence of a trough-wall crack in the circumferential weld between the outlet nozzle and the outlet connecting pipe. Nozzle size is 24 in (NP 600) with a thickness of 47 mm (1.85 in). The exchanger was in service since 2012. Like for the E416 WHB discussed in previous section, this unit is part of a preventive replacement and repair plan. In 2002 during the out of service period of this asset, a transition piece was installed at the outlet nozzle to facilitate the field installation. The equipment configuration with indication of outlet nozzle is shown in Figure 11. The modification procedure was qualified and PWHT in the furnace for the entire equipment as well as step cooling test on a welding test coupon were performed. Regular turnaround inspection was carried out in 2018 by applying TOFD, PAUT and UT backscatter with no relevant indications.

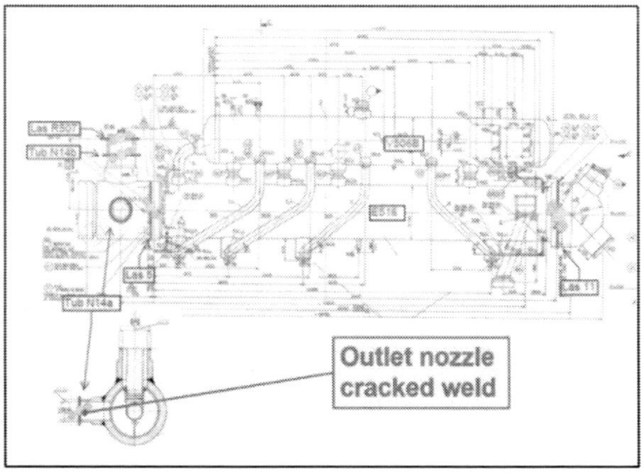

Figure 11. E516: GA drawing with indication of outlet welded connection that failed.

Failure investigation on a sample taken from the nozzle which include the crack surface point out that a combination of intergranular and transcrystalline crack surfaces were present together with micro voids on the intergranular surfaces. High hardness value as high as 390 HV was measured indicating that nitriding occurred at the inner surface as shown in Figure 12. The fact that no indications were found during 2018 inspection indicates that a fast crack growth occurred. The analysis of pressure vs. temperature operating condition since the installation also pointed out that the equipment operated in an unsafe area during transient plant trips with respect to the MPT curve that has been developed after the failure of R306 ammonia converter. Nitrided layer at the inner surface also played a role in the crack initiation. All the above, combined to some uncertainty on the PWHT effectiveness as no internal temperature measurements points were present during welding on the shop as well as at site, led to the conclusion that HAC developed due to a combination of factors that involve nitriding, operating conditions, and uncontrolled fabrication process.

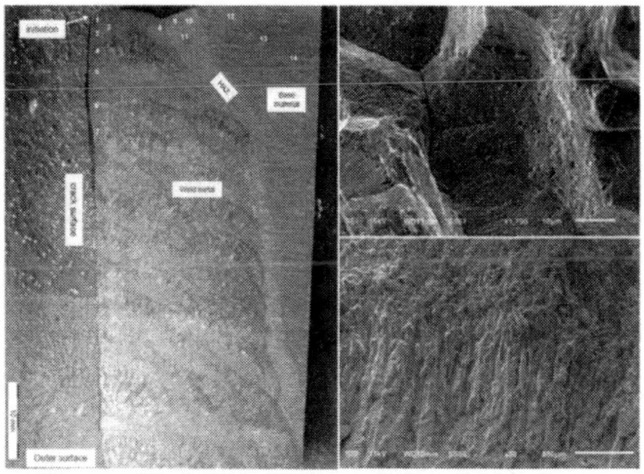

Figure 12. E516: Macro of the cracked area with hardness measurement point and SEM images of crack surface.

Conclusions

Main pieces of equipment in the ammonia synthesis loop where high thickness and low alloy steel welded construction is combined with high hydrogen partial pressure and high operating temperature are subjected to HAC. Technical standards and recommended practices available nowadays provide best practices for construction of equipment in low alloy steels (1¼Cr – Mo, 2¼Cr – Mo, 3Cr – Mo and the V-modified variants) to mitigate the hydrogen trapping effect in the material lattice. Unproper PWHT procedure resulting in non-acceptable hardness levels as well as the detrimental effect of other degradation mechanism such as nitriding proved to have a considerable effect on reducing the resistance to HAC of these pressure equipment. A detailed analysis of the operating conditions and the definition of specific MPT curves considering the amount of available hydrogen together with the geometry of the equipment and operating loads have been identified as the best approach to mitigate the risk of HAC in combination with a proper inspection plan involving advanced NDT techniques although the recent experiences within Yara gave evidence that this degradation could be difficult to properly predict and prevent with in-service inspection. For these reasons the Yara's internal procedure have been updated to increase the quality control on the manufacturing and repair action, such as strong prescription PWHT and hardness check. Inspection plan and recommendation on inspection techniques have been updated accordingly and MPT curves have also been developed for some specific assets. As this degradation mechanism could potentially give severe process safety issues, an extensive campaign for assessing the risk of HAC of all the asset in the ammonia units together with the review of the start-up/shut-down procedure and definition of master MPT diagram at unit level is ongoing for all Yara plants.

References

[1] *Damage Mechanisms Affecting Fixed Equipment in the Refining Industry*, API RP 571.

[2] *Materials and Fabrication of 21/4Cr-1Mo, 21/4Cr-1Mo-1/4V, 3Cr-1Mo, and 3Cr-1Mo-1/4V Steel Heavy Wall Pressure Vessels for High-temperature, High-pressure Hydrogen Service*, API RP 934-A.

[3] *Fabrication Considerations for Vanadium-Modified Cr-Mo Steel Heavy Wall Pressure Vessels*, API RP 934-B.

[4] *Materials and Fabrication of 11/4Cr-1/2Mo Steel Heavy Wall Pressure Vessels for High-pressure Hydrogen Service Operating at or Below 825 °F (440 °C)*, API RP 934-C.

[5] *Technical Report on the Materials and Fabrication Issues of 11/4Cr-1/2Mo and 1Cr-1/2Mo Steel Pressure Vessels*, API RP 934-D.

[6] *Recommended Practice for Materials and Fabrication of 1¼Cr-½Mo Steel Pressure Vessels for Service above 825 °F (440 °C)* , API RP 934-E.

[7] *Impact of Hydrogen Embrittlement on Minimum Pressurization Temperature for Thick-wall Cr-Mo Steel Reactors in High-pressure H2 Service—Initial Technical Basis for RP 934-F, API RP 934-F (Part1).*

[8] *The Effects of Hydrogen for Establishing a Minimum Pressurization Temperature (MPT) for Heavy Wall Steel Reactor Vessels*, API RP 934-F (Part4).

[9] *Recommendations for Establishing the Minimum Pressurization Temperature (MPT) for Equipment*, WRC Bulletin 562.

[10] *Fitness For Service*, API 579-1/ASME FFS-1.

[11] *Risk-Based inspection Methodology*, API RP 581.

[12] Bosselaar et al. "Failure of a Hot Walled Ammonia Synthesis Converter at Yara Sluiskil", AIChE Ammonia Symposium (2021).

Titanium and Zirconium Corrosion Benefits and Reactivity Concerns in Ammonia and Nitric Acid Process Equipment

Titanium and zirconium metals are used for ammonia, nitric acid and ammonium nitrate process equipment because of their excellent corrosion resistance to aggressive oxidizing environments. However, these metals can be reactive in certain conditions, introducing potentially unexpected hazards. This paper discusses the advantages and disadvantages of using titanium and zirconium as a material for process equipment. . Case histories will be shared where reactions have occurred along with ways to mitigate failures of titanium and zirconium equipment.

Daniel J. Benac, P.E., CFEI, Dr. Milo Kral, and Dorothy Shaffer
Baker Engineering and Risk Consultants, Inc. (BakerRisk®)

Introduction

Titanium and zirconium metals are used for ammonia, nitric acid and ammonium nitrate process equipment because of their excellent corrosion resistance to aggressive oxidizing environments. However, these metals can also be reactive in certain conditions and have some unique design and fabrication requirements.

Advantages and disadvantages of using titanium and zirconium as a material for process equipment are discussed in this paper. Examples of case histories are shared where reactions have occurred, and ways to mitigate potential failures of titanium and zirconium equipment are presented.

Stainless Steel Utilization in Nitric Acid and Ammonium Nitrate Service

Certain stainless steels such as 304L have been the workhorse material in nitric acid and ammonium nitrate plants. Selected stainless steels have been used for pipe reactors, spargers, nitric acid cooler condensers, nitric acid heaters, and piping. Even though stainless steels have been used extensively, the accelerated corrosion rates in more severe services have resulted in replacement of equipment at regular scheduled intervals or have required maintenance during unscheduled shutdowns. These added maintenance costs have provided the incentive for companies to consider using more corrosion-resistant materials, such as zirconium and titanium. Although stainless steels corrode faster than titanium or zirconium, stainless steel alloys are not considered reactive metals; that is, stainless alloys are not known to easily ignite or explode under certain conditions, and they have a much lower initial cost.

Stainless steel corrodes at various rates, depending on the nitric acid concentration and alloy, as illustrated for standard Type 304 in very hot nitric acid (190 °C/374 °F) in Figure 1. The corrosion rate generally increases as the acid concentration increases. Type 304 stainless steel has lower corrosion rates for 10% nitric acid concentrations or less, but the corrosion rate is known to accelerate rapidly above about 10%. Special grades of stainless, high chrome alloys and duplex materials have been developed to improve corrosion rates. For example, 304L Nitric Acid Grade (NAG) with higher nickel and very low carbon and low silicon, phosphorus, and sulfur residuals is extremely resistant to intergranular corrosion after sensitization. However, in cases including reboiling attack in cooler condensers and other equipment, titanium and zirconium are notably superior.

At about 50% concentration in boiling nitric acid, the Type 304 stainless steel corrosion rate is as great as 1,000 mils per year (25 mm/year). (1 mil = 0.001 in) This rate is about 10 times that of titanium and about 200 times that of zirconium under these conditions.

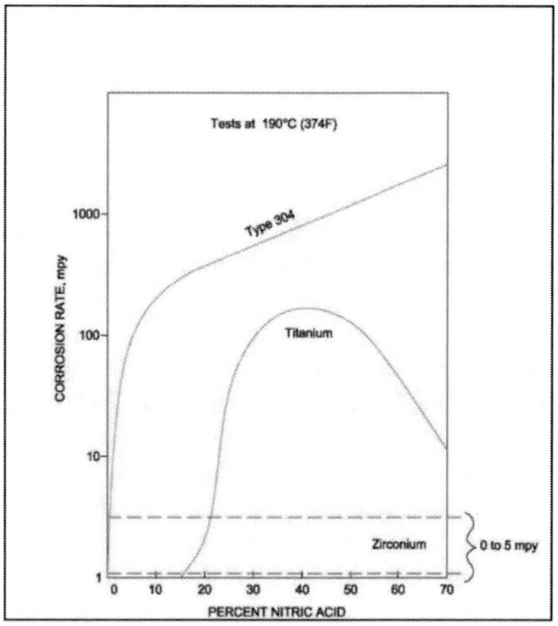

Figure 1. Corrosion plot comparing Type 304 Stainless Steel to Titanium and Zirconium in Hot Nitric Acid [1]

Case Study: Use of Titanium and Zirconium

A nitric acid reboiler was part of a nylon production plant. The temperature of the nitric acid at the inlet of the reboiler was 77°C (171°F), and the reboiler was heated on the shell side by 1.2 MPa (175 psig) steam at 187 °C (370 °F). The previous Type 304L stainless steel tubes were failing about every five years, so a decision was made to move to Titanium Grade 2 tubes, as it had worked well in other parts of the distillation column. [2]

After six months operating with the new tubes, a hydrotest of the reboiler indicated numerous titanium tubes had developed through-wall leaks in the middle of the bundle, due to Nitric Acid Reboiling. The tubes showed little corrosion near the ends.

Although titanium had worked well in the milder conditions of the column, the localized conditions in the reboiler had exposed the nitric acid to higher temperatures such that the nitric acid would boil; see **Error! Reference source not found.**. Titanium is unsuitable for boiling nitric acid in the 30+% concentration range. The tubes showed very little corrosion at each end where acid temperatures and concentrations were acceptable to titanium, but unacceptably high corrosion rates where incipient boiling of the nitric acid occurred. The titanium bundle was replaced by Zirconium tubes of Grade 702 (UNS R60702/ASTM B523) tubes of ASTM B523.

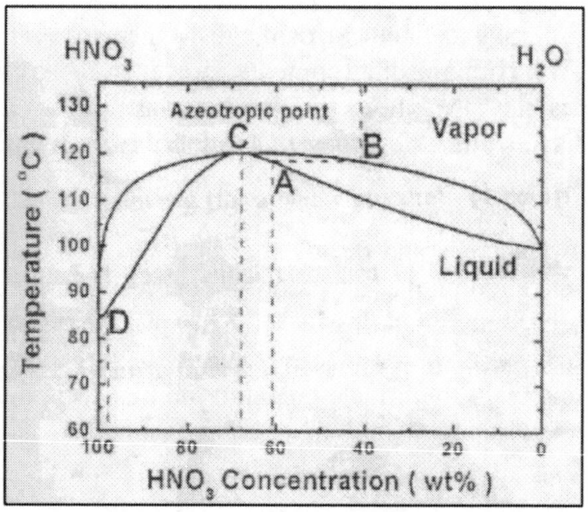

Figure 2. Nitric Acid Boiling Curve

Reactive Metals Applications in NitricAcid and Ammonium Nitrate Processes

The two principal applications of zirconium and titanium are in nitric acid and ammonium nitrate process units. Since zirconium and titanium have superior corrosion resistance in aggressive oxidizing environments, as compared to stainless or duplex stainless steels, they are often considered as replacements in applications where a stainless steel alloy has been attacked in the past. The excellent corrosion resistance of zirconium in nitric acid, even in the presence of most impurities, has been recognized for more than thirty years.

Titanium and zirconium are used in condensers, exchangers, tubes, piping, spargers, and other equipment.

The advantages of using titanium and zirconium as a material of construction include:

- Reduced corrosion rates.
- Reduced need for replacement or frequency of maintenance.
- Reduced shutdown or turnaround costs.
- Improved reliability of equipment before a replacement is required.

The potential disadvantages of using titanium and zirconium for equipment include:

- Requirement for materials' specific design considerations.
- Higher initial cost.
- Increased fire and explosion risk if subjected to adverse conditions.
- Requirement for additional hazard analysis, consequence assessment, and operational procedures (particularly process control and shutdown procedures).
- Need for commensurate instrumentation (per PHA study and SIL assignments) and DCS programmed actions.
- Need for additional staff training on welding, cutting, and handling titanium and zirconium.
- Need for training on metal fire response.
- Special disposal requirements to alert reactive hazards.

Although they have good general corrosion resistance, these alloys can be prone to certain life-limiting failure mechanisms. One concern for these alloys is the pick-up of hydrogen/formation of hydrides.

Case Study: Titanium Hydride Formation

A titanium 3Al-2.5V float was added to an ammonia unit. [3] The titanium float system is a magnetic level indicator that consists of a chamber and a magnet-equipped titanium float that raises and lowers depending on changes in the fluid level. Erroneous level readings and problems with the float had been reported to site supervision. The titanium float, shown in Figure 3 was removed for investigation and was found cracked, as shown in Figure 4.

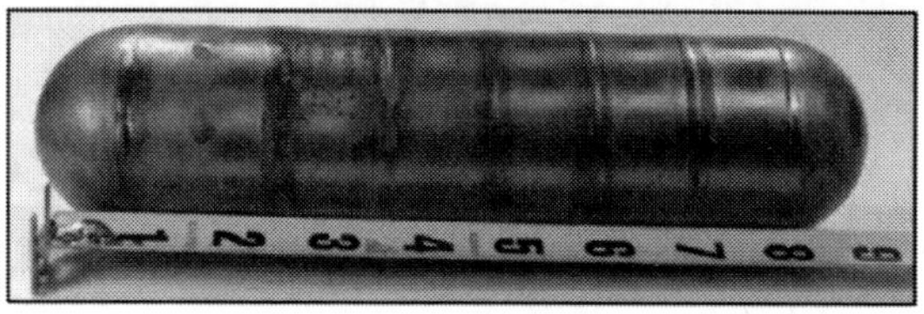

Figure 3. The titanium float

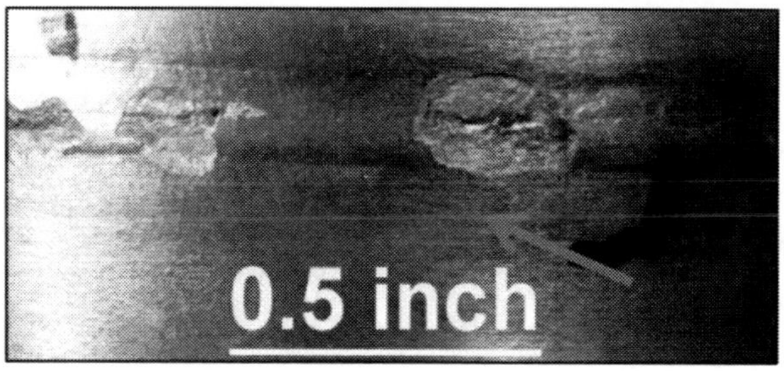

Figure 4. Close-up of the weld showing cracks

The typical operating temperature was 271 °F (133 °C), and the typical pressure was 400 psig

(2.8 MPa). The principal process medium included: Water 10.4%, Hydrogen 54.7%, Nitrogen 18.3%, and CO_2 15.9%.

Metallurgical analysis determined that the cracking and spalling of the titanium float was due to the excessive pick-up of hydrogen. The gas analysis of the weld region showed absorption of excessive amounts of hydrogen at the weld and heat affected zone. The weight percent of hydrogen was as much as 0.479% in the weld compared to 0.015 % in the non-cracked sheet metal.

Hydrogen, when present above specification, can form hydrides in the metal and cause metal embrittlement, hardening, cracking, and spalling.

The most likely source of the hydrogen was due to operation in a hydrogen, steam, and ammonia environment. The areas where the most hydrogen was absorbed were at the weld and weld heat affected zones (HAZ), likely due to residual stresses. These regions were also harder. Figure 5 shows the affected weld and HAZ.

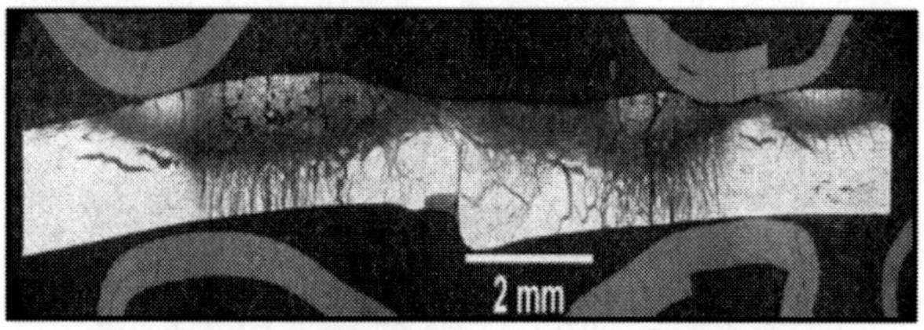

Figure 5. Cross section through the weld and HAZ showing the cracking and hydrides

Reactivity Issues with Titanium and Zirconium

Investment in titanium and especially zirconium can be justified by anticipation of longer life/fewer replacements of components. However, industry fires and reactivity failures in both zirconium and titanium equipment and piping in nitric acid, ammonium nitrate, and other industries raises precautions when fabricating and disposing of the metal.

Fire Concerns

Nearly all metals will burn in air under certain conditions, some burning rapidly. Certain metals, including titanium and zirconium, are referred to as combustible metals [4] because of the ease of ignition of thin sections, fine particles, or molten metal, followed by an extremely aggressive reactive metal fire. Temperatures produced by burning metals are generally much higher than temperatures generated by burning flammable organic liquids. Some metals, including titanium and zirconium, can continue to burn in carbon dioxide, nitrogen, or steam atmospheres, making fire extinguishment difficult. Metal fires are generally denoted as Class D fires, and specialized firefighting equipment must be employed.

Pipes, castings, and other bulk pieces of titanium and zirconium are not usually fire hazards unless they are involved in a large external fire, but they may contribute to Ammonium Nitrate (AN) Explosions under highly abnormal process conditions.

Titanium is practically only a significant fire hazard when finely divided or exposed to high temperatures, as described in NFPA 481. [5] Heat generated while machining, grinding, and drilling titanium may be sufficient to ignite small pieces, so water-based fluids should be used for cooling. Fine particles should be removed regularly and stored wet in a metal container. Carbon dioxide, foam, and dry chemical fire extinguishers are not effective in fighting titanium fires. Special equipment for Class D fires has been developed, including, for example, the Met-L-X system by Ansul. [6]

Case Study: Titanium on Fire!

In a report, a major fire was described that occurred at BASF Ludwigshafen works during startup. [7] There were no injuries. The fire started in a heat exchanger with a titanium tube bundle. Three water cannons were directed at the burning, partially open heat exchanger, and an initial attempt was made to extinguish the fire using a

Class B or C-powder extinguishing vehicle. During the initial phase, 12 cubic meters per minute (2750 gpm) of water was used. The intense heat ignited surrounding cabling, control units, and wooden objects, requiring secondary firefighting efforts. The water application washed most of the titanium tubes out of the exchanger and extinguished them. In 45 minutes, approximately 500 m^3 (132,000 gal) of quench water collected in the plant.

Zirconium Reactions

Zirconium, like titanium, is a combustible metal, with some differences in physical properties. Again, finely divided zirconium is easily ignited and zirconium dust can even spontaneously combust. Similar to titanium, large pieces of zirconium typically are difficult to ignite unless the oxide surface is removed or damaged. Zirconium burns with little smoke generation. Details of handling zirconium are discussed in NFPA 482. [8]

Like titanium, finely divided zirconium can react. An example of this is a localized high energy reaction that occurred at an nitric acid plant while plugging tubes in a nitric acid cooler condenser with a zirconium tubesheet. Events are as follows:

Case Study: Zirconium Reaction

On May 30, 2007, a localized explosion occurred at a nitric acid plant while plugging tubes in a nitric acid cooler condenser. [9] The shell and tube heat exchanger was out of service at the time, and plugs were being hammered into leaking tubes through a zirconium-clad stainless steel tube sheet. The incident occurred while installing the tenth plug of the day following the typical plugging procedure. Upon impact of the large hammer, a white-yellow flash was observed along with a very loud noise. Three of the personnel standing close to the exchanger immediately experienced ruptured eardrums. The zirconium cladding separated from the tube sheet over most of its area, and cladding parts blew off the tube sheet, landing as far as 119 feet (36.3 m) away. The event was not localized on the tube sheet near the plug impact area, but also occurred in the upper part of the tube sheet, as shown in Figure 6.

A gap was found between the tube sheet and the zirconium cladding at a typical width of about 0.05 inch (1.3 mm). A black powder-like material was found deposited in a flow pattern inside some of the tubes and on the exterior of some of the newly installed plugs. The deposit pattern indicated that a combustion event had occurred, and that the deposits were from the release of high-pressure gas/particle "exhaust" through failure locations. Analysis of the powder showed it to be zirconium oxide and zirconium hydride nitride, products of oxidation reactions between zirconium and nitric acid.

Figure 6. Debonded tubesheet and failed Nitric Acid Exchanger

Factors that contributed to this event include:

- improper welding on the zirconium tubesheet,
- a reactive zirconium surface, and
- the use of hammer impact to drive plugs to seal leaking tubes.

Recommendations to reduce the risk of this type of event recurring include:

- Using welding methods to avoid debonding of the zirconium cladding.
- Conduct periodic inspections of the clad tubesheets to ensure that they have not debonded.
- Consider alternative methods of driving plugs

into zirconium-clad tubesheets other than by hammer impact to reactive metals.

Other Potential Failure Causes

Other failures include hydrogen fluoride (HF) attack in nitric acid plants, in process locations where condensation occurs. While HF is uncommon in ambient air, refrigerants are often used in processing plants for cooling. These may contain fluoride-containing chemicals that will react across a nitric acid converter to form fluoride compounds that are highly corrosive to Zirconium. Note that caution should be taken to monitor for refrigerant or other chemical leaks where air compressor intakes can introduce fluoride compounds into the process. Chemical uptake into process air should be considered during the site's process hazard analysis, including contaminants in raw materials.

Failures may also occur due to mechanical design and fabrication issues where differing properties for titanium and zirconium (including expansion coefficients, modulus of elasticity, weld characteristics, and hardness values) are not taken into consideration. See Appendix A for some general guidelines.

Summary

The improved corrosion resistance of titanium and zirconium over stainless steels makes these materials attractive options for equipment and piping. However, there have been enough documented incidents involving the reactivity of titanium and zirconium to warrant careful consideration and added controls when these materials are used. Special procedures and added controls should be seriously considered if titanium or zirconium are to be used in plants under certain conditions, for example:

- Avoid thin materials, such as packing, where a greater surface area to weight ratio promotes localized reactions and fires are prone to be sustained.
- Carefully avoid corrosion or conditions that would promote dust or formation of gritty corrosion deposits.
- In areas where oxidizing conditions occur (such as ammonium nitrate spargers or piping low points), avoid accumulation of corrosion products by ensuring grit and dust can be reliably evacuated.
- Take care to check for localized conditions that might invite failures.

Zirconium and titanium are very useful, but all materials have their limitations, as illustrated by these incidents. Good design and fabrication procedures are essential to ensure the significant investment in these specialized materials results in safe production over many years of operation.

Acknowledgements

The shared information from the 2020-2023 Ammonia & Fertilizer Joint Industry Program (JIP) facilitated by BakerRisk has contributed to the content of this paper. [10] The JIP is a successful collaboration among a significant group of ammonia, nitric acid, and ammonium nitrate producers to share materials and corrosion knowledge based on investigations following incidents.

References

1) T.-L. Yau, "Zirconium for Nitric Acid Solutions," *Industrial Applications of Titanium and Zirconium*, ASTM STP 917, ASTM 1986, p 57.

2) M.M. James, "Effect of Process Conditions on the Corrosion Performance of Reactive Metals," Paper No. 3454, Corrosion 2003.

3) N. Cherolis, D. J. Benac, D. Shaffer, J. Redman, "The Value of Investigating and Trending Minor Failures to Prevent Major Incidents" Journal of Failure Analysis and Prevention, June 2018, Vol. 18.

4) *Fire Protection Handbook,* A.E. Cote, Ed., 18th Ed, NFPA, Section 4/Chapter 16, "Metals."

5) NFPA 481: Standard for the Production, Processing, Handling, and Storage of Titanium," 2000 edition.

6) ANSUL Website: https://www.ansul.com/

7) Cabelka, U., "Titanium Fire – A Seldom Incident in the Chemical Industry," BASF AG: Works Fire Depart: Reports: Titanium fire, 1999.

8) NFPA 482: Standard for the Production, Processing, Handling, and Storage of Zirconium," 1996 edition.

9) "Terra Industries Inc. Nitric Acid Cooler Condenser Explosion," Presentation at ANNA 2007 meeting, Park City, Utah.

10) 2023 Ammonia & Fertilizer JIP BakerRisk Project No. 01-07250-004-20 Identification of Damage Mechanisms and Mitigation Options for the Nitric Acid and Ammonium Nitrate plants, April, 2024.

Appendix A - General Guidelines when using Titanium and Zirconium

Below are twelve practices to prevent titanium tube-bundle fires. [A1]

1. Hot work policy and permit procedure.
2. Prohibit hot work on equipment containing combustible metal tube bundles.
3. Mount permanent signs on equipment containing combustible metal tube bundles.
4. When significant hazard may be present, conduct a hot work hazard analysis.
5. As part of the hot work permit process, identify susceptible combustible metals.
6. When combustible metals are present, include a checklist for combustible metals on the hot work permit.
7. Train personnel who issue and perform hot work, on hot work fire prevention, including causes and hazards of metal fires.
8. Consider cold cutting procedures as an alternative to hot work.
9. Manufacturers advise that individual titanium tubes can be welded, or torch cut, but there is a severe fire hazard associated with hot work performed on tube bundles.
10. Equipment designers should consider issues with combustible metal equipment over the whole life cycle.
11. Include hot work as potential causes of high temperature deviations in heat exchanger HAZOPs.
12. Ensure that the hot work policy is actually implemented: required management commitment and tireless vigilance at all levels.

Generally, for titanium and zirconium equipment, to minimize the risk of equipment failure, or fire or explosion, the following guidelines should be considered (guidelines apply to both zirconium and titanium):

- PHAs should address the potential increased risks of using titanium and zirconium equipment.
- Proper fabrication practices and approved weld procedures should be in place for welding on new and existing equipment in order to avoid weld contamination and possible brittle fracture.
- Zirconium may at times be bonded or welded to another material, such as stainless steel. These welds may appear to be visually acceptable, but due to the formation of iron-zirconium constituents, the corrosion behavior and mechanical properties can be degraded. [A2] Fusion welds and bonds to non-zirconium alloys should be avoided in potentially corrosive conditions.
- Stainless steel equipment that is changed in order to use titanium or zirconium usually requires a redesign for the specific characteristics of the new materials and to reduce cost. Often, the large corrosion allowances that are necessary in stainless applications are reduced in the redesign, resulting in thinner component specifications. These thinner tubes and walls may be more subject to physical wear, such as from vibration or baffle cutting.
- Accelerated metal loss rate can occur due to erosion-corrosion, high flow rate, or turbulent conditions, compared to static corrosion rate predictions. Such locations should be monitored.
- The presence of finely divided zirconium should be avoided, since it may easily or even spontaneously ignite in air.
- Storing of fine turnings and chips should be minimized because of potential reactivity. If collected, chips and turnings should be kept in non-flammable containers, stored in isolated areas, kept separate from other flammables, and covered with water or oil.
- Do not dry grind zirconium to remove scale or corrosion. A water-based lubricant should be used if grinding must be performed, to quench any sparks. [A3]
- Surfaces should be cleaned of any fine zirconium dust that may have collected.
- Training should be performed on reactive metal fire prevention and firefighting (Class D fires). Fire prevention equipment should be in

place when working on titanium and zirconium equipment. If a fire starts, do not apply water or ABC or BC-type fire extinguishers. Instead, use dry sand or a Class D fire extinguisher. Additional extinguishers may be needed for other types of fires in the area.

- Avoid operations that expose zirconium to high temperatures (such as welding or flame cutting).
- Avoid exposing zirconium to strong oxidizers (such as fuming nitric acid, nitrogen tetroxide, or liquid oxygen).
- Avoid exposing zirconium to decomposing AN or the reaction products.
- Consider the possible hazards of reactive or pyrophoric zirconium surfaces or corroded regions. Zirconium operated under chemical reducing or anaerobic conditions are especially suspect. This may include opening a vessel or pipe to the air. Possible reactive surfaces should be steamed to reduce their reactivity.
- Stagnant, deadlegs, and low-flow rate locations, that can also cause deposits to form and allow for a concentration of material, should be monitored for high corrosion rates.

Maintenance Practices

Below are some maintenance items for consideration when working with zirconium and titanium:

- Guidelines should be in place for maintenance teams on repair and handling of titanium equipment, including end-of-life disposal fire hazards.
- When zirconium and titanium surfaces are broken up into fine particles from maintenance practices such as grinding and sand blasting, care should be taken in handling the dust and waste products. These particles may be easily ignited or pyrophoric.
- Welding and cutting are possible ignition sources. Hot work permits for zirconium and titanium equipment should include considering these hazards and good practices for minimizing them.
- Contamination from copper and iron particles can result in localized corrosion. AN plants have minimal copper and brass present. To minimize iron particle contamination, it is recommended that tools, such as wire brushes, grinding stones, and files, be dedicated to zirconium use.
- Decommissioning of zirconium and titanium equipment can be hazardous. There have been reports of decommissioned equipment being destructively dismantled and catching on fire. Vendors handling titanium and zirconium should be aware of possible ignition and fire conditions.

References

A1. Mahnken, G., "Watch Out for Titanium Tube-Bundle Fires," Chem. Engineering Progress, April 2000, p. 47.

A2. D. Holmes, Corrosion Resistance and Mechanical Properties of Zirconium-Titanium Fusion Welds'" Corrosion Paper 6210, 2006.

A3. Titanium: A Technical Guide, ASM International 1988, .p. 90.

Repeated Secondary Reformer Waste Heat Boiler (101-C) Tube Rupture and Following Effects

The Ammonia 1A Plant experienced significant failure events on the Secondary Reformer Waste Heat Boiler (101-C) where the tube ruptured during two successive start-up attempts. In both cases, these incidents led to boiler feed water (BFW) flooding in the high-temperature Shift (HTS) catalyst bed. Total replacement of the HTS catalyst was required after the first event, whereas efforts to more slowly heat up following the second event were succesful in restoring the HTS catalyst performance.

This paper describes the failure analysis of the Secondary Reformer Waste Heat Boiler (101-C), the sequence of events of the 2 incidents, investigation and repair of the damage of the equipment, and suggests corrective actions during plant start-ups to prevent re-occurrence in the future.

Panggy Aji Pratama, Fitri Kurniawan, Digna Jatiningsih

PT Petrokimia Gresik, Indonesia

Introduction

Between May and July of 2023, the ammonia 1A plant experienced two occurrences of tube failures in the Secondary Reformer Waste Heat boiler (101-C). In both cases, the HTS catalyst bed was flooded with boiler feed water. In the first instance, a subsequent re-start attempt with an uncontrolled HTS bed pre-heat rate compromised the HTS catalyst, requiring replacement and repair of vessel internals. This paper discusses prevention and recovery from such events in the future.

Description of Facility

PT Petrokimia Gresik operates an expansive fertilizer and related products facility in East Java, Indonesia.

The ammonia 1A plant is the first ammonia plant of PT Petrokimia Gresik based on the low-energy natural gas reforming process offered and licensed by the M.W. Kellogg company and has been operating since 1994. The Ammonia plant design is based on producing 1350 MTPD ammonia.

Ammonia 1A Plant Front End Description

The raw synthesis gas is produced from natural gas in four major steps:

- Compression, preheating, and desulfurization of the natural gas feed.
- Steam Reforming of the hydrocarbons in the feed to an intermediate level in the primary reformer (101-B) furnace.

- Autothermal reforming of the primary reformer effluent in the secondary reformer (103-D), where the methane is reduced to a low level while introducing sufficient process air to provide the nitrogen required for ammonia synthesis reaction.
- Conversion of carbon monoxide and steam in the reformed gas to carbon dioxide and hydrogen in the high-temperature shift reactor (104-D1) and low-temperature shift reactor (104-D2).

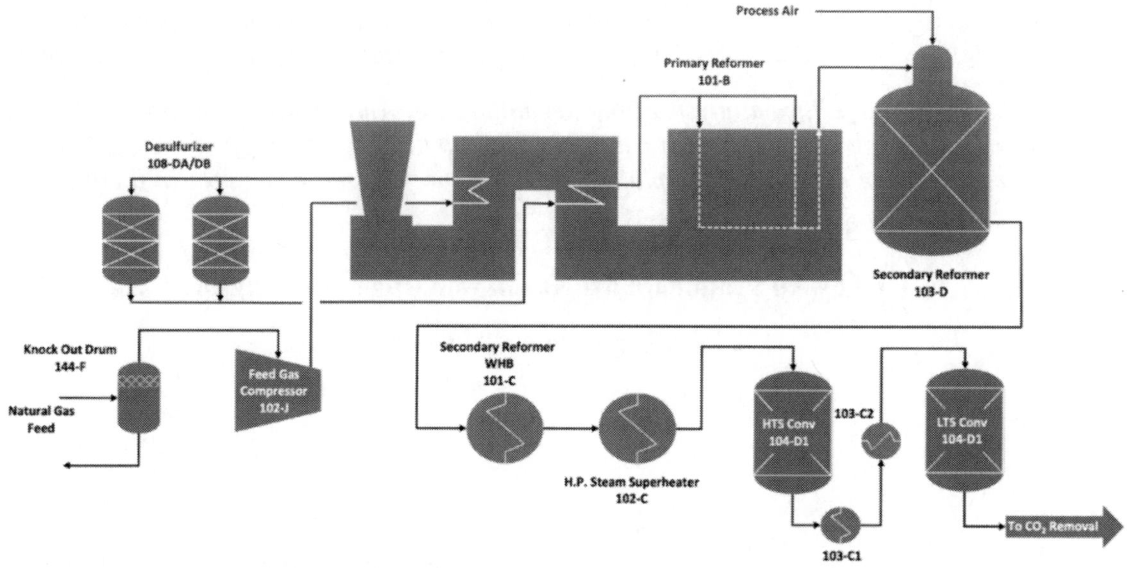

Figure 1. Ammonia Front End Process Flow Diagram

Desulfurizer (108-DA/DB)

The natural gas for feed is compressed to 44,7 kg/cm^{2}g (635 psig) in a turbine-driven feed gas compressor (102-J). The natural gas feed then flows through the feed preheat coil of the 101-B primary reformer to the inlet of the 108-DA/DB desulfurizer. The heated feed gas then enters the desulfurization vessel 108-DA & 108-DB, at the outlet sulfur content will be reduced to less than 0,01 ppm.

Primary reformer (101-B)

The desulfurized feed is mixed with medium pressure (M.S.) steam in a ratio S/C of 3.2:1, the temperature after preheating around 620 ºC (1148 ºF) and developed a reformed gas temperature near 810 ºC (1490 ºF) at the outlet of the catalyst tube.

Secondary reformer (103-D)

The partially reformed gas stream from 101-B enters the 103-D plenum horizontally and is combined with preheated air streams from 101-J through the internal tube in the plenum. The flow continues downward, exits the secondary reformer and passes directly to the secondary reformer waste heat boiler 101-C. The partially cooled gas then passes through the H.P. steam superheater 102-C.

High Temperature Shift Converter (104-D1)

The process stream from 102-C enters the top of the HTS 104-D1 through an internal gas distributor and flows down through the catalyst bed. The reaction in the HTS 104-D1 is exothermic, and the process stream leaves the vessel at a temperature of about 430 ºC (806 ºF). From 104-D1, the process stream is cooled to 200 ºC (392 ºF) by flowing in series through the shell side of the HTS effluent waste heat boiler 103-C1/103-C2.

Steam System Description

High-pressure (H.S.) steam in the Ammonia 1A Plant is generated by a series of shell and tube steam generators inside the process (Figure 2). Preheated boiler feed water is pumped from the deaerator (101-U) to the steam drum (101-F) by the boiler feed water pump (104-J or 104-JA). The steam generator consists of HTS Effluent Waste Heat Boiler (103-C1/C2), Ammonia Converter Effluent Waste Heat Boiler (123-C1/C2), and a Secondary Reformer Waste Heat Boiler (101-C) will maintain the plant in steam balance. The H.P. steam with a pressure 123 kg/cm^{2}g (1749 psig) and a temperature of 510 °C (950 °F).

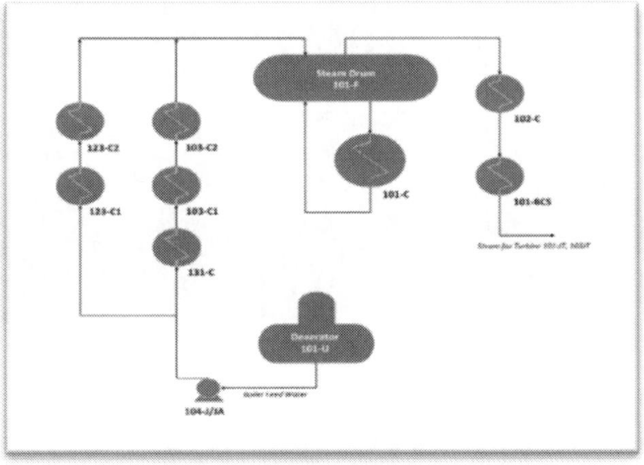

Figure 2. High-pressure (H.S.) Steam Generation Diagram

Water is withdrawn from the steam drum (101-F) via gravity and thermally circulates through the Secondary Reformer Waste Heat Boiler (101-C), a vertical single pass shell and tube-type exchanger.

The partially cooled gas then passes through the H.P. Steam Superheater (102-C). The 102-C is a vertical, single-pass shell and tube exchanger. Process gas flows through the tubes and superheats the H.P. steam.

Chronology of Events

- **On May 30, 2023**, the Ammonia plant shut down because of a governor control problem on process air compressor turbine 101-JT.
- **On May 31, 2023**, the start-up process began with the primary reformer 101-B firing. During the start-up process, the operator observed high-pressure (H.S.) steam pressure decrease rapidly from 110 kg/cm^{2}g (1564 psig) to 45 kg/cm^{2}g (640 psig) within less than five minutes and the operator decided to shut down the total ammonia plant. On the morning of June 1, 2023, the operator found several relief valves at the front end area popping, The maintenance team did a re-setting of those relief valves to make sure the setting pressure of those relief valves as per design data.
- **On June 2, 2023**, the start-up process began with steam heating of the primary reformer 101-B. During the start-up process, the operator observed some anomalies and unusual conditions in the ammonia plant :
  - Water carried over at the front end stack in a very high quantity, something the operator had never seen before.
  - The temperature indicator downstream of 101-C showed a low value after the reaction at secondary reformer 103-D occurs.
  - The supervisor then shut down the total ammonia plant complex to find the problem.
- **On June 2, 2023**, a leak-test procedure was conducted on 101-C, 102-C, and some other heat exchangers to find the water at the front-end stack. After the leak test activity was completed, the operation team found that 101-C had an exchanger tube leakage problem, as evidenced by several drain points near 101-C that contained boiler water characteristics. Maintenance activities were carried out for 12 days to replace the 101-C tube bundle (Tube Bundle A) with the spare tube bundle (Tube Bundle B).

Figure 3. Water carryover at front end stack

- **On June 14, 2023** start-up activity begins after the successful replacement 101-C tube bundle. A problem arose on June 16, 2023, when process gas was introduced to the high-temperature shift converter 104-D1.The pressure differential indicator across the catalyst bed achieved 15,21 kg/cm^{2}g (216 psig), and the CO slip at the outlet of the HTS reached more than 5% even with low front end rate. The ammonia plant shut down again to overcome the problem for ten days. It was determined that the HTS catalyst would require replacement.
- **On June 26, 2023, the** ammonia plant re-started again after replacing the HTS catalyst.
- **On June 27, 2023**, after process air introduction to secondary reformer 103-D, the operation team faced the same problem. The high-pressure (H.S.) steam pressure decreased rapidly, and the operation team decided to shut down the total ammonia plant to prevent the same problem at high-temperature shift converter 104-D1 from happening again.
- Because the same symptoms were observed as the previous shutdown event, another leak test procedure on 101-C and 102-C was conducted to find the problem, and Petrokimia Gresik team found tube leakage on 101-C like the previous event.
- Maintenance activities were conducted for 14 days to replace the 101-C tube bundle B with the repaired 101-C tube bundle A.
- **On July 11, 2023**, following repair of the 101-C, the ammonia plant was successfully re-started and production commenced on July 14, 2023, with good results.

Major Inspection findings during the shutdown events

- **Tube rupture on the 101-C tube (Tube Bundle A)**

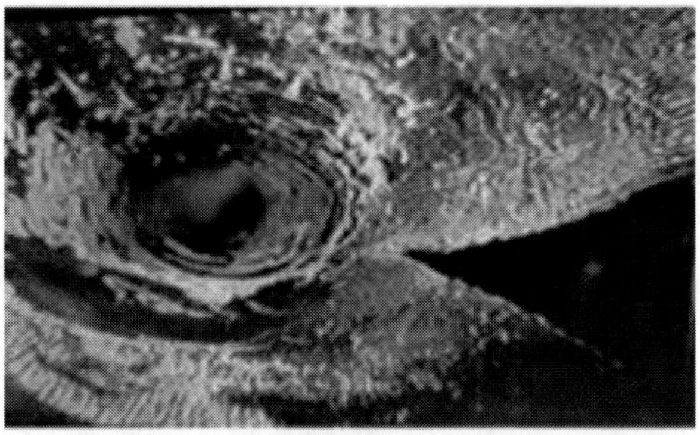

Figure 4. Tube rupture on a tube of bundle A 101-C

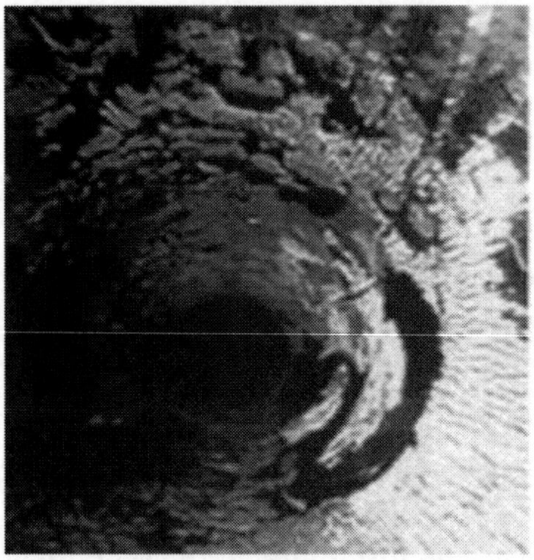

Figure 5. Scaling on a tube of bundle A 101-C

Figure 6. Tube rupture location on bundle A

- Upon completing an initial inspection, it was determined that there was damage to one tube of tube bundle A. Based on boroscope checking on bundle A of 101-C, 1 (one) tube to be found had a rupture with characteristic "fishmouth" and bulging as seen in Figure 4.
- Based on preliminary findings by PT performed by Petrokimia Gresik's team, this "fishmouth" shape, thin-lipped rupture, and bulging of the tube indicate tube bundle A of 101-C had experienced short-term overheating phenomena.
- The majority of tubes from bundle A of 101-C had scale on the "ID" side of the tube wall that inhibited heat transfer from the hot process gas to the BFW. Laboratory analysis revealed that most of the scaling compositions were CaO, MgO, and phosphate.

- **Condition inside HTS 104-D1 converter**

As a result of the 101-C Tube Bundle A failure, the 104-D1 HTS bed temperature dropped to 110 ºC (230 ºF) as per Figure 8. This indicates a significant amount of BFW from the 101-C WHB had carried over to the 104-D1 HTS catalyst bed.

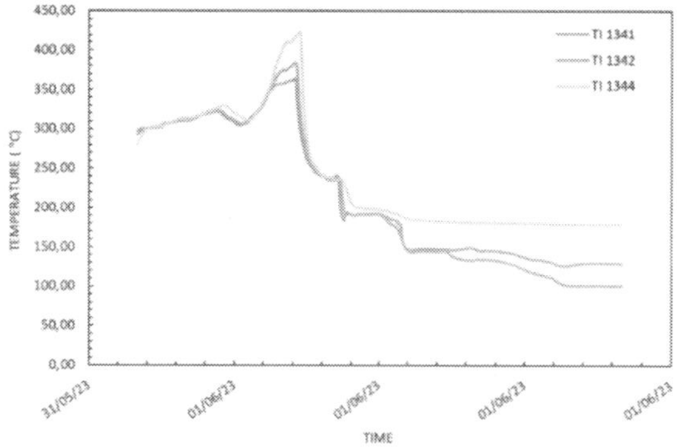

Figure 7. Temperature profile chart of HTS converter after Tube Bundle A Failure

Figure 8. The internal condition of the 104-D1 HTS converter

- The internal condition of 104-D1 HTS converter showed terrible condition of the catalyst and grating on the top catalyst bed.
- The catalyst at the top part of the 104-D1 converter had changed from tablet shape to fine mesh due to being submerged in boiler feed water (BFW) and rapidly heated without first performing a controlled dry-out process.
- Some grating from the top of the catalyst bed was missing and found at the bottom and middle part of the 104-D1 converter during the unloading catalyst activity.

- **Tube rupture on the 101-C tube (Tube Bundle B)**

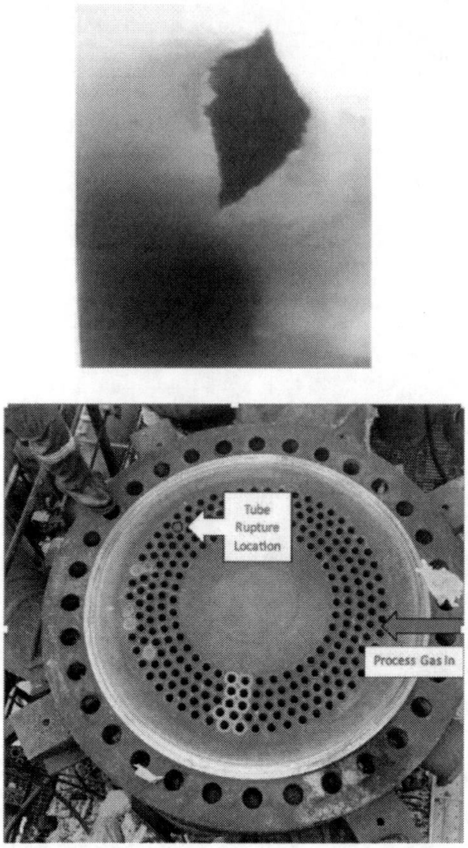

Figure 9. Tube rupture on a tube Bundle B 101-C

- Based on the boroscope camera checking on bundle B of 101-C, 1 (one) tube was found to have a rupture with the same characteristic "fishmouth" and bulging like tube bundle A as seen in figure 9.
- Based on these symptoms, tube bundle B of 101-C also experienced short-term overheating phenomena like tube bundle A.

- **Condition inside HTS 104-D1 converter after second tube rupture**

As a result of 101-C Tube Bundle B failure, the 104-D1 HTS bed temperature dropped to 110 °C (230 °F) as per Figure 10. It indicates a significant amount of BFW from 101-C WHB had carried over to the 104-D1 HTS catalyst bed.

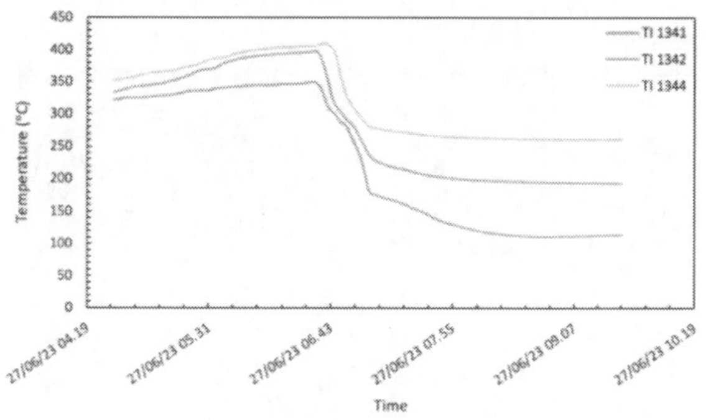

Figure 10. Chart Temperature profile of HTS converter after Tube Bundle B Failure.

Figure 11. The internal condition of the 104-D1 HTS converter

The internal condition of the 104-D1 HTS converter showed good condition of the catalyst and internal part of the 104-D1 HTS converter. No abnormalities to be found on the catalyst bed even after the catalyst was submerged in boiler water.

Evaluation of the Problems

Based on evaluation from the Petrokimia Gresik team, the problem's root cause was short-team overheating during plant start-up activity. Several potential root cause possibilities that led to the short-term overheating phenomena at 101-C WHB were considered:

1. **Not Enough BFW flow on Tube side 101-C / Low-low level steam drum 101-F**

Based on data from the Distributed Control System (DCS) record, there was no indication of a low level at the steam drum, with all six (6) level transmitters showing normal values at the event

Tube Bundle A and Tube Bundle B leakage. It is concluded that the low-low-level steam drum was not the root cause of the tube bundle A and B damage.

2. Tube Side Overpressure

Based on data records from the DCS at the onset of the Tube Bundle A and Tube Bundle B leakage events, the H.P. Steam pressure indication, which represents tube side pressure, only reached a maximum value of 118 kg/cm^{2}g (1678 psig) , which is within the normal range. It is concluded that the tube side overpressure was not the root cause of the tube bundles A and B got damage.

3. Corrosion Problem

Based on visual inspection and damage characteristics at tube bundle A and B, there was no sign of corrosion inside or outside of the failed tube . It was, however, revealed that tube bundle B was stored in the open area from 2018 to 2023 without adequate preservation. This possibly became one factor that contributed to tube bundle B damage incurred during the start-up process.

4. Process Gas Temperature at Shell Side Exceeding the maximum limit value

From the DCS historical data, the maximum temperatures at the outlet of the secondary reformer (103-D) to 101-C WHB at the onset of tube bundle A and B leakage only reached 913 °C (1675 °F) and 918 °C (1684 °F).

Based on the ammonia plant operating manual from the M.W. Kellogg, the 101-C WHB inlet temperature TI-1334 shall not exceed the maximum temperature of 1100 °C (2012 °F).

From the data above, it is concluded that in the event tube bundles A and B had failed, the shell side process gas temperature did not exceed the maximum design temperature of 101-C WHB.

5. Damage on Internal Part 101-C WHB

Based on an internal check of tube bundle A and tube bundle B, there was no sign of damage at the inlet gas distributor or impact bar, and all internal parts were in very good condition. It is concluded that damaged internals was not the root cause of the problem.

6. Scaling Problem on Inner Tube of 101-C WHB

There was a thick scaling deposit at "ID" side of tube bundle A during the event tube bundle A got damaged. This scaling deposit was found in almost every tube bundle A 101-C tube. Petrokimia Gresik's team believed that the scaling problem was one of the main factors that contributed to the failure of tube bundle A 101-C.

Later, it was found that several months before, there was a problem at the water treatment section that slipped some of the minerals into the boiler to feed the ammonia plant.

At tube bundle B, there was no sign of a scaling problem at the tube wall, so scaling was not the root cause of the failure of tube bundle B.

7. Temperature Rate Increase at inlet 101-C WHB Exceeding Design Value

TUBE BUNDLE A

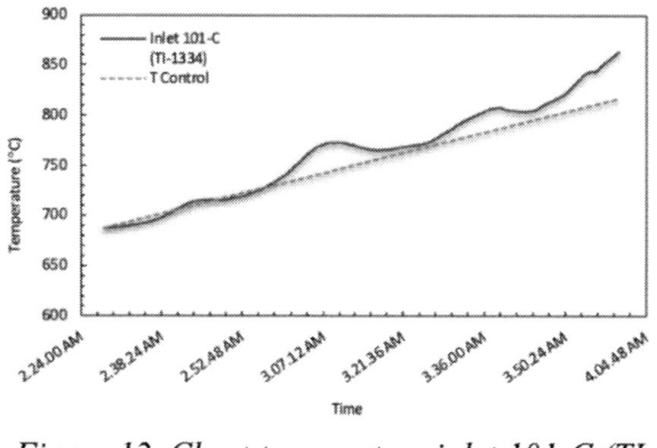

Figure 12. Chart temperature inlet 101-C (TI-1334) during Feeding Air June 1, 2023

During the start-up process on June 1, 2023, as shown in Figure 12 above, the temperature increase rate at the inlet 101-C WHB TI-1334 exceeds the maximum control rate of temperature rise of 85 C°/hr (185 F°/hr).

Based on the Ammonia Plant operating manual from M.W. Kellogg, the temperature increase at inlet 101-C WHB should be limited to a maximum of 85 C°/hr (185 F°/hr).

It was also found from visual inspection that the 101-C tube failure was located at the bottom part of 101-C, where the temperature and heat flux is at the highest value.

As a result of this event, the vaporization rate of boiler water at 101-C WHB is much higher than the maximum allowable design vaporization. This phenomenon, combined with the deposit problem at "ID" side of tube wall 101-C WHB, eventually led to short-term overheating damage to the 101-C WHB boiler tube for tube bundle A.

TUBE BUNDLE B

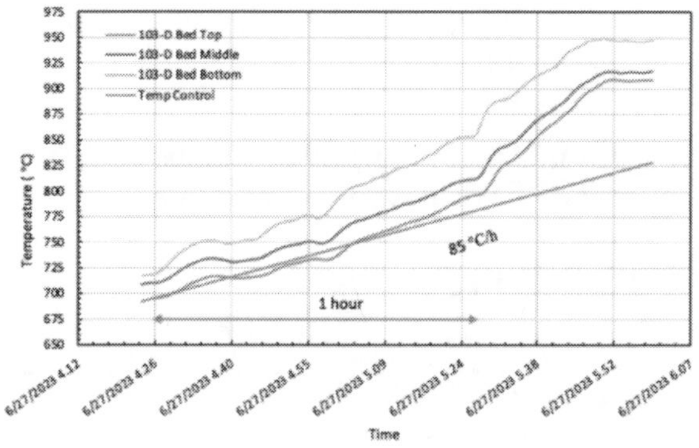

Figure 13. Chart temperature catalyst bed 103-D during Feeding Air June 27, 2023

During the tube bundle B failure event, the 101-C WHB upstream temperature, TI-1334, showed an inaccurate indication and was thus unavailable for making a determination. To determine the root cause problem of failure 101-C tube bundle B, we compared the bed temperature of the secondary reformer 103-D observed during the failure events of tube bundle A and tube bundle B. Based on DCS recorded data, the bed temperature rate increase of the 103-D secondary reformer preceding the events of failure of tube bundle A and B were similar, with the temperature rate increase at the event tube bundle B failure being slightly higher. It is concluded that the root cause of the problem of Tube Bundle B failure is the temperature rate increase at the inlet of the 101-C WHB exceeding the design value, leading to short-term overheating damage.

First Event 104-D1 HTS Converter Submerged in BFW

The high differential pressure (D.P.) problem and high CO slip observed at the outlet of the 104-D1 HTS converter root cause problems will be analyzed as follows:

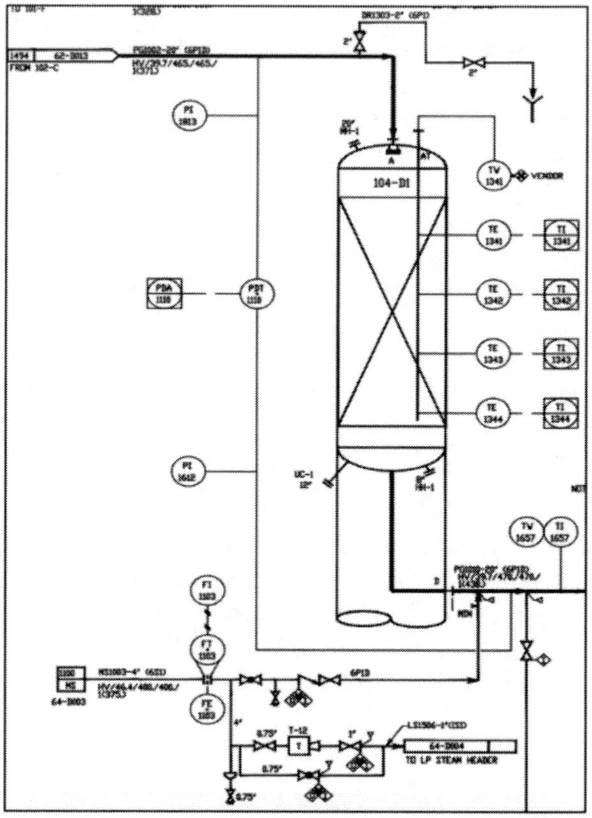

Figure 14. Heating Method 104-D1 HTS Converter

- During the attempted re-start on June 14, 2023, the 104-D1 HTS catalyst required heating-up to increase the bed temperature. This heating-up procedure uses an injection of M.S. steam from the bottom of the 104-D1 HTS converter by a 4-inch pipe as seen in Figure 14.

- Because of the prior leakage problem at 101-C, the catalyst had been submerged in boiler water, and the plant operator was unaware of that phenomenon. M.S. steam was directly injected from the bottom of the 104-D1 HTS converter. This steam injection rapidly vaporized the embedded moisture in the soaked catalyst, and as a result of this activity, broke the catalyst into a fine mesh that created a high-differential pressure (D.P.) at the 104-D1 HTS converter during the subsequent plant start-up attempt.
- The damaged catalyst was then unloaded and replaced to resolve the problem. Catalyst replacement was carried out nine days later, after Petrokimia Gresik team contacted some local catalyst sources.

Corrective Action Plan

1. Repair Procedure of Bundle A 101-C Waste Heat Boiler

Following Tube Bundle B's failure at the start-up on June 27, 2023, the team decided to repair and re-install the damaged tube bundle A. A tube plug was fabricated using Inconel material and AWS ER NiCrMo-3 with a diameter of 1,6 mm (0,06 inch) as an electrode.

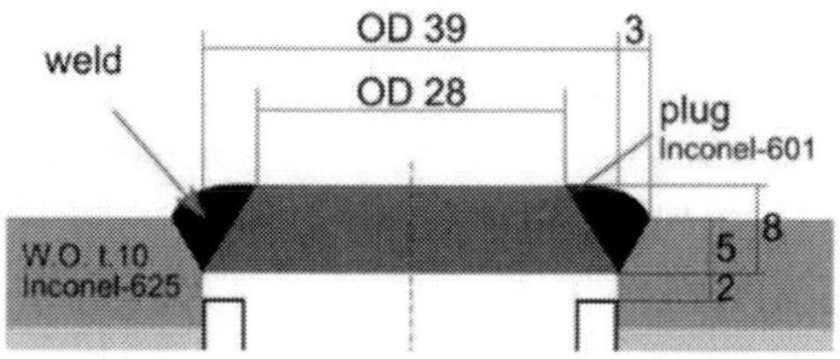

Figure 15. Tube Plug Configuration

As per Figure 15, the Inconel plug has a thickness of 25 mm (0,98 inch) and was attached inside the damaged tube. After the welding process was complete, the tube bundle was then pneumatically tested with a pressure of 0,5 kg/cm^{2}g (7 psig) to verify the welding quality.

A new problem arose when we found some tube-to-tube sheet joints adjacent to the plugged tube had a leakage or crack after the pneumatic test as per Figure 16. Based on the evaluation from the Petrokimia Gresik team, the problem occurred because of the high heat input from the welding process, which caused propagated cracks at the tube-to-tube sheet joint.

Figure 16. Crack at tube-to-tube sheet joint after pneumatic test

The corrective action to overcome those problems was removing the damaged tube-to-tube sheet welding up to the Inconel layer to the depth of 9 mm (0,35 inch), then rewelding again using the same electrode.

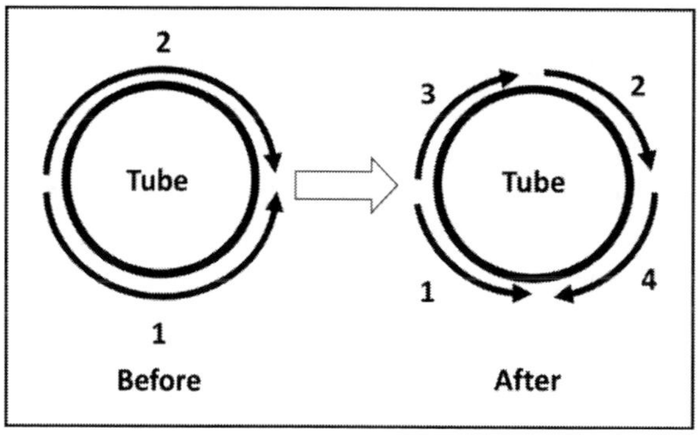

Figure 17. Modified Welding Process

As per Figure 17, the welding step was modified to minimize heat input to tube-to-tube sheet joints that affected adjacent welding joints. Initially, we conducted a two-step process to weld the tube plug inside the tube. After those problems arose, the welding process changed to four-step welding as per Figure 17, with a crossing welding sequence.

The tube bundle then had a pneumatic test again, and the repaired work had a satisfactory result.

2. Mechanical cleaning of Tube Bundle A

Mechanical cleaning was done using a hydro blaster/high-pressure water cleaning at 2460 kg/cm^{2}g (35.000 psig) to remove all deposits at the tube wall.

Improvements during plant start-up and shutdown

Shut down

Leak test procedure of 101-C WHB, 102-C Steam Superheater, 103-C1/C2 HTS Effluent BFW Preheater/Steam Generator shall be carried out after plant shutdown, to make sure all the exchangers are in good condition and to prevent 104-D1 HTS catalyst being submerged by boiler feed water (BFW).

Start-up

1. During the start-up process, strict adherence to existing Standard Operating Procedures (SOPs) is important, and the temperature rate increase at the inlet of the 101-C WHB shall not exceed the design value. A new temperature rate indicator at the inlet of the 101-C WHB was installed to prevent similar failures from taking place in the future. Figure 19 below shows the new temperature rate indicator installed in the DCS that the operator is required to monitor and control during the start-up process.

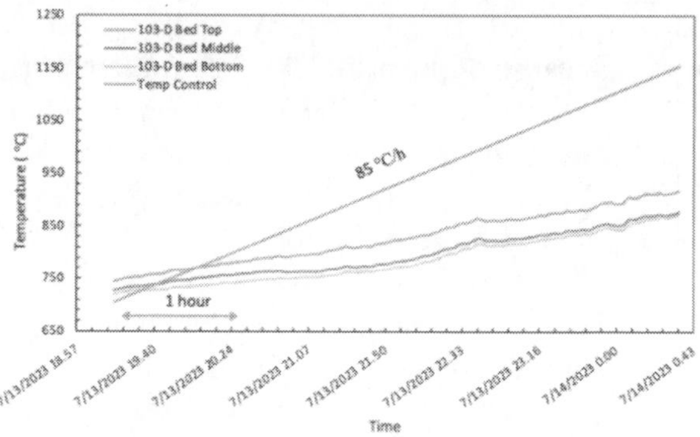

Figure 18. Chart temperature catalyst bed 103-D during Feeding Air July 13, 2023

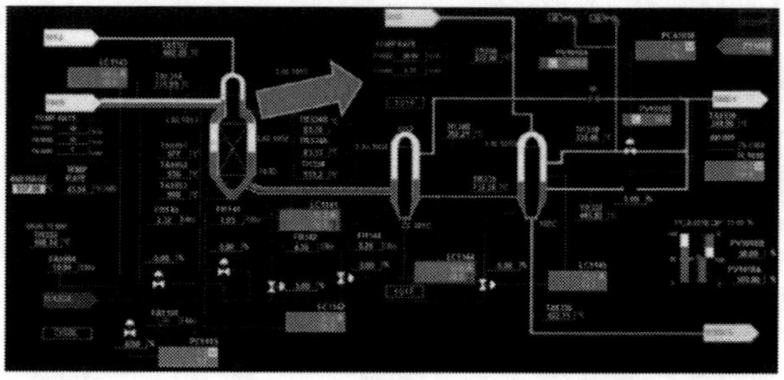

Figure 19. Temperature rate monitor on display DCS

2. To prevent a similar failure at the 104-D1 HTS catalyst before steam sparge flow is introduced from the bottom catalyst vessel, a few additional checklists will be carried out:
   - Make sure there is no condensate at the inlet and outlet 104-D1 HTS converter: several drain points around the 104-D1 HTS converter shall be open until all condensate discharges from the system.
   - Sparger steam shall be heated to reach a minimum of 50ºC (122 ºF) higher than the saturated temperature at the corresponding pressure.

3. If there is any indication of the 104-D1 HTS converter being submerged by Boiler Feed Water (BFW), some additional step/inspection will be carried out to determine whether the catalyst replacement is required or and to prevent unnecessary additional work during the start-up process:

- Visual inspection of catalyst bed under nitrogen condition from top manhole to determine catalyst condition. If there is any sign the catalyst has broken or become fine mesh, then catalyst replacement is required.
- During the 104-D1 HTS catalyst heating up, the temperature rate increases by 15-20°C/hr (59 – 68 °F/hr) shall be maintained to avoid excessive water evaporation in the catalyst bed.

Summary and Conclusion

After finishing all the maintenance activities, the start-up process was carried out from July 12-14, 2023. The temperature rate increase at the inlet of the 101-C WHB was controlled below 85 C°/hr (185 F°/hr) as per Figure 18 below, and 101-C WHB ran satisfactorily. The performance of the 104-D1 HTS catalyst also can be recovered after being submerged by BFW by completing a controlled heat up to gradually dry out the catalyst. The pressure drop of 104-D1 HTS catalyst only reached 0,23 kg/cm^2 (3,2 psi) compared to the first event that reached 15,21 kg/cm^2 (216 psi) with an uncontrolled heat-up rate from the steam sparing. The CO slip at outlet 104-D1 HTS was also maintained at 2,04% compared to the design value of 3,65%.

Recent findings about Flow Accelerated Corrosion (FAC) in aqua ammonia loops

Flow Accelerated Corrosion is a well-known degradation mechanism that may affect several systems in chemical plants like boiler feed water loops, water treatment system, acidic solutions piping - e.g., sulfuric acid, amine system etc. Carbon and low alloy steels show the poorest resistance to this degradation, but stainless steels and higher alloys can be affected as well. Flow conditions, temperature, and fluid chemistry – for instance pH – are the main parameter affecting the severity of the degradation. In the last years at Yara, severe corrosion has been observed in equipment and pipelines in the so-called aqua ammonia loop. Those findings challenge the current understanding of FAC and a direct correlation with severe corrosion rate and high pH level seems to be present. In the paper the findings will be presented, a possible correlation between FAC and pH in aqua ammonia will be discussed and possible mitigation measures will be introduced.

Stefano Cavalleri, Giulio Paci, Roy van Lier
Yara International ASA

Rodney Babwah
Yara Trinidad Ltd.

Introduction

In recent years, extensive and severe internal corrosion has been observed on carbon steel equipment and lines in aqua ammonia service at several Yara facilities across the globe. In all cases, the damage has a distinct flow-accelerated corrosion (FAC) signature. FAC is a specific form of erosion-corrosion and it is primarily driven by the corrosion process aided by chemical dissolution and mass transfer. In practice, there may be some contribution from the mechanical factors that lead to a removal of corroded scallops on material surface that can flow out in case of high velocity process fluid. The damages occurred in the aqua-ammonia loops in the ammonia plant, which have been recently found, challenge the current understanding of FAC. Inspection results and analysis of recent findings suggest that for high pH levels i.e. above 9 , which is a common condition in ammonia-water mixtures containing up to 25% of ammonia, carbon steel becomes more susceptible to FAC than at the medium pH range. As is well-known from the steam generation system in the power industry, FAC is affected by many parameters, like material composition, temperature, pH, dissolved oxygen content, pipe geometry and flow velocity. Evidence showed that the maximum FAC rate appears at ~ 150 °C (300 °F). As a result, FAC caused rupture of large, medium, and small diameter pipelines carrying either single phase (single-phase FAC) or two phase wet steam flow (two-phase FAC).

Failure at Yara Trinidad

The ammonia stripper (T-3411) located in the ammonia recovery section at Yara TRINGEN II ammonia plant experienced severe corrosion and related leakages over the time. This has resulted in the need for recurrent replacement of this component and material changes in the attempt to permanently fix the issue. On April 28th, 2022, a loss of containment event in the ammonia stripper T-3411 of TRINGEN II plant occurred. The ammonia recovery unit was quickly stopped and isolated by plant operations. No one was affected by the ammonia loss and the main plant remained online. The liquid distributor was dislodged from the tower through the rupture opening arising from the incident. This was located some 12 m (40 feet) from the tower location. The event could potentially have caused physical injury to the personnel in the vicinity of the failed equipment. A release of ammonia to the environment also occurred in the period until the unit was stopped. The plant operations team responded to the event by first identifying the area in which the incident occurred via the DCS and then promptly shutting the unit down. Investigation into this incident commenced on May 4th, 2022. Due to this serious LoC event site management decided to perform a RCPS (Root Cause Problem Solving) to identify the main causes of the failures and define possible mitigation actions. The investigation team was built including local team from process, operation, maintenance, and inspection department. Experts from Yara central support team were also involved in the investigation as well as an independent third-party company. The mandate for the team and the goal of the investigation was:

1. Determine and document the sequence of events, as well as all other information, circumstances, and conditions that are considered to have contributed directly and indirectly to the incident.
2. Review and comment on emergency response actions and performance.
3. Look into technical, organizational, and human aspects of importance to the incident.
4. Identify non-conformity on regulations and company requirements related to the incident.
5. Identify preventive actions and recommend corrective actions for the site.

Description of the process

The purge gas station is a closed loop circuit in TRINGEN II plant where the ammonia is absorbed from a gaseous phase into the circulating water at the absorption side. Off gases are vented to other process loop sections. The ammonia is then released from the circulating water in a stripper column on the desorption side. Purge gas from the ammonia synthesis via V-3414 - 35.4 bar (513.4 psi) and 4°C (39 °F) - and purge gases from TRINGEN 1 plant - 36.2 bar (525.0 psi) and 13°C (55 °F) - are guided to the entrance of heat exchanger E-3431 A/B/C/D. In E-3431, ammonia is absorbed into the circulating water. The absorption heat is removed by a cooling water heat exchanger. E-3431 B/C/D are heat exchangers; E-3431 A is a packed bed vessel. The relatively ammonia free off gases (helium purge gas) are vented towards the furnace H-3401, and towards an injection point between E-3409 and E-3410. The aqueous ammonia solution is heated in inlet-outlet-exchanger E-3434 and brought into ammonia stripper T-3411 (Figure 1), where the solution is heated further to 212 °C (414 °F) with stripper reboiler E-3426 (41 bar steam). At this high temperature, ammonia and water vapor are released to top condenser E-3427. The water phase is condensed and circulated back from T-3411 to E-3431 D, cooled through heat exchangers E-3434 and E-3425.

Failure event summary

The rupture on the shell of the ammonia stripper occurred immediately beneath the middle flange and to the left (West) of the liquid feed nozzle N2/N11 (Figure 2).

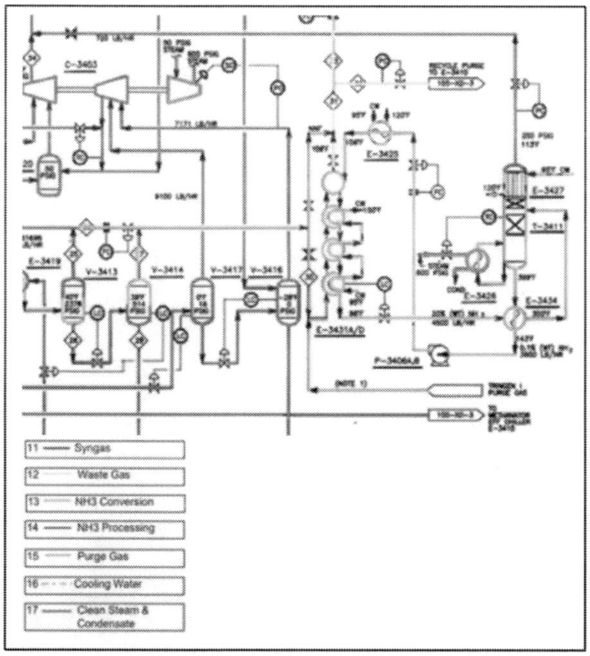

Figure 1. Integrity loop of purge gas in TRINGEN II Plant.

At the instant of rupture, the distributor tray (below the nozzle N2) was propelled outwards, possibly colliding with the internally projected liquid feed line N2. The liquid feed line N2 (2in Sch. 160 pipe) was found to be significantly thinned and was warped/twisted consistent with impact from the distributor tray. This led to the decision to remove and replace the line. The removed section was subsequently sent to an independent laboratory for a metallurgical analysis on June 1st, 2022. Evidence suggested that the 2 inches feed line corroded externally by flow accelerated corrosion, to such an extent that the line became perforated between the 10-12 o'clock positions, just before the elbow as shown in Figure 3. As the 2 inches feed line was only 7.4 inches away from the shell ID, it was likely that the shell was subjected to liquid impingement corrosion by the incoming feed.

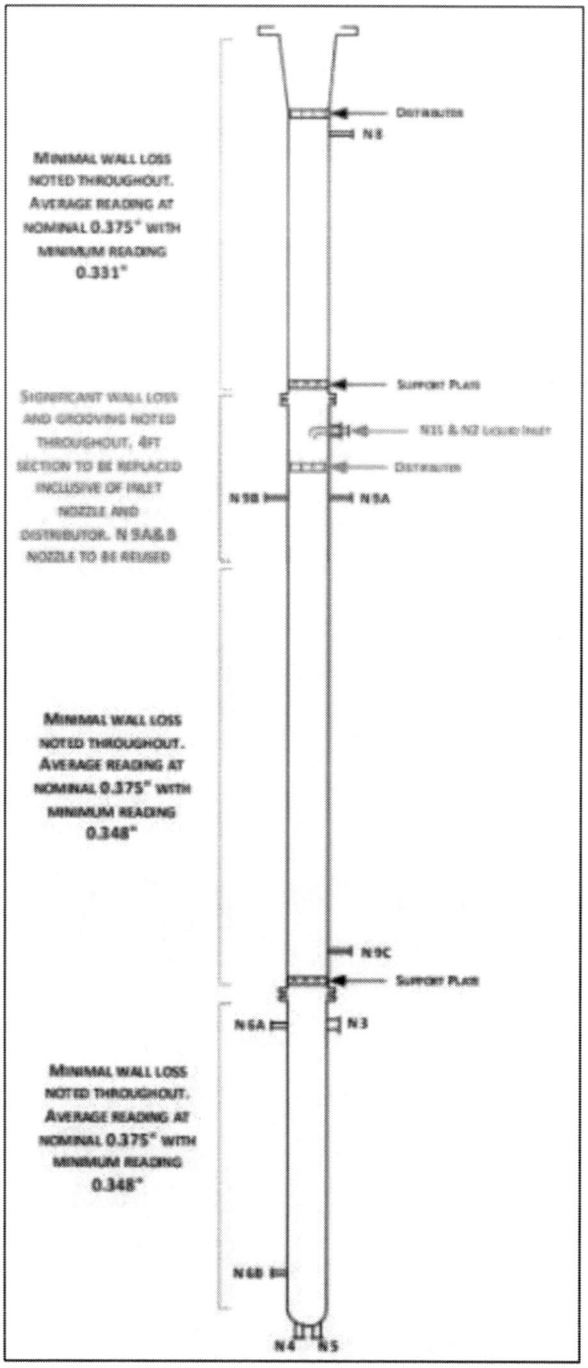

Figure 2. Tower inspection summary with red highlight injection point and perforated shell area.

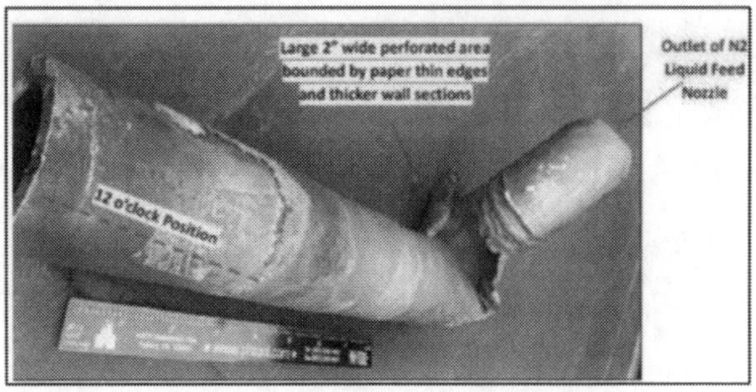

Figure 3. Perforated liquid feed nozzle.

The shell corroded relatively quickly in the vicinity of the impingement until it suddenly ruptured under normal operating conditions. The likely sequence of the event that occurred was a small leak (weep) under the insulation at first (unidentified) and as the thinned-out area grew, a rupture driven by the internal pressure of the vessel occurred as shown in Figure 4.

Figure 4. Failed section of NH_3 stripping tower.

It is worth highlighting that some symptoms indicating possible loss of containment were identified during previous inspections performed on T-3411 the since 2017. On August 12th, 2017, for instance borescope inspection was carried out and black deposits were noticed at three locations along the column as reported in Figure 5.

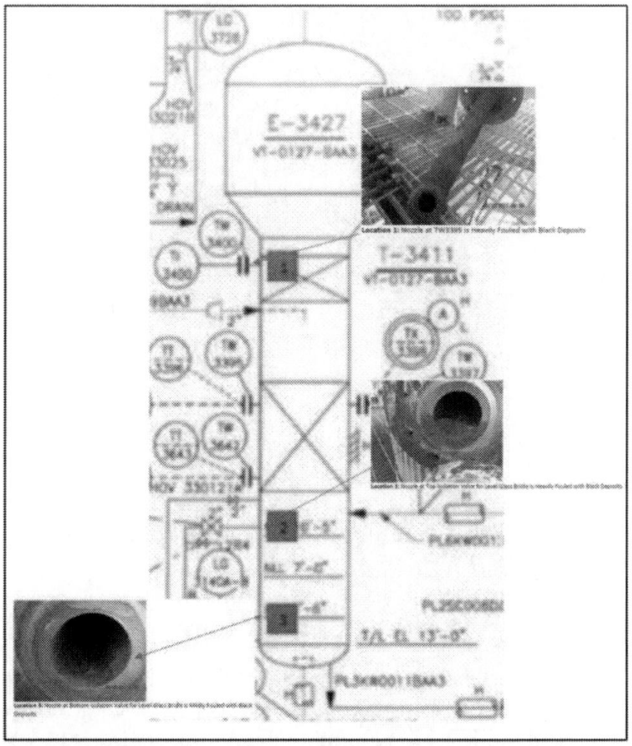

Figure 5. Results of borescope inspection on the nozzle of T-3411 carried out in 2017.

On August 17th, 2017, the complete column including the overhead exchanger was thoroughly inspected. The T-3411 was found to have considerable corrosion, confirming the deposit identified during previous borescope inspection. Severe corrosion/pitting was observed on the upper section of column (section right under the exchanger) with a minimum measured wall thickness of 2.9 mm (0,11 in) vs. a nominal thickness of 9.5 mm (0.374 in) as shown in Figure 6. In addition to the wall loss there was also significant corrosion of the distributor and its support as well as the packing support plate (Figure 7). The two lower sections of the column were in a much better condition with measured minimum wall thickness 9.1 mm (0.358 in) with no sign of pitting. The distributor and packing support plate in the middle section were also in a good condition. Nozzles N9A, N9B and N11, however showed areas of corrosion on the gasket surfaces and were replaced.

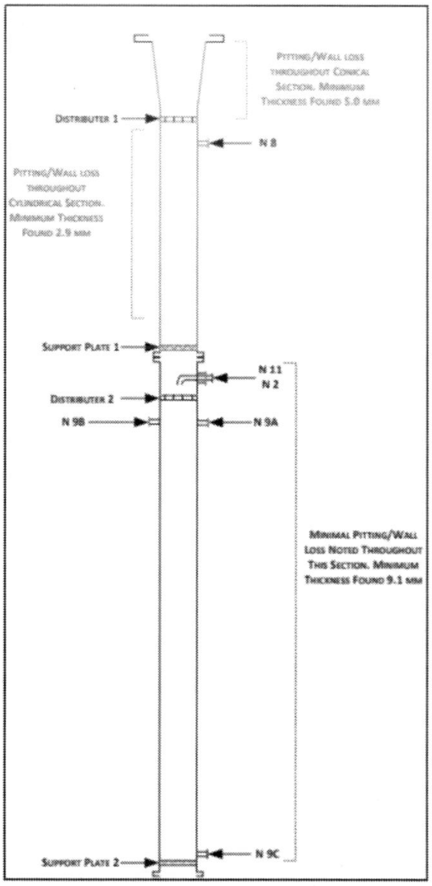

Figure 6. Corrosion in the upper part of the tower, as found in 2017.

The heat exchanger in the top E-3427 was also found to have corroded on the carbon steel sections of the bundle (baffles, tie rods and tube sheets) as well as packing area around bundle external. Corrosion was also noted on shell internals although the minimum measured thickness was still acceptable. The visible tube sheet of the E-3427 showed numerous pinholes (about two hundred). Due to the extent of holes on the tube sheet of the exchanger and possible thinning of the tube sheet, this exchanger was marked for replacement (to be noted that this exchanger was replaced in 2003 where only the tubes were upgraded to 316 SS).

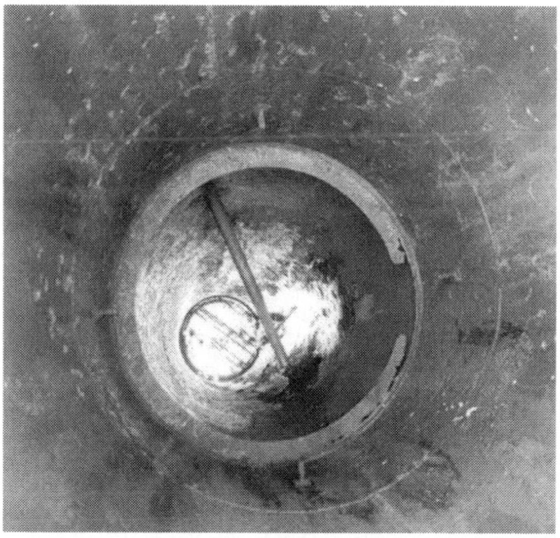

Figure 7. Corroded tray support.

On November 11th, 2021, the E-3427 and the upper section of tower was removed due to heavy ammonia leak observed during operation. Subsequent UT measurements done on both upper and middle sections led to the decision to remove the middle section for further analysis. A leak occurred through a weephole section behind the upper insulation ring. An external sleeve that was installed in 2017 for additional wall reinforcement was seal welded onto the insulation ring. The other side of the insulation ring was only partially welded to the tower leaving two open areas in which the leaking ammonia vented. The failure of the tower occurred on the shell just beneath the support ring for the distributor and channeled through the sleeve and out through the weephole. The support ring was heavily corroded and thinned out with perforations noted. Repairs to the leaking area (which was possible from the inside only) were unsuccessful due to the thinning level of the shell. The support plate for rings was missing and found at the top of the middle section of the tower in a badly corroded state. Notable bulge in tower internal were also identified in an area close to support plate. Heavy black deposits were also noted at the internals including rings and walls as already noticed back to 2017. All the findings are shown in Figure 8.

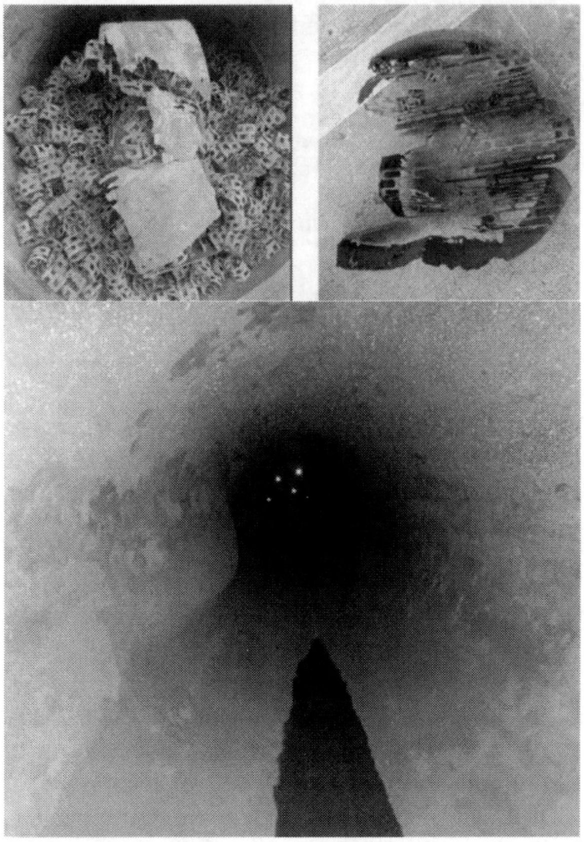

Figure 8. Corrosion Corroded support ring and bulged tower shell.

The failure analysis performed by the laboratory that determined that most likely cause of the failure was Flow Accelerated Corrosion. Investigations consisted of visual examination of the removed injection point, ultrasonic thickness evaluation of the injection quill, stereoscopic analysis, and microstructural analysis. From the visual inspection of the inlet nozzle piping, it was noted that there were several features such as grooving (scalloping), wall loss, and thru-wall perforations. This was the most severe along the top surface of the line (in the vicinity of the outlet elbow welded joint), all occurred on the OD of the line, and along the straight run of piping directly upstream of the outlet elbow. This confirmed the external initiation of the corrosion, most likely influenced by the process, occurred over the time in operation, resulting in regions of excessively thin wall sections, and thru-wall perforations in some instances. Ultrasonic thickness checks were performed along the failed 2 inches of liquid feed line N2 to quantify the extent of wall loss experienced. These checks were performed at 4 cardinal points. Readings were taken every one inch longitudinally along the sample, starting from the flanged end. It was noted that the major wall loss area corresponds to the area exposed to external flow of process liquid down the tower. The presence of a small hole in the vicinity of the elbow welded joint was observed on the inlet pipe. While the thru-wall perforation was not directly in line with the direction of the vessel shell rupture, the presence of this thru-wall perforation, combined with the extent of excessively thinned areas, strongly suggest several thru-wall perforations previously existed along the 'paper thin' regions of the line. These thru-wall perforations would have grown larger over time and most likely caused increasingly worse liquid impingement corrosion to the adjacent shell. Although severe wall thinning was noted along piping, the HAZ on both sides of the weld appeared to be preferentially attacked. The weld microstructure was typical for the material and process with no visible findings, and the HAZs were well defined. The elbow experienced significant wall thinning near the weld and its microstructure was typical of a forged carbon steel fitting: fine, refined grain structure with small pearlite colonies within a ferrite matrix. The microstructure was unaffected on the OD near the perforated areas, which is typical of Flow Accelerated Corrosion (FAC) as it is a chemical dissolution process (Figure 9).

Figure 9. Micro of the folded pipe on the left and adjacent section to the right.

Corrosion mechanisms within NH_3 Stripping Tower (T-3411)

From the work done by the investigation team, it has been concluded that there would have been multiple corrosion mechanisms working simultaneously on the T-3411 tower. The combined effect of these gives an explanation to the high rate of corrosion experienced and manifested over the years. The three (3) key forms of corrosion identified are:

1. Primary mechanism: Carbon Steel corrosion in Low/No Oxygen Environment.
2. Secondary mechanism: Flow Accelerated Corrosion.
3. Secondary Mechanism: Galvanic Corrosion.

The primary mechanism has always existed and is a function of the carbon steel material in the service conditions. An inhibitor would normally be used to mitigate this effect but there are no historically information about inhibitor usage in this service in the plant. Conditions 2 and 3 can be seen as having the effect of accelerating corrosion rates (either generally or in a localized manner) above and beyond what is normal for condition 1 in the tower. Looking at the primary driver, there is a basic form of corrosion that exists on the carbon steel for this equipment which is a function of the specific environment of the tower (anaerobic, temperature, pH) and produces as the by-product of corrosion various forms of iron oxide (FeO, Fe_2O_3 and Fe_3O_4). The black deposit observed historically coming from the tower are essentially two (2) of these products (FeO and Fe_3O_4). Fe_2O_3 is essentially rust (reddish brown powder) seen on steel exposed to the atmosphere. The most important of the oxides formed is Fe_3O_4 (magnetite) which normally provides an adherent layer that that would protect the carbon steel substrate. However, in the specific environment of the tower, within a temperature and pH range, the magnetite layer can be unstable and removable (see solubility graph Figure 10). The traditional inhibitor (sodium dichromate) for this service, would perform the role of providing a stable protective magnetite layer across the unstable temperature range. Unfortunately, sodium dichromate has been associated with human health issues and is not in use anymore.

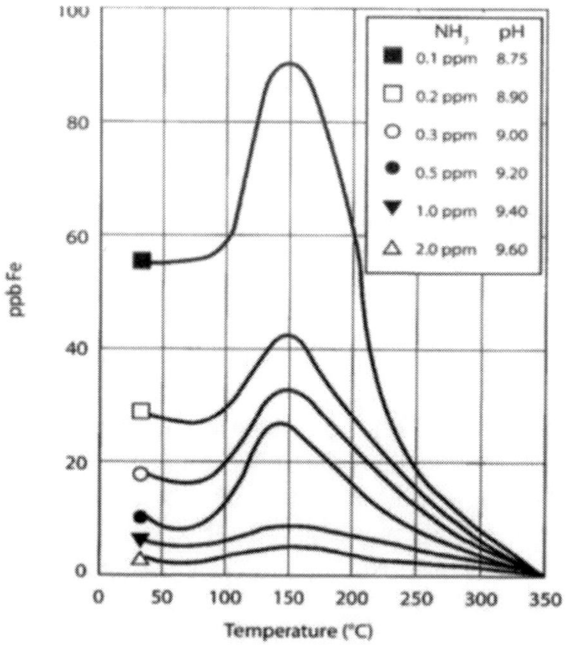

Figure 10. Solubility of magnetite in function of temperature and pH.

It is therefore safe to conclude that the basic corrosion mechanism has always been present within the tower. Manifestation of this has been the presence of 'black stuff' historically noted on inspection of the tower and found within the packing and in the stripper bottoms pumps. In addition to the basic corrosion mechanism, a special form of erosion-corrosion also exists within the tower known as Flow Accelerated Corrosion. This acts to increase the corrosion rate on the carbon steel surfaces and may evidence in a localized manner. The conditions required to support Flow Accelerated Corrosion (FAC) are present in the tower. This is essentially an unstable protective layer (magnetite) and sufficient flow rate for the fluid to remove the unstable layer. The mechanism of FAC is to remove the protective layer from the carbon steel, thereby exposing the substrate material to further attack. Therefore, if the layer cannot be regenerated fast enough (no oxygen), it leads to degradation of the base material resulting in higher corrosion rates. The loss of material on the top outer portion of the inlet flow pipe has the unique appearance (scalloping) of FAC. So, is the loss of material on the vessel wall which led to the eventual failure via rupture of the vessel wall due to flow impinging on the wall from the thinned-out perforation in the inlet piping.

In response to the occurrence of the corrosion mechanism, there has been a shift towards the use of more noble material (stainless steel) in parts of the tower (Exchanger E-3427, packing rings etc.). While this change has had the intended beneficial effect, the presence of stainless steel alongside carbon steel in the environment could set up a galvanic cell which would have the dual impact of keeping the cathodic material (stainless steel) sound but also increasing the corrosion rate on the anodic material (carbon steel). This theory has been however discarded and not really taken into consideration.

Yara independent failure analysis in the tower shell

With the aim to validate the results of the investigation done by a dedicated team, another failure analysis was conducted by another independent laboratory and in a piece of tower shell affected by impingement corrosion. The lab received a part of a shell from the upper section of the ammonia stripper column (T-3411). Before cleaning, the internal surface has shown a smooth appearance covered by brown and orange deposits. Beneath the deposits, a uniform black scale was covering the surface. After cleaning, wide pits were present on most of the surface. No directional pattern was observed. At low magnification, the internal surface exhibited a smooth appearance. At higher magnification, the appearance was rough, and the surface was covered with smaller pits with a penetration depth from a few microns up to ≈ 60 µm. No directional pattern was observed. The deposit layer on the internal surface is not thick (< 200 µm). The base material microstructure consisted of perlite with ferrite. It was typical for a carbon steel such as A 516 Gr. 70: no irregularities were observed. - EDX analysis of deposits showed the presence of corrosion products of carbon steel.

Based on the results of the investigation, it appeared that the received part was subjected to erosion-corrosion. Erosion-corrosion refers to material loss due to a synergetic effect of corrosion and erosion. Erosive effects can accelerate corrosion processes due to the removal of semi-protective corrosion products. First, a corrosive effect was observed on the failed part: the main wall thickness loss was located on the internal surface. On the macroscopic scale, an irregular surface with pits and valleys without a clear indication of directionality. - At higher magnification, the surface exhibited a rougher, attacked morphology with superficial pits. Corrosion products of carbon steel were covering the internal surface, also at the attacked areas. However, there was evidence that flow conditions also affected the corrosion rate. The deposit layer was thin, which indicates that corrosion products were likely to be washed away from the surface, exposing once again the fresh metal surface to further corrosion. In the report discussion it was pointed out that the corrosion rate of carbon steel in contact with ammonia solutions is expected to be low: Carbon steel is fully acceptable for ammonia service. In the absence of atmospheric oxygen, aqueous solutions of ammonium hydroxide have little action on unalloyed and low-alloy steels. (DECHEMA, 2003), (Craig, 2002), (DILLON, 1994) (Outokumpu, 2022) (Rabald, 1988). Carbon steel in contact with uninhibited ammonia-water solution at 100°C exhibit corrosion rate below 0.001 mm/y (Technology, 2000).

At the end of the discussion the same key points raised from the investigation team were also highlighted with the following statement: "*However, ammonia-water solutions are not completely inert towards carbon steels, especially at*

higher temperatures. The corrosion attack is expected to form non-protective magnetite scales. High flow rates can cause these scales to be continuously removed leading to an accelerated attack. Conventionally in advanced ammonia absorption systems, a corrosion inhibitor in the form of sodium chromate (Na_2CrO_4) or sodium dichromate ($Na_2Cr_2O_7$) is added to the working fluid to inhibit the working fluid from reacting with the steel (Technology, 2000), (LANXESS Elastomers B.V., n.d.). Sodium chromate is effective for operating temperatures up to about 200° C.
While uninhibited ammonia-water is not inert towards carbon steels, the corrosion rate is still expected to be limited at current operating temperatures (up to 154 °C). This is in accordance with the observations of the plant that the attack mainly occurred at the locations subjected to the highest flow rates".

Comparison with other Yara's strippers

After two investigations and thorough assessment there were solid basis and confirmation about the mechanism involved in the corrosion of the tower, the team decided to get a gauge of what the experience has been for others, information was sought from other operations within Yara to compare data. In summary, the following was feedback from the three (3) operations sites:

Sluiskil – Black deposits which they have identified as magnetite have been a feature of the operations. There has been no problem experienced with wall thinning. The packing was changed to 304SS in 1998. One picture was provided showing a hole formed on the top section of the inlet pipe (Figure 11). This was seen in the 2020 turnaround.

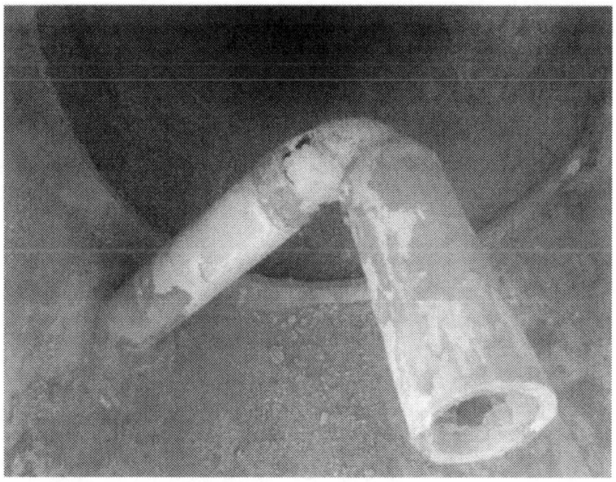

Figure 11. Hole in the inlet tower nozzle.

Babrala – Babrala has similarly seen evidence of black deposits as a part of the operations. The trays are routinely cleaned during stops. No experience with wall thinning or unusual rates of corrosion noted. A picture provided of the lower tower shows an even layered distribution of magnetite around the perimeter of the walls (Figure 12). The internals of the tower (trays, inlet pipe etc.) are all made from Stainless Steel.

Figure 12. Left lower tower section – right upper tower section.

Belle Plaine – Belle Plaine's experience is somewhat closer to Trinidad in that they have experienced one case of an inlet nozzle hole on the top section, wall thinning and loss of containment through a leak (weep) from the wall of the tower. It seems that the main area of thinning may have been because of liquid impingement also. This is based on the proximity of the thinned area to the nozzle. The shell was patched using pads to increase the thickness and the inlet nozzle was changed to Stainless Steel to mitigate the risk of a further leak. Flow Accelerated Corrosion (FAC) was identified as the likely cause of the leak on the nozzle.

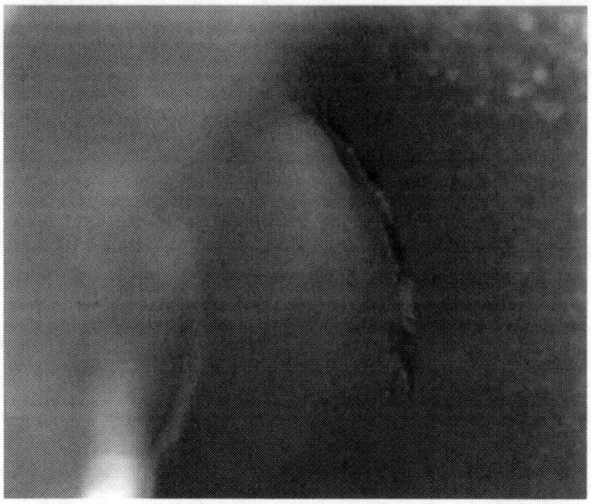

Figure 13. Detail of the inlet nozzle in Belle Plaine tower.

Conclusion

The investigation team were able to establish that there are multiple degradation forces at work within the T-3411 stripper tower at Trinidad. The first (magnetite dissolution) is a function of the process (pH, temperature, oxygen levels) and material of construction. The second (Flow Accelerated Corrosion) is driven by the flow rates (general or localized). The third (galvanic corrosion) is a function of the combination of materials within the vessel and the fluids contained.
These mechanisms have been reviewed and a comparison of the performance of some strippers at other Yara sites was done. From the comparison between the recovery strippers at some other Yara sites, it was clear that they all experienced magnetite dissolution which manifested itself in the form of black powdery residue. This was not surprising as can be expected to be present in such systems in the absence of an inhibitor to mitigate this outcome. They all had carbon steel as the material of construction for the main structure and stainless steel as the material of construction for much of the wetted internal components (packing, trays etc.). The corrosion experience has been variable across the compared strippers (from no events to multiple significant events). These results seemed to suggest that galvanic corrosion might not be the major driving force in determining the outcomes as far as corrosion is concerned though it has been considered as one of the forces at work in the process. Process conditions (pH, oxygen levels, flow rates and temperature) all have an impact on the magnetite stability and there were material differences among the plants. Given that the examined strippers all manifested evidence of magnetite instability by way of black deposits being present in the systems, the level in the internal flow regime may be the most feasible explanation for the outcomes experienced.

Considering this, focus needs to be placed on two (2) key areas:

1. Flow Distribution – It is critical that the distribution of flow in the columns is in accordance with design intent. Any compromises in this area would lead potentially to distortions in flow such as channeling which places any carbon steel surface affected at risk of increased rates of corrosion.
2. Material Selection – Given the fact that the key corrosion risk exists towards carbon steel, it needs to be determined whether in fact a change to a less ‘at risk’ material for the vessel shell is not the best long-term solution to this problem. Most of the internals have been shifted over the years to stainless steel and these appear to be unimpacted. The question therefore of making the entire vessel into one material arises which reduces any galvanic impact on the process also.

Based on the corrosion events experienced in the T-3411 stripper, it is evident that while longer terms solutions are being determined, shorter term actions are needed to mitigate the risk of repetitive failure and lend to a better understanding both corrosion mechanism and actual corrosion rates experienced. These conclusions have been further enhanced from observations made during a recent stop of the unit for inspection and installation of additional monitoring mechanisms. Although FAC mechanism is quite known in the boiler feed water and steam generation system, more focus should be insighted in the aqua ammonia system, including detailed corrosion model and possible application of new filming forming substances to mitigate this phenomenon..

References

[1] B. Dooley, H. Lister, "Flow-Accelerated Corrosion in Steam Generating Plants, PowerPlant Chemistry, 20(4) 2018, 194-244.

[2] M. Appl, Ammonia – Principles and Industrial Practice (Weinheim, Germany: Wiley-VCH), 1999.

[3] M. Davies, Materials Selector for Hazardous Chemicals – MS6: Ammonia and Caustic (St. Louis, MO, USA: MTI), 2004.

[4] B. Trémilon, La Chimie en Solvants Non-Aqueux (Paris, France: Presses Universitaires), 1971.

[5] Dillon, C. P., Corrosion Control in the Chemical Process Industries, Natl Assn of Corrosion, 1994

Unprecedented Catalyst Overheating due to Feed Preheat Coil Failure

On March 18th, 2023, Ammonia Kujang 1A faced operational disruptions due to a vacuum system issue, causing speed fluctuation in all compressors and subsequent shutdown of the back-end section. Control measures stabilized the Air and Natural Gas Compressors, but overheating in the HTS and LTS catalyst bed increased the temperature up to 1010 °C [1850 °F] and 515 °C [959 °F], respectively prompting a total plant shutdown while no interlock was triggered. Inspection revealed damages, including coil fracture and catalyst breakdown due to overheating caused by air exposure and water submersion. Maintenance was performed, but catalyst replacement wasn't possible. This paper provides a detailed discussion of the observed symptoms during the event, efforts to reactivate catalysts, and share insights aimed at improving the interlock system to effectively mitigate the risk of feed preheat coil failure.

Tatang Arind Pradibta, Robert Sarjaka, Ariffianto
PT Pupuk Kujang, Indonesia

Introduction

Pupuk Kujang is a subsidiary of PT Pupuk Indonesia (Persero), located in Karawang, West Java, Indonesia, operating two ammonia production facilities. Ammonia Kujang 1A, operational since 1978, utilizes MW Kellogg (KBR) Technology with a capacity of 1000 MTPD. Similarly, Ammonia Kujang 1B, operational since 2005, also employs KBR Technology with the same capacity. Amid operational challenges since 1978, Ammonia Kujang 1A underwent a retrofit in 1994 to enhance energy efficiency and extend the overall service life of the plant.

As a typical KBR Technology licensed ammonia plant, Ammonia Kujang 1A utilizes a top-fired Primary Reformer (101-B) to carry out a hydrocarbon cracking reaction to break down natural gas into synthesis gas. The flue gas from the radiant box primary reformer area with a temperature of ± 1100 °C [2012 °F] then flows through the Convection Section before finally being released through a chimney. In the Convection Section area, there are eight (8) coils to recover heat from flue gas, including Feed Preheater Coil (MK-M) that is designed to increase the temperature of natural gas to 390 °C [734 °F], ensuring optimal conditions for the hydrotreating reaction on the Hydrotreater Reactor (101-D).

Routine checks are conducted every 18 months to ensure the reliability of the heater coils, which includes assessing wall thickness and monitoring corrosion in the manifold and U-bend areas. Between 1994-2020, 7 out of 8 coils were replaced, with the Feed Preheater Coil (MK-M) remaining unchanged. Figure 1 provides an illustration of the convection section coil configuration.

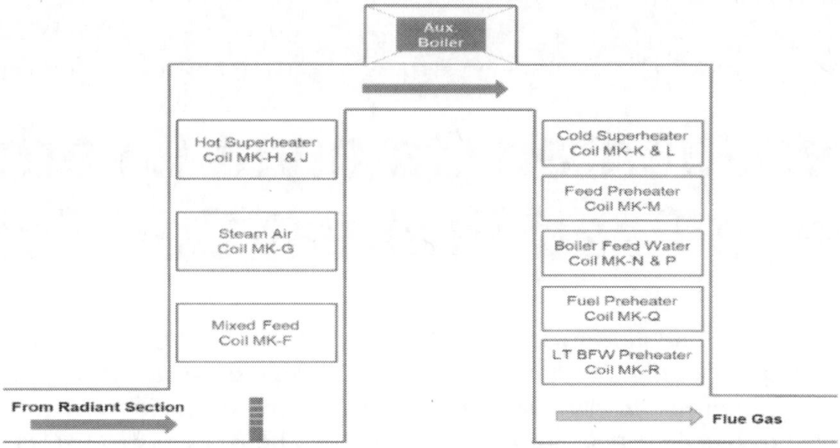

Figure 1. Diagram of Primary Reformer Convection Section – Coil Configuration

Failure Event

On March 18th 2023, when operating at full capacity, Ammonia Kujang 1A experienced a sudden surge in vacuum system pressure, escalating from 100 to 170 mmH_2O [3.93 to 6.69 inH_2O], accompanied by the activation of the compressor's field alarm. This sudden increase in vacuum system pressure led to fluctuations and uncontrolled speed in all compressors. In anticipation of potential equipment damage, the Syngas Compressor (103-J) and Refrigerant Compressor (105-J) were promptly shut down, and the production rate was adjusted to 90%. The syngas was vented to PCV-4 (inlet 103-J), revealing an anomaly wherein the opening of PCV-4 is only 10%, deviating significantly from the expected standard opening of 65-70% at a 90% operational rate.

Fifteen minutes after the vacuum trouble, Air Compressor (101-J) and Natural Gas Compressor (102-J) were successfully controlled, and the flow of natural gas (FRC-1) and air (FRC-3) were stabilized according to the standard ratio. However, there was a rapid temperature increase in Secondary Reformer catalyst bed up to 1,366 °C [2491°F], and extremely high temperature of HTS and LTS catalyst bed up to 1,010 °C [1850 °F] and 515 °C [959 °F], no interlock was triggered.

Considering the extreme increase in catalyst bed temperature, the decision was taken to initiate a total shutdown of the facility for a thorough analysis of the issue. During the shutdown process, a high-level alarm was triggered in the HTS Effluent KO Drum (117-F), necessitating the implementation of manual drainage in the field. The condensate flow rate in 117-F was exceptionally high, and visually, the condensate exhibited a dark brown color as depicted in Figure 2. Laboratory analysis confirmed that the condensate from 117-F contained a PO_4 content of 16 ppm. The HTS-LTS configuration diagram and trend of temperature during the event are depicted in Figure 3 and Figure 4 respectively.

Figure 2. Condensate Sample from 117-F

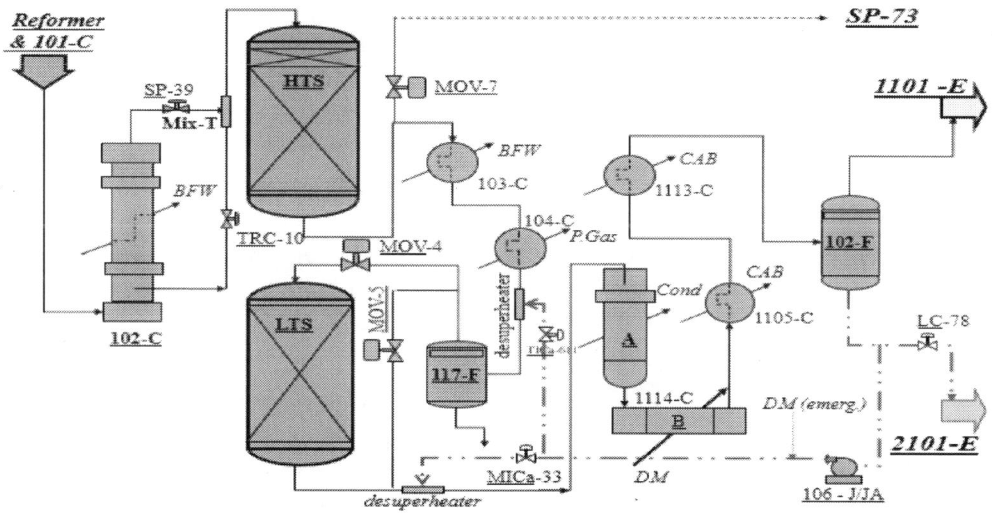

Figure 3. Detail Diagram of HTS – LTS Configuration

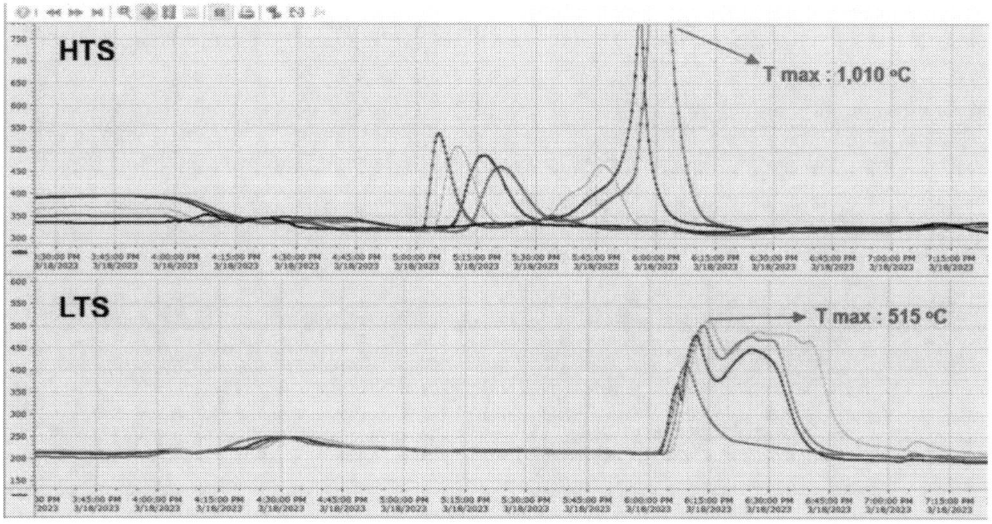

Figure 4. Trend of HTS-LTS Temperature During Failure Event

The temperatures in the HTS and LTS exhibited a gradual decline; however, a sudden temperature increase occurred in the Methanator (106-D) catalyst bed, rising from 290 °C [554 °F] to 320 °C [608 °F]. Immediate isolation of the 106-D catalyst bed was executed, and syngas was vented through PCV-5 (inlet methanator). Subsequently, the production rate was reduced to 50%, and a strange sound was observed in the convection section of the Primary reformer, as reported by field operator. Concurrently, it was observed that the outlet temperature of the HTS and LTS bed was closely equal to its inlet temperature. To analyze this issue, a comprehensive laboratory analysis was conducted on the gas inlets of the primary reformer and secondary reformer, revealing trace amounts of CH_4. Following the shutdown of the Air Compressor (101-J), laboratory analysis of the natural gas at the primary reformer's inlet indicated a CH_4 content of 19.77% and a strange sound remained observed in the convection section.

Findings

Primary Shift Effluent Waste Heat Boiler (103-C) Leak

Figure 5 illustrate a highly vigorous water flow in the gas-side drain of 103-C, providing valid evidence of a leaking tube. The leakage in 103-C was observed during a high-pressure test (HP Test) in the BFW system at a pressure of 40 kg/cm^{2}g [569 psig]. Inspection of the internal part of 103-C revealed a local metal loss of approximately +/- 1 mm [0.04 inch] on the upper tubesheet in the gas inlet-side, as depicted in Figure 6. Additionally, one tube was found to be fractured.

Figure 5. Leak on 103-C during HP Test

Figure 6. Metal Loss on Tubesheet in the Gas Inlet-Side

Fracture of the Feed Preheater Coil (MK-M)

Detecting leaks in the convection section coils poses inherent challenges due to the confined space, stacked coil configurations, and the presence of several coils in the form of fin tubes. In March 2023, the process of identifying gas leaks in the convection section spanned up to five days, employing a method that pressurized each coil alternately with nitrogen and low pressure steam at a pressure of 8 kg/cm^{2}g [114 psig]. The disassembly of the convection section walls facilitated inspection access. Subsequently, it was revealed that a tube in the feed preheater coil (MK-M) had fractured, as depicted in Figure 7.

Figure 7. Fractured Fin Tube on Feed Preheater Coil (MK-M)

The fracture in the feed preheater coil (MK-M) resulted in the majority of the feed gas being discharged within the MK-M coil, while the remaining feed gas could not enter the primary reformer tube due to lower gas pressure compared to steam pressure. This scenario caused an anomaly during shutdown, wherein the gas inlet analysis of the primary reformer indicated trace amounts of CH_4. Despite the feed gas being unable to enter the primary reformer, the Low Gas to Air Ratio interlock (BPS-3) remained inactive, as the Feed Gas Flowmeter (FRC-1) and Feed Gas Low-

Flow Alarm (LFA-18) sensors were positioned upstream of the feed preheater coil, as depicted in Figure 8.

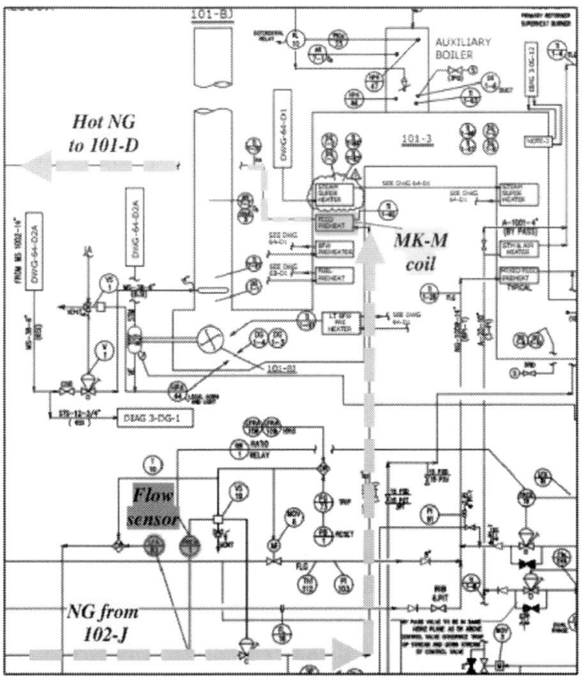

Figure 8. FRC-1 and LFA-83 sensors on Upstream of Feed Preheater Coil (MK-M)

HTS Catalyst Breakage

During internal inspection of 103-C, a significant amount of reddish-brown metal powder was found on the gas inlet cover. A thorough inspection of the HTS outlet pipe revealed the accumulation of metal powder in MOV-7 (valve venting system) and HTS Effluent KO Drum (117-F), as illustrated in Figure 9. The substantial buildup of metal powder on 117-F led to the blockage and damage of the demister within 117-F.

In Table 1, the results of the metal powder analysis are presented, indicating that the analyzed substance is identified as HTS catalyst powder. These findings suggest that the HTS catalyst was exposed to air and immersed in Boiler Feed Water from 103-C, subsequently leading to catalyst breakdown due to an unprecedented overheating surpassing the catalyst's sintering temperature. In November 2023, the HTS catalyst was replaced, revealing a significant portion of the catalyst had disintegrated into powder, and some parts of the catalyst had fused with the bottom screen, as depicted in Figure 10. Examination of the HTS vessel structure, made of carbon steel SA 516 Gr 70 with a thickness of 90 mm [3.5 inch], using ultrasonic and thickness tests showed no significant metal loss on the inner wall of the vessel and no defect on metal joints or weldings, confirming the HTS vessel's suitability and safety for normal operations.

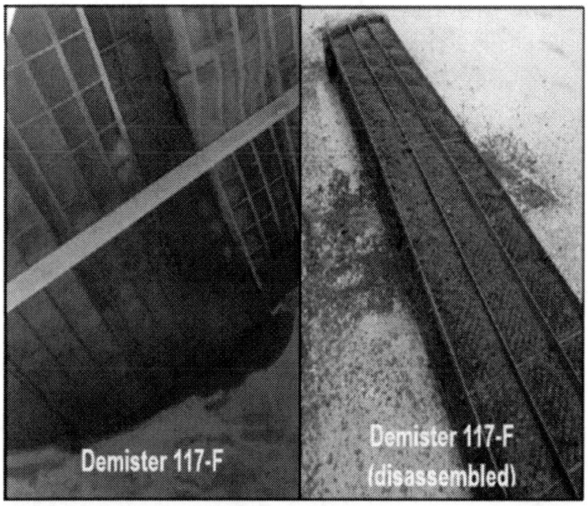

Figure 9. Metal Powder Accumulation in 117-F

Figure 10. HTS Catalyst Melt and Fused with Bottom Screen

Parameter	HTS Catalyst	Metal Powder
Fe_2O_3	86 - 92 %	86.58 %
Cr_2O_3	6 - 10 %	6.76 %
Al_2O_3	N/A	2.23 %

Table 1. Laboratory Analysis of Metal Powder Found on 103-C Cover

Failure Analysis

The failure of Feed Preheat Coil (MK-M) resulted from a disruption in the vacuum system, subsequently influencing the speed of all compressors. The sequence of events diagram is presented in Figure 11.

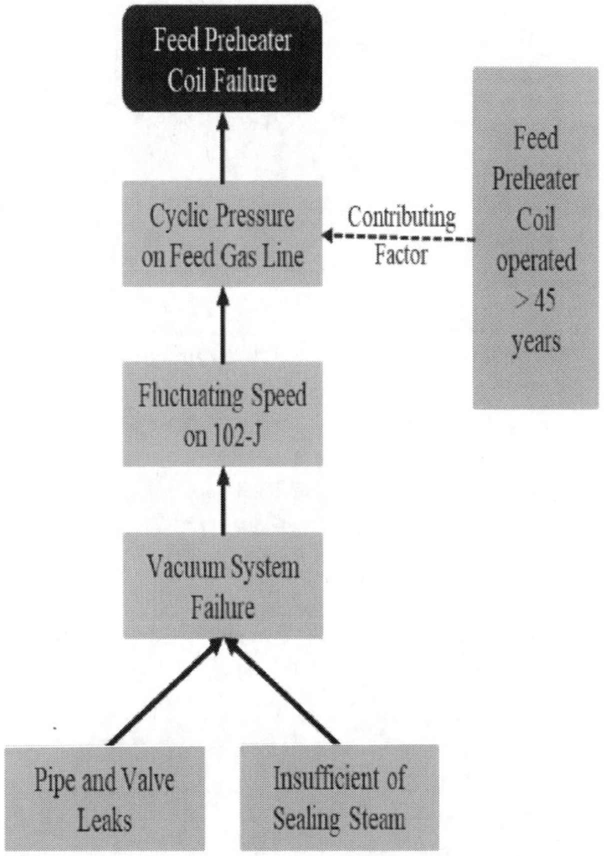

Figure 11. The Sequence of Event Diagram of Feed Preheater Coil Failure

Leaks on Piping and Valve of Vacuum System

Based on historical records, there have been several incidents of leaks in the vacuum system pipes, especially in the condensing turbine exhaust channel of cooling water pump (2209-JBT). If a leak occurs, ambient air will be sucked into the pipe, resulting in an increase in vacuum in the steam condenser which will result in a decrease in the performance of the condensing turbine and potentially cause turbine speed instability (hunting).

The inspection findings revealed a leak originating from the T-shaped junction connecting the turbine casing to the condensate pot, as depicted in Figure 12. This leakage is attributed to pipe thinning resulting from prolonged usage and poses challenges in detection as it is typically concealed by insulation. Additionally, a leakage was also detected in the insulation of the condensate pot venting valve (point C).

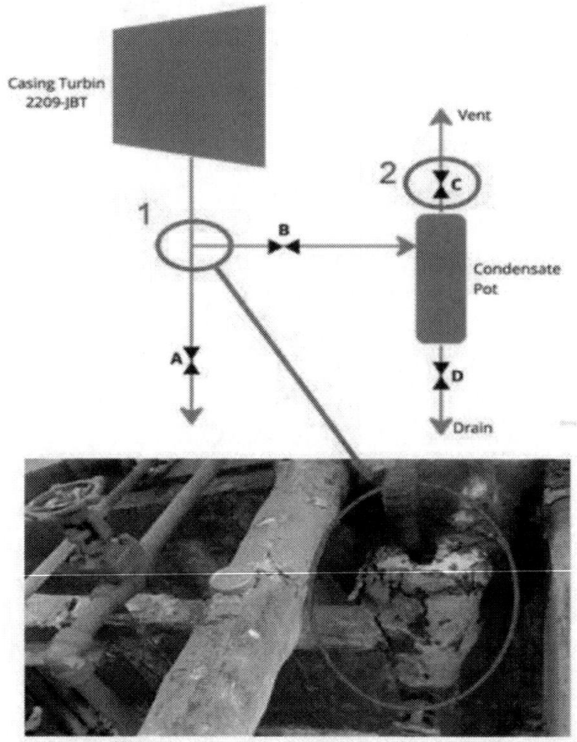

Figure 12. Leaks on Vacuum System

Insufficient of Sealing Steam Pressure

Condensing turbines are equipped with steam sealing mechanisms designed to prevent air ingress into the turbine vacuum chamber. Various factors, including turbine dimensions, seal configuration, and operational demands, determine the requisite steam sealing pressure, typically ranging between 0.1 to 1.0 bar absolute [1.45 to 14.5 psi].

Despite a thorough review of the 2209-JBT turbine technical documentation, no specific guidelines regarding steam sealing pressure were identified. Consequently, steam sealing adjustments have been manually conducted to date by operating the block valve, relying solely on visual assessments of which adequacy is determined by observing steam emissions. However, the pressure gauge affixed to the 2209-JBT turbine's steam sealing mechanism exhibits visual distortion, as illustrated in Figure 13, introducing uncertainty regarding sealing efficacy. Inadequate sealing steam pressure could potentially contribute to vacuum pressure escalation.

Figure 13. Pressure Gauge of Sealing Steam on 2209-JBT

Contributing Factor :
The Extended Usage of Aging Fin Tube Coil

Since the commencement of operations of Ammonia Kujang 1A, the Feed Preheat Coil (MK-M) has not undergone replacement and has thus far functioned for over 45 years. Operating such an aging fin tube coil entails significant risk. The current predictive maintenance practices involve periodic thickness measurements and penetrant tests conducted at every major overhaul, with the most recent thickness measurement, conducted during 2020 Major Overhaul, yielding satisfactory results, as indicated in Table 2, with the lowest thickness of 5.4 mm [0.21 Inch] while the minimum thickness limit is 1.8 mm [0.07 Inch]. It should be noted that thickness measurements are only carried out on the U-bend part and not on the fin tube due to limited maintenance access.

Elbow	ACTUAL THICKNESS (mm)								
	ROW. 1			ROW. 2			ROW. 3		
No.	a	b	c	a	b	c	a	b	c
1	6,18	5,85	5,90	6,40	5,80	5,74	6,30	6,53	6,30
2	6,00	5,83	5,94	5,77	5,80	5,87	6,62	5,61	6,18
3	5,88	6,01	5,91	5,91	5,54	5,70	5,39	6,12	5,92
4	6,01	6,11	6,15	6,10	6,02	6,16	5,92	6,08	6,16
5	6,25	5,98	6,78	5,82	5,94	6,13	5,60	5,77	5,85
6	5,80	5,92	5,87	5,82	6,08	6,06	5,95	6,25	6,30
7	5,62	6,20	6,11	5,88	5,78	5,92	6,27	6,18	6,25
8	6,02	5,80	5,95	6,42	6,77	6,51	5,92	5,64	6,25

Table 2. Thickness Measurement of U-bend Feed Preheat Coil (MK-M)

Catalyst Reactivation

The HTS and LTS catalysts were exposed to air and submerged in Boiler Feed Water, leading to an unprecedented overheating surpassing the catalysts' sintering temperature due to oxidation. Consequently, some HTS catalysts disintegrated into powder, which was detected in several downstream equipment units. Despite relatively short operational periods for both HTS and LTS catalysts, with each being in use since March 2022 (12 months) and September 2020 (29 months), respectively, both are expected to be inactive due to extensive oxidation. Ideally, replacing the HTS and LTS catalyst is imperative to ensure the continued operation of Ammonia Kujang 1A, given the confirmed damage to the HTS catalyst. However, as of March 2023, there is an insufficient stock of readily available HTS and LTS catalysts, necessitating repairs to be conducted without catalyst replacement.

HTS Catalyst Reactivation

The reactivation of the HTS catalyst involves a gradual heating process to dry out the catalyst bed, followed by a reduction stage using the process gas outlet from the reformer. In Ammonia Kujang 1A, the heating up process is conducted by using steam sparger facility, as depicted in Figure 14. During the heating-up stage in April 2023, a significant condensate flow was observed in the HTS drain. To facilitate the flow, an additional drain was implemented by opening the TRC-11 flange (103-C bypass).
The detailed stages of the HTS catalyst reactivation are as follows :

1. Do proper N_2 purging to sweep off any remnant oxygen inside HTS reactor. While doing N_2 purging, open all low point drain at and around HTS reactor to ensure smooth flow.
2. Start HTS heating up with rate 10 °C/hr [50 °F/hr] from ambient temperature until the temperature just below or at the condensation temperature (90-100°C) [194-212 °F].
3. Hold this temperature until HTS catalyst is dry and monitor the water flow released in the drain. The dry catalyst is indicated by stable/very slow increment of water flow in drain.
4. Once the catalyst is indicated as dried, continue heating up at 10°C/hr [50 °F/hr] rate until temperature in between 175-200 °C [347-392 °F].
5. Continue heating again at 25 °C/hr [77 °F/hr] until temperature 285-300°C [545-572 °F] as preparation for feed gas introduction to Primary Reformer.
6. Conduct HTS reduction. HTS reduction will happen immediately once feed gas is introduced to Primary Reformer. During HTS reduction period, S/C ratio at Primary Reformer inlet should be controlled at 15 – 20 (mol/mol). HTS reduction will be indicated by heat wave passed thru HTS bed reactor. When HTS reduction is completed, all bed temperatures should be near the inlet temperature after feed gas is introduced to Primary Reformer.
7. Once HTS catalyst is fully reduced, S/C ratio at Primary Reformer inlet could decreased gradually to 6 – 8 (mol/mol). Monitor bed temperature profile at HTS.
8. Secondary Reformer top bed temperature could be increased. Once the top of the secondary reformer catalyst bed has been brought to approximately 650 °C [1202 °F], air may be introduced with combustion automatically occurring.
9. Closely monitor on HTS bed temperature profile as well as dP. If no abnormality is observed, S/C ratio and plant load could be normalized.
10. Once plant operation has reached steady state, gas analysis at inlet and outlet reactor could be done to confirm catalyst performance.

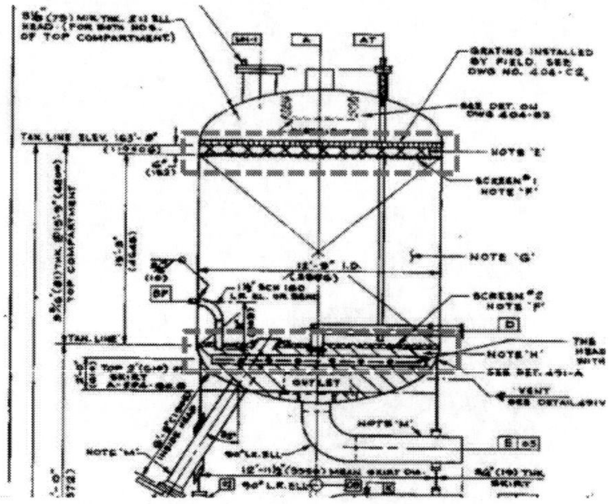

Figure 14. Top and Bottom Steam Sparger Facility on HTS Ammonia Kujang 1A

LTS Catalyst Reactivation

The reactivation of the LTS catalyst involves a gradual heating process to dry out the catalyst bed, followed by a reduction stage using hydrogen gas. In Ammonia Kujang 1A, the heating up process is conducted by circulating nitrogen using Air Compressor (101-J). During the heating-up stage in April 2023, a significant condensate flow was observed in the 101-J Interstage Cooler (130-JC) drain. The detailed stages of the LTS catalyst reactivation are as follows :

1. Do proper N_2 purging to sweep off any remnant oxygen inside HTS reactor, open all low point drain around LTS reactor to ensure smooth flow.
2. Establish normal and stable N_2 circulation for LTS unit. Once N_2 circulation circulation is well established (a minimum space velocity of 200 hr-1 to ensure good flow distribution across the catalyst bed), monitor dP at LTS and compare the measured dP with historical data. If no significant dP increment is observed, N_2 circulation/LTS start up circulation could be maintained and continued as preparation for LTS heating up.
3. Start LTS heating up with rate 10°C/hr [50 °F/hr] from ambient temperature until the temperature just below or at the condensation temperature (90-100°C) [194-212 °F].
4. Hold this temperature until LTS catalyst is dry and monitor the water flow released in the drain. The dry catalyst is indicated by stable/very slow increment of water flow in drain.
5. During this period, calibrate the H_2 flow meter by introducing a small flow of H_2 into the carrier gas for a brief period, about 60-90 seconds. Sample the carrier gas containing H_2 at the inlet sample point and obtain laboratory analysis for percent H_2. Compare the lab results with the meter setting to check the meter correction factor.
6. Once the catalyst is indicated as dried, continue heating up at 10°C/hr [50 °F/hr] rate until temperature in between 175 – 200 °C [347-392 °F].
7. Reduce the reactor inlet temperature to 165-180°C [329-356 °F]. When this temperature is achieved, hold for a certain time to allow for homogenous temperatures distributed across the entire bed.
8. Start to introduce H_2 into the N_2 to achieve 0.5-1.0% of H_2 concentration in inlet gas, maintain inlet temperature at 165-180°C [329-356 °F] and take samples of LTS inlet and outlet gas and analyze for percent H_2.
9. As the reduction has been confirmed to proceed, indicated by the outlet H_2 concentration is essentially nil and there is a measurable temperature rise in the bed, increase the inlet H_2 concentration to 1.0 – 1.5% gradually while maintaining the inlet temperature at 165-180 °C [329-356 °F].
10. At these conditions, a peak temperature of approximately 220 °C [428 °F] will move through the bed as the catalyst is reduced. The maximum temperature in the bed should be limited to 230 °C [446 °F] during the reduction.
11. During the reduction, the reaction zone moves down through the bed. When the reaction zone approaches the bottom of catalyst bed, H_2 breakthrough can be expected.
12. At the first sign of the reduction close to completion, increase the inlet temperature to 213-218°C [415-424 °F at the rate of 10-15°C/hr [50-59 °F/hr] when the inlet and outlet H_2 concentration differ less than 0.3% and the temperature in the lower portion of the bed has fallen below 205 °C [401 °F].
13. When all bed temperatures are 200 °C [392 °F] or higher and the inlet and outlet H_2 concentration difference is within 0.3%, slowly raise the inlet H_2 to 3.5-4.0%.
14. Continue with 3.5-4.0 mole-% inlet H_2 and observe the bed temperatures. After 1 (one) hour during which the bed temperatures have not increased and the inlet and outlet H_2 concentration difference remains within

0.3%, increase the inlet H_2 concentration to 5-6%.

15. The reduction is considered complete when no further H_2 is consumed, and all bed temperatures are stable.
16. After completion of the reduction, cut the feed of H_2-rich gas into the carrier gas, place the LTS reactor to online mode and purge H_2 contained gas in the start-up circuit with nitrogen.
17. Prepare the LTS reactor to be placed online. Before doing so, check the HTS outlet gas composition to ensure it is safe for LTS. Once confirmed, bring the LTS reactor online and closely monitor the LTS bed temperature profile and differential pressure (dP).
18. Once LTS operation is normal and stable, gas analysis at LTS inlet and outlet should be done to confirm LTS catalyst performance.

Table 3 illustrates the comparison of the HTS-LTS outlet analysis before and after the incident. Despite a slight increase in CO slip observed at the HTS and LTS outlets, reaching 2.8% and 0.28% respectively, this elevation does not result in a temperature rise in the methanator catalyst bed.

Parameter	Before	After	EoR Standard
Rate of Production	100 %	100 %	100 %
CO outlet HTS	1.97 %	2.8 %	2.68 %
CO outlet LTS	0.12 %	0.28 %	0.17 %
ΔP HTS Bed	0.26 bar [3.7 psig]	0.46 bar [6.7 psig]	< 1 bar [< 14.5psig]
ΔP LTS Bed	0.67 bar [9.7 psig]	0.76 bar [11 psig]	< 1 bar [< 14.5psig]

Table 3. Comparison of HTS & LTS outlet Analysis and ΔP Before/After Reactivation

Conclusion

In ammonia plants that have a natural gas flowmeter on the upstream side of the preheater coil, such as the Ammonia Kujang 1A plant, there exists a scenario of significant operational failure that involves the failure of the Feed Preheater Coil unit. The risk of feed preheater coil failure has not been adequately mitigated by the existing interlock system, and the potential for coil failure is substantial, especially in aging plants. The failure of the feed preheater coil unit led to extreme overheating of the HTS-LTS catalysts and resulted in an 18-day shutdown of Ammonia Kujang 1A.

To prevent similar incidents in the future, several programs need to be implemented :

a. Addition of a Low Flow Feed Gas sensor at downstream of the Feed Preheat Coil.
b. Feed Preheater Coil Replacement
c. Ensure all pressure gauge indications in the vacuum system can be read properly and regularly maintained
d. Carry out routine checks of piping in the vacuum system
e. Review existing interlock system for potential unmitigated risk

The incident of HTS-LTS catalyst overheating due to air exposure and water submersion occurred for the first time at Ammonia Kujang 1A or similar plants. Nevertheless, the reactivation efforts of the HTS-LTS catalysts have been successfully carried out, allowing the plant to operate at full capacity for over 7 months without the need for catalyst replacement.

Once through ammonia converter for revamp of a fertilizer facility

Koch Fertilizer collaborated with Uhde and Johnson Matthey to develop a revamp solution that increased ammonia production capacity of a Koch Fertilizer facility. A legacy Kellogg ammonia plant was revamped by integrating Uhde's once through ammonia converter upstream of the existing ammonia synthesis loop. This paper describes the once through ammonia converter concept with some key details around design, installation, commissioning, and operation.

Rahul Solunke[1], John Brightling[2] and Dirk Meurer[3]
[1]Koch Fertilizer, [2]Johnson Matthey and [3]Uhde

Introduction

To help achieve the revamp capacity, process experts from Uhde developed an innovative cartridge insert for an existing pressure shell based on latest design principles. Use of the existing pressure shell saved costs and time. Much of the installation work was done while the existing plant was running which minimized the turnaround activities. Covid pandemic posed a significant challenge requiring elevated attention to personnel health. Another challenge was to keep the pyrophoric converter catalyst under inert Nitrogen atmosphere during catalyst loading and all startup activities. Uhde and Koch Fertilizer together worked on a HAZOP study to ensure safe operation with respect to personnel, process and the environment.

A significant novel challenge of the revamp scheme was the very low operating pressure of Uhde's once through ammonia synthesis converter - about 960 psig (66 barg). Johnson Matthey's testing on their catalyst KATALCO 74-1 confirmed acceptable performance at such low pressure which made this revamp option technically viable. Johnson Matthey's on-site support for catalyst reduction also significantly helped in commissioning of the new Uhde once through ammonia converter.

The joint solution from Uhde and Johnson Matthey along with some other major revamp scope like reformer tube assembly replacement (reharp), high pressure synthesis loop's reversal and upgrades to CO2 removal process increased the ammonia production capacity by about 20%. The site needed to learn and adapt to the interactions between Uhde's new once through converter (low pressure) and existing two old loop converters (high pressure), which they quickly did. Since its commissioning in September 2022, both Uhde's once through converter and Johnson Matthey 's catalyst have run reliably and met revamp project expectations. The site has seen a significant boost in gas efficiency improvement as well.

Pre- and Post-Revamp Equipment Layout

Figure 1 and figure 2 below show the plant layout before and after revamp. As seen in figure 2, once through converter was installed between the low case and high case of the syngas compressor (103-J). The loop was also reversed which moved the ammonia chillers and 106-F ammonia separation drum from upstream to downstream of the high-pressure loop converters A and B. This change in location of the ammonia chillers required to add an additional step to remove the water from incoming makeup gas. Typically, such drying is done by either a molecular sieves unit or an ammonia wash unit. Koch Fertilizer adopted to go with ammonia wash unit which took some of the liquid ammonia from 106-F for washing the moisture out from syngas at the discharge of the low case of the syngas compressor. Once through converter required its own ammonia chiller (119-CA) and ammonia separation drum (126-F).

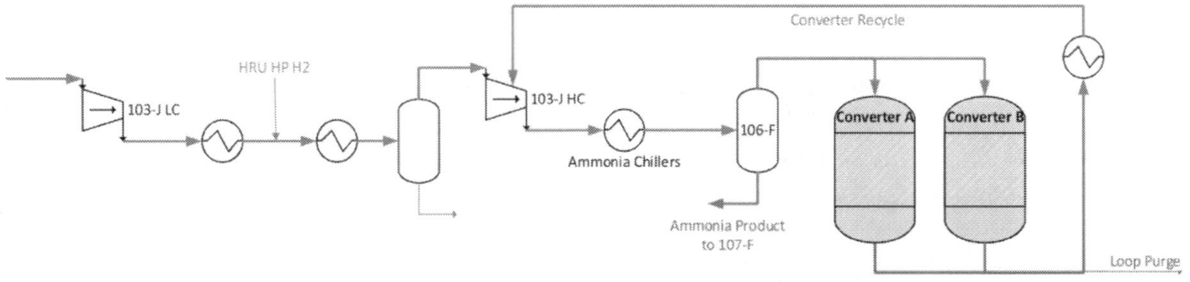

Figure 1. Pre-revamp layout

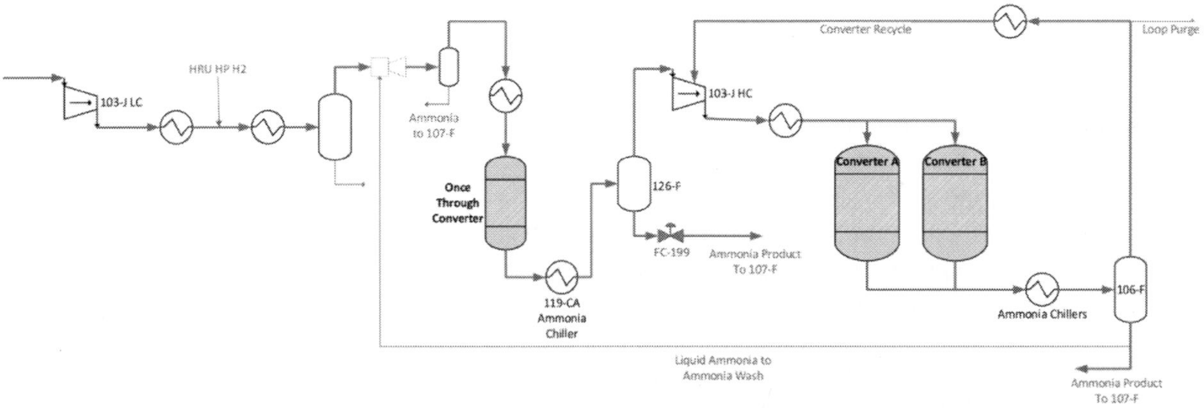

Figure 2. Post-revamp layout

Dual Pressure Process

Uhde has developed the dual pressure ammonia process as a way to design and build large plants with low project execution risks. The considerations leading to its development are discussed in more detail in [1] and [2]. Today, the five largest ammonia plants in the world use this process, and another is under construction. The process is schematically shown in Figure 3.

Compared to the conventional process, in the dual pressure process a once-through ammonia synthesis is inserted upstream of the loop synthesis, operating at the outlet pressure of the low-pressure (LP) casing of the syngas compressor. The ammonia produced there is separated, and the remaining synthesis gas is sent into the conventional synthesis loop. The desired effects are:

- The syngas compressor load is lowered;
- Synthesis loop equipment and piping sizes are designed for approximately two thirds only of the total production.

For large plants, both contributes to lowering the scale-up risk to an acceptable level. The dual pressure process is not completely new but quite conventional with the only major difference being the newly introduced once-through synthesis.

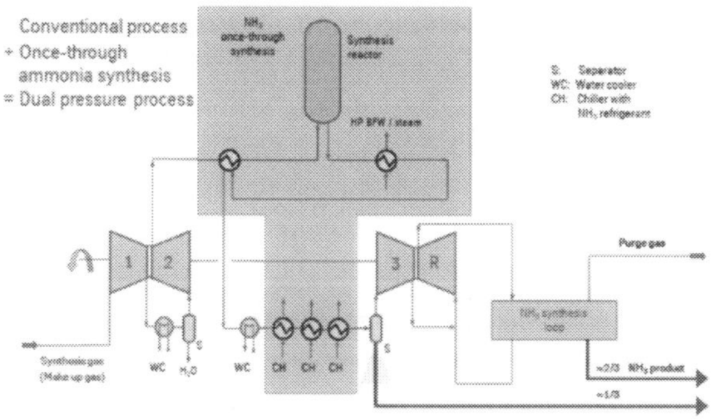

Figure 3. Simplified scheme of thyssenkrupp Uhde's dual pressure ammonia synthesis

In the project in focus in this paper, the dual pressure process has been applied to an existing plant for a capacity increase. This is the second project of this kind; a similar revamp has been finished in 2007, where Uhde has realized a capacity increase by 30% in a 1,000 t/d ammonia plant from 1973 at Duslo, Sala, Slovakia [3].

Cartridge Design

The ammonia converter cartridge used in this project is in some points different to those used in new plants. The reason being the fact that an existing pressure shell was chosen to be re-used by Koch Fertilizer in order to lower the cost of the project. This led to a very narrow slim design plus the requirement not to exceed the shell design temperature, a heat exchanger is included to cool the effluent gas down to 252°C (486°F). This heat exchanger is installed at the bottom of the cartridge. In addition, the cartridge is designed with two radial catalyst beds and an internal heat exchanger located in the center of the first catalyst bed. To keep the design slim, the gas from the main inlet is routed to the bottom heat exchanger through the gap between the pressure shell and the cartridge. This also ensures that the limited design temperature of the shell is not exceeded. The gas then passes the bottom heat exchanger on shell side and is routed towards a central pipe upwards to the internal heat exchanger and then through the first catalyst bed. For control of the catalyst bed inlet temperatures two other gas inlets are available to bypass the internal heat exchanger and both heat exchangers. The cartridge has a modular design. The internal heat exchanger and the part with the first catalyst bed can be removed separately to get access to the second catalyst bed for loading and unloading of catalyst. Refer to Figure 4 for a sketch.

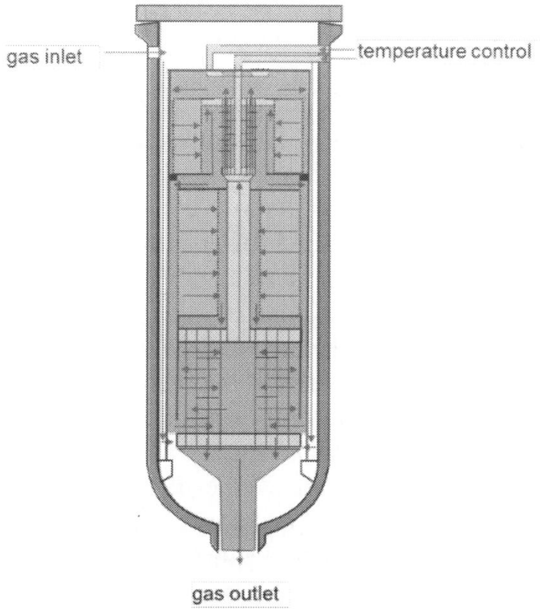

Figure 4. Sketch of the new converter cartridge. Orange: catalyst beds.

Advantages of OT Synthesis Design for Construction

Often in a revamp for a significant capacity increase, there are many equipment items which must be replaced or modified.

One of the advantages of the dual pressure synthesis is that the existing loop equipment can remain unchanged. The once-through synthesis is installed upstream of it with very few tie-in points. It can be installed completely while the plant is in operation, and only a short shutdown is needed to finish the tie-ins.

Catalyst Testing for Low Pressure Operation

Koch Fertilizer's once through reactor design needed to be implemented at the unusually low operating pressure of 960 psig (66barg), This is believed to be the lowest operating pressure for commercial ammonia synthesis in any production worldwide highlighting the innovative challenge of Koch Fertilizer's revamp project. As pressure is decreased the rate of the ammonia synthesis reaction declines – hence high catalyst activity is essential the catalyst selected for the design was KATALCO 74-1A.

As the pressure of operation was lower than that where KATALCO 74-1A catalyst had proven operation before additional test data was generated by Johnson Matthey to provide assurance that of the performance over a lower pressure range, as this was a lower pressure than had been in commercial operation before.

The highly active ammonia synthesis catalyst, KATALCO 74-1 series was developed by ICI in the 1980s specifically for operation at low pressure with the AMV and LCA processes designed for 1160psig, 80 Barg synthesis . The catalyst is magnetite iron-based and contains the same promoters as found in conventional synthesis catalysts (Al_2O_3, K_2O and CaO), plus a small amount of cobalt as CoO. The higher activity is the result of the inclusion of cobalt and optimization of promoters to give the best results under low-pressure conditions. Handling of the finished catalyst is the same as conventional synthesis catalysts; the cobalt concentration is too low to present any hazards during catalyst charging, operation or discharge. Reduction also follows the same general procedure as conventional catalysts, with the advantage that the reduction temperature of KATALCO 74-1A is significantly lower than with standard catalysts.

Coincidentally in parallel but connected directly to this project the University of Cambridge, UK also tested Johnson Matthey ammonia synthesis catalyst at pressures down as low as 20 bar as is envisaged for some possible future Green ammonia projects that could for example be based pressurised H_2 available using Proton Exchange Membrane (PEM) electrolysis [4]. The university's testing independently supports the finding that KATALCO 74-1A offered significantly increased activity over conventional catalyst.

This increase in activity and an increased ease of reduction are achieved by the incorporation of cobalt oxide as a promotor with the effect of increasing the rates of nitrogen adsorption and ammonia desorption from the surface of the catalyst, hence increasing the rate of the overall synthesis reaction.

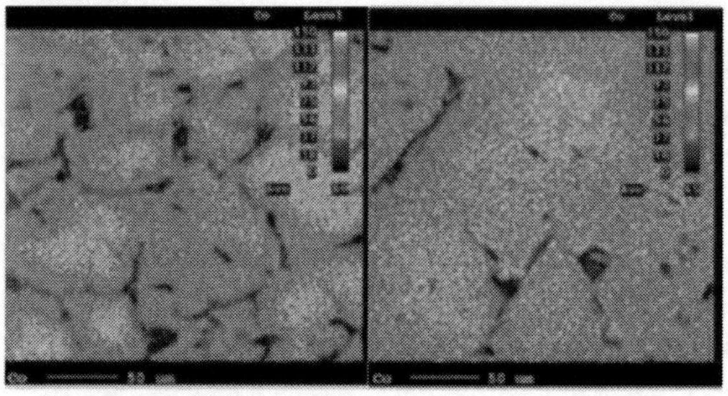

Figure 5. EPMA (Electron probe micro-analyzer) images show Co promotion in magnetite oxidic and reduced (left and right respectively)

As can be seen from Figure 5, cobalt is present and distributed throughout the catalyst structure, rather than being located at discrete points within the lattice or particle. The incorporation of cobalt into the iron lattice distorts the structure of the fused iron catalyst, generating layers of cobalt spinels which typically produce smaller iron crystallites on reduction. This is a significant factor in generating the high activity and stability of the ammonia synthesis catalyst performance over time. The magnetite structure is a Fe3+ and F2+ matrix meaning some of the Fe3+ sites can be substituted by Alumina. Alumina is recognized as a strong refractory structural promotor which reduces sintering rate compared to wustite-based Fe2+ catalysts which cannot contain alumina.

The conclusion of the University of Cambridge paper [5] is that such high activity catalyst could be a practical alternate to the sensitive and much more expensive Ru-catalyst that are often proposed for lower pressure service.

These findings are supported by industrial experience of KATALCO 74-1 catalyst, in 2002 [6] the catalyst was used in a complete converter retrofit to replace an earlier retrofitted Ru-based catalyst in the final 4th bed 'booster KAAP converter' [7]. This duty being a most demanding bed due to the combination of high inlet ammonia and lowest operation temperature.In operation in this booster converter the catalyst not only easily achieved design rates of over 18% but lasted in-service for 21 years until the plant closed due to it being unable to secure an affordable gas supply. Hence a long running commercial reference has demonstrated that high activity Fe-based catalysts are a practical and robust solution compared to Ru catalyst which is too expensive and sensitive.

Once Through Converter Catalyst Loading

The catalyst was preloaded in the new converter using a filling hopper, Figure 6a from which multiple hoses each fitted with Uhde / Johnson Matthey proprietary RSK distributors, Figure 6b to ensure dense even loading of the catalyst beds charging KATALCO 74-1A oxidic catalyst to the 2nd lower bed and KATALCO 74-1AR, pre-reduced to the 1st upper bed.

Figure 6a. KATALCO 74-1 hopper loading

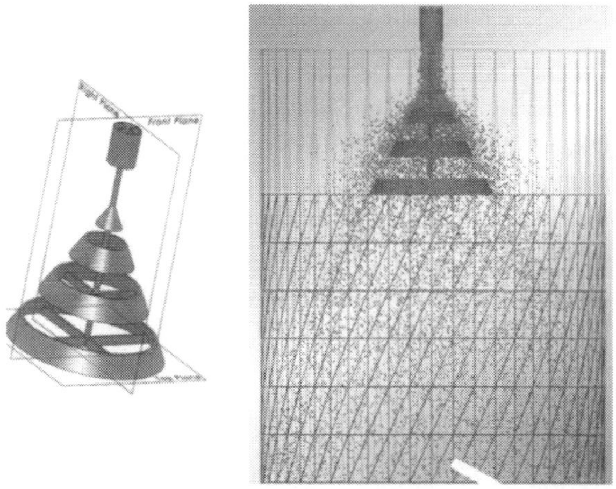

Figure 6b. Design of Uhde RSK loading device

To ensure minimum plant downtime and an efficent project Koch Fertilizer decided to install the cartridge and charge the catalyst before the shutdown keeping the vessel under nitrogen atmosphere and leave it standing isolated alone, until shutdown. During the shutdown tie-ins were made to connect to the plant ready for commissioning and the catalyst reduction. The time during which the reactor stood isolated under nitrogen atmosphere was approx. 50 days

Once Through Converter Catalyst Reduction

Koch Fertilizer installed Johnson Matthey's catalyst in Uhde's once through ammonia converter. For reduction Johnson Matthey's Technical engineers were on site to assist and monitor the reduction. Catalyst came in the oxidized form and needed to be reduced to activate it for the ammonia synthesis reaction. The reduction of the catalyst was done with the synthesis gas by following all the guidelines and procedures prescribed by Johnson Matthey. The procedure required monitoring of several operational parameters. The three major items to watch for were (1) isolation of ammonia chillers during early part of the reduction, (2) heating ramp rate of the top catalyst bed prior to reaching the target temperature for reduction and (3) water content in the syngas at the converter outlet during the entire reduction process.

As the Once through converter had an ammonia chiller downstream of the converter, water released during the reduction could freeze inside this chiller if it was left in operation potentially leading to tube rupture and a lengthy outage. Therefore, for the early part of the reduction, catalyst reduction procedure required complete isolation of this chiller. It was advised to open the chiller isolation valves only after ammonia concentration in the liquid collected in ammonia separation drum reached about 25%. At such concentration level, all risk of water freezing can be safely said to be eliminated.

A 56 C/hr. (100 F/hr.,) heating ramp rate was used before reaching the reduction target temperature which ensured reasonable speed for heating of the catalyst bed while making sure that converter piping, nozzles, and catalyst itself did not experience any unreasonable thermal stresses.

Once the reduction target temperature was reached, a 1000 ppm limit on the water content in the outlet syngas ensured full control over the dynamics between completion of the reduction and onset of ammonia synthesis reaction. Ammonia synthesis is an exothermic process. Reducing the catalyst too fast could kick-start the ammonia synthesis reaction in a large portion of the catalyst bed too early leading to large exotherm which could be difficult to manage while the reduction process is still in-progress. If temperatures inside the bed become too hot while the reduction is still ongoing, higher levels of water/steam evolved could react with the catalyst leading to its re-oxidization. Repetitive reduction-oxidization could damage the pore structure of the catalyst leading to significant catalyst activity loss.

Good gas distribution during reduction is particularly important because uneven flow through the catalyst bed will give uneven reduction, and therefore uneven distribution of the ammonia synthesis reaction during the reduction period.

Catalyst that is reduced at a low flow rate will then be exposed to a higher water content than catalyst reduced at a higher flow rate. This could lead to differences in catalyst activity. The damage caused by high water concentrations is more severe at high temperature.

Operational and Safety Challenges in Catalyst Reduction

Ammonia wash strainer failures – As part of the revamp, the ammonia synthesis loop was reversed which moved ammonia refrigeration chillers from upstream of the high-pressure loop converters to a downstream location. This change in chillers layout required an addition of ammonia wash system between low pressure case and high pressure case of the syngas compressor in order to ensure that synthesis gas going to the loop converters is fully dry. During startup of the plant, site noticed that strainers on ammonia wash system were getting clogged up quickly. To make the matter worse, revamp did not add a parallel strainer which meant that the strainer must be cleaned online, or site should stop all the startup activities to clean the strainer. Unfortunately, site had to trip the once through converter few times during the reduction process until site devised a strategy to safely backwash the strainer to a tote while keeping the once through converter online.

The ammonia wash upstream of the once through converter ensured that synthesis gas stays dry, and it also added some ammonia to the synthesis gas going to the ammonia synthesis converters. This carryover ammonia was sufficient to avoid freezing risk in chiller downstream of the once through converter when the reduction was in progress. However, whenever ammonia wash was not available due to strainer plugging issues the freeze risk in the chiller was elevated. To mitigate this issue, the chiller's refrig ammonia flow was controlled in such a way that chiller temperature does not drop below the water freezing temperature.

Strainer plugging was determined to be due to inadequate cleaning procedures during the commissioning of the revamp. Plugging frequency eventually dropped to several months.

Water carryover to high pressure loop converters – When the once through converter catalyst was being reduced the high pressure loop converters were fully online. The ammonia separation drum (126-F in figure 2) was the only location where water from the reduction process could be separated out before the syngas goes to the high-pressure loop converters. It is well known that high pressure loop converters cannot tolerate more than 10ppmv of oxygen compounds in the inlet syngas. Therefore, it was very important to monitor the temperature of ammonia separator drum. Chiller refrig ammonia flow was controlled in such a way that ammonia separator drum's temperature was not too cold to cause any freeze risk but at the same time not too hot to cause too much water carryover to the high-pressure loop converters.

Higher inerts to high pressure loop converters – After once through converter catalyst's reduction was complete, site started to divert all the synthesis gas to go through the once through converter which was the normal flow path per the revamp design. After this flow change, site experienced a trip of the high-pressure loop converters. At first, it was feared that reduction was not fully complete and water from reduction could have landed on the high-pressure loop converter catalyst causing complete loss of the synthesis reaction. However, a close look at the temperature of the ammonia separation drum (126-F in figure 2) revealed that water carryover capable of extinguishing the high-pressure loop converters was technically impossible. This closer look at the data also revealed that as the site started to divert more and more syngas to go through the once through converter, the loop inerts started to increase. However, the purge flow had not been adjusted to keep the loop inerts in check, meaning the loop converters eventually tripped on high loop inerts. The amount of loop inerts was not unusual for the converters but the rate at which they increased was unusual. This was a significant learning for the plant operations. In a normal loop without the once through converter, there is no reason for any increase in the loop inerts at constant purge as long as the front end is stable. However, when the synthesis gas starts to go through the once through converter, ammonia is produced in that reactor which is separated in a downstream ammonia separator drum. So, the synthesis gas leaving ammonia separator drum and going to the high-pressure loop converters contains a higher concentration of inerts than the makeup gas at the suction of the syngas compressor. Plant operations quickly learned this process change and started to adjust the purge flow as synthesis gas from discharge of the syngas compressor was getting diverted to fully go through the once through converter. This resulted in a good control over the loop inerts which ensured stable operation going forward.

Ammonia separator liquid level control valve failure – When the catalyst reduction was almost complete, plant experienced a trip due to the failure of the level control valve (FC-199 in figure 2) on ammonia separator drum (126-F in figure 2). Since the reduction was almost complete, this valve failure did not cause any major setback on the reduction procedure. However, site had to get the plant back online by running on the bypass valve and manually adjusting that valve to control the liquid level until a new valve came onsite. The second new valve also failed after few weeks of operation. It was later determined that the control valve's metallurgy was not suited for the combination of pressure, temperature, and material composition inside the ammonia separation drum (water+ammonia during reduction). After upgrading the metallurgy, site did not see any control valve failure.

Performance in Normal Operation

After the once through converter catalyst was fully reduced, site met the revamp production and efficiency targets. Apart from once through ammonia synthesis converter, the plant revamp also included several other major changes like reformer tube reharp, synthesis loop reversal and CO2 removal upgrades. After all these changes, plant's ammonia production has increased by about 20%. Once though converter's share in this increased capacity is about 50%. Once through converter operates at an intermediate syngas pressure (about 62 barg, 900 psig) which results in lower syngas compressor power requirement per ton of ammonia produced which ultimately results in better gas efficiency. After all revamp changes, site observed about 5% increase in efficiency (i.e. natural gas consumed per ton of ammonia produced). Once through converter's share in this efficiency improvement was significant.

With a recent data set, a full performance analysis by Johnson Matthey showed the catalyst to be operating as expected and that the inlet conditions to the converter meant it was operating at optimal conditions needed for maximum conversion rate.

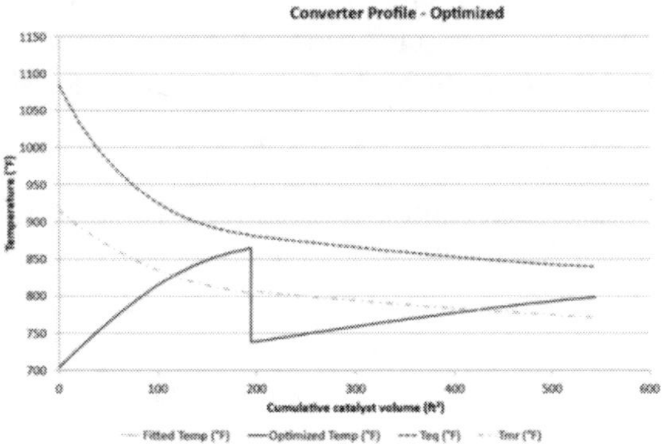

Figure 5. JM's performance evaluation of once through catalyst beds

Conclusion

It has been more than two years since the new once through converter became fully operational. It is maintaining its production capacity without any reliability issues. It is too early to comment on long-term reliability of the converter or long-term performance of the catalyst. The project had its own challenges and site learned some lessons too. But, overall, Uhde's once through ammonia converter with Johnson Matthey's KATALCO 74-1 catalyst has proved to be a good revamp change for a Koch Fertilizer facility.

References

[1] D. Lippmann, J. Larsen, "Extending Ammonia Plant Capacity Limits with Proven Technology", Nitrogen Conference 2002, Doha, Qatar, 2002

[2] K. Noelker, C. Meissner, "4,700 mtpd Single-Train Ammonia Plant based on Proven Technology", AIChE Safety in Ammonia Plants Symposium, 2016

[3] D. Lippmann, "Successful Execution of Ammonia Plant Revamps in Eastern Europe and Russia", Nitrogen + Syngas Conference, Moscow, Russia, 2008

[4] An overview of water electrolysis technologies for green hydrogen production S. Shiva Kumar , Hankwon Lim, https://doi.org/10.1016/j.egyr.2022.10.127

[5] Low Temperature and Pressure Single-Vessel Integrated Ammonia Synthesis and Separation using Commercial KATALCO Catalyst, Collin Smith, Laura Torrente-Murciano, Peer Reviewed July 2022, Department of Chemical Engineering and Biotechnology, University of Cambridge. Murciano_Oct22 (cam.ac.uk)

[6] Twenty Years of Low Pressure Ammonia Synthesis, C W Hooper, Synetix - AIChE Ammonia Safety Manual 2002

[7] First two 1,850 MTPD Grassroots Ammonia Plants, R B Strait, KBR; I Welch, PCS – AIChE Ammonia Safety Manual 2000

Challenges Experienced in the Longest Run of HTS Catalyst

A High-Temperature Shift (HTS) catalyst charge experienced high differential pressure (dP) across the bed within a few months of installation. The high-temperature issue was resolved by skimming the top hold-down layer and replacing it with alumina balls. Operating challenges experienced and the root cause of the hold-down layer issue are discussed in this paper as part of the "Lessons Learned". Close monitoring of catalyst bed operating conditions, periodic catalyst performance evaluations, and optimization of the catalyst bed operation resulted in a run of sixteen years, one of the longest in the global ammonia industry. A great spirit of teamwork between plant operations and catalyst suppliers contributed to improved safety (not handling catalysts in frequent catalyst changeouts), reliability, and energy-efficient operation of a 53-year-old ammonia plant.

Kok How Chew, Taylor Archer, and Teck Huai Ng
Clariant

Venkat Pattabathula
SVP Chemical Plant Services

Nik Amin
Incitec Pivot

Stephen Maule
Sedgman Prudentia

Introduction

Incitec Pivot Limited (IPL) operated an ammonia plant designed by JF Pritchard with an initial nameplate capacity of 600 MTPD. The plant had been upgraded to about 900 MTPD since its commissioning in the late 1960s. The unique features of this ammonia plant were:

- A relatively low-pressure (450 psig, 32 bar) front-end
- A high-pressure back-end (2600 psig, 182 bar)
- A medium-pressure steam system (400 psig, 28 bar, 750 °F, 400 °C).
- A closed-loop refrigeration system and a jet engine drove a reaction turbine, which drove the synthesis gas (syngas) compressor.

The plant is located at Gibson Island (GI) in the suburbs of Brisbane City on the East Coast of Australia.

Manufacturing at the site ceased in early 2023 after being unable to secure an economically viable long-term gas supply.

Background

The temperature of the high-temperature shift (HTS) converter inlet was controlled with quench water supplied from condensed process water. A large knockout (KO) pot, including demister, was installed upstream of the high-temperature shift in the 1990s due to frequent condensate carryover to the catalyst bed.

The front-end flowsheet of the plant is shown in Figure 1.

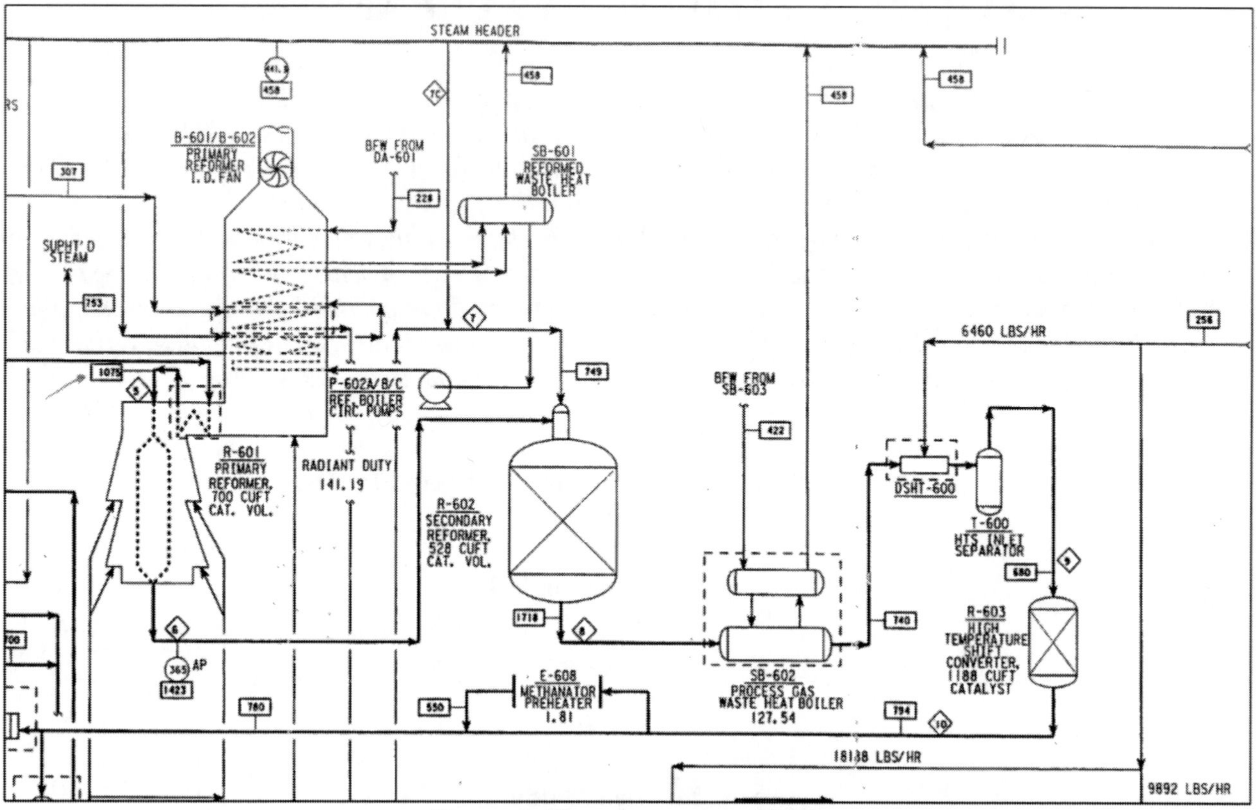

Figure 1. Flowsheet around HTS

Historical Performance

Before the 2007 turnaround, the HTS catalyst was typically changed every 4 to 5 years. From February 2002 to November 2003, the plant operated with a particular catalyst batch for 22 months.

Throughout this period, Gibson Island (GI) faced challenges with the catalyst's performance, which did not meet operational efficiency and effectiveness expectations and required changeout within a turnaround cycle. Despite efforts to optimize performance, the catalyst's performance fell short of the desired standards, necessitating further exploration into catalyst technologies.

GI utilized another catalyst batch from December 2004 to March 2007 for 28 months. However, the performance of this batch also failed to meet expectations fully. Issues such as pressure drop anomalies, suboptimal reaction rates, or other operational inefficiencies have persisted, highlighting the need for a more robust and reliable catalyst solution.

These experiences underscore the importance of catalyst performance in the operation of the HTS reactor system. Recognizing the limitations of the existing catalyst batches, GI loaded ShiftMax 120 in April 2007. This replacement aimed to address the shortcomings observed with previous catalyst batches and enhance the overall performance and efficiency of the HTS reactor.

2007 Turnaround

The HTS catalyst was replaced in 2007 after four years in operation. The catalyst was loaded with a Dypor hold-down layer. The plant was restarted in March 2007.

Reaction Profile Comparison: ShiftMax 120 vs. Competitor

The start-of-run exotherm is a critical parameter in catalyst performance evaluation, reflecting the catalyst's ability to initiate and sustain chemical reactions at elevated temperatures. In the context of HTS applications, a higher start-of-run exotherm indicates superior catalytic activity and reactivity, facilitating the conversion of carbon monoxide and water vapour into carbon dioxide and hydrogen. This enhanced reaction activity translates into improved process efficiency and productivity.

In our technical assessment, as shown in Chart 1, ShiftMax 120 demonstrated a unique exotherm distribution pattern within the catalyst bed, presenting a strategic advantage over competitor catalysts previously selected by GI.

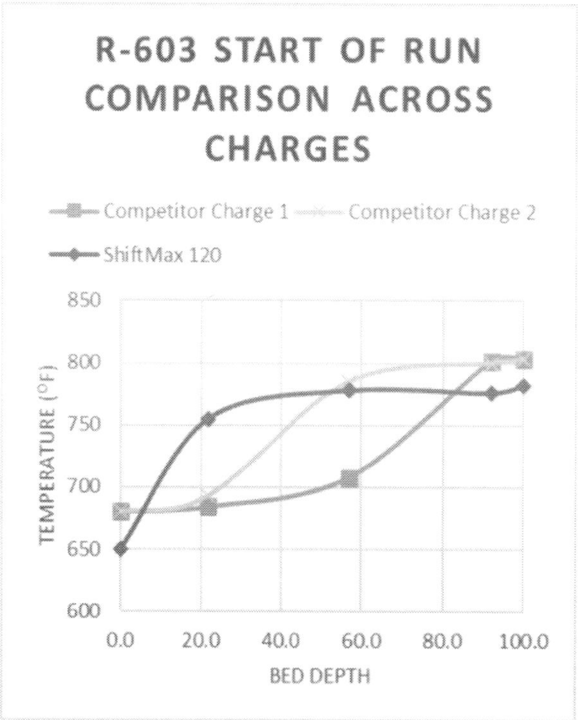

Chart 1. Start of Run comparison across various Catalyst Charges

While competitor catalysts previously installed by GI exhibited a less steep exotherm span from 20% to 90% of the bed for a complete reaction, ShiftMax 120 achieves a complete exotherm within the first 25% of the bed. This steeper reaction exotherm indicates higher activity, ensuring that a significant portion of the catalyst bed remains reactive, facilitating efficient polishing and accommodating higher shift reaction requirements.

High differential pressure across HTS

Immediately after it was brought online, a step increase was noticed in pressure drop across the reactor.

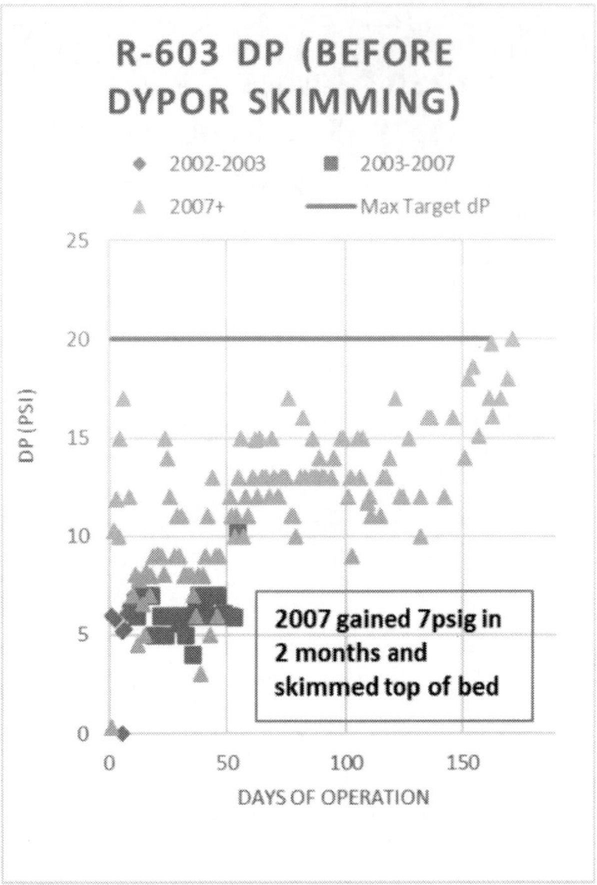

Chart 2. Differential pressure over catalyst bed (before Dypor Skimming)

Within six months of operation, the differential pressure went up to 20 psi, almost twice the differential pressure differential pressure expected, based on previous operating experience. Data from pressure gauges at different locations on the vessel confirmed that the differential pressure was mainly on the top of the bed.

Several team meetings were held with members from IPL, Clariant, the catalyst supplier, and catalyst loading contractor Contract Resources. Based on these discussions, it was decided to shut down the HTS while keeping the primary and secondary reformers in operation. This plant could block the HTS and isolate it from the reformers. Risk assessments and a job safety environmental analysis (JSEA) were conducted to ensure an inert entry was safe. A life support system was available next to the reactor during inert entry.

The catalyst reactor was opened after a cool-down operation, and entry was made into the vessel. When the bed was opened, the Dypor hold-down material was found ball-milled on the top of the bed, lying on top of the hold-down screen. Contract Resources vacuumed the broken Dypor material, removed the screen and vacuumed 2-3 inches (5-7.6 cm) of HTS catalyst from the reactor. After that, the support screen was placed back on top of the catalyst, and 2" low silica-alumina balls were loaded on top of the support screen. When the HTS returned to operation, the differential pressure was restored to a normal range and stayed that way for many years. The increased differential pressure was discussed before shutting the plant down. According to IPL reliability team hydraulic studies with the process licensor, the root cause of the milling of Dypor was higher operating rates, causing gas to swirl in the top of the reactor, including the top portion of the catalyst bed.

Opening the vessel shortly after the high-pressure drop prevented further impact on the HTS catalyst from broken particles and dust penetrating deep into the bed.

New Operating Practices

Catalyst producers improved HTS catalyst over the years by adding a stabilizer to the catalyst formulation. Clariant has included copper to increase the low-temperature activity, prevent over-reduction of the Fe phase, and chromium to inhibit Iron HTS crystallite growth (slows thermal sintering), allowing the operator to achieve longer run-length at stable conditions. This improved formulation has allowed the HTS catalyst to operate at a much lower (626 °F, 330°C) temperature than before (680 °F, 360°C) while maintaining higher catalyst activity.

ShiftMax® 120 is an HTS catalyst that combines high activity and thermal stability with extreme robustness, enabling it to withstand some boiler leakages. The catalyst also effectively prevents Fischer-Tropsch by-product formation at low steam-to-gas conditions.

Today, ShiftMax® 120, with an improved production route, is also available as ShiftMax® 120 HCF, which offers the same outstanding performance but contains essentially no hexavalent chromium, which complies with EU REACH regulations.

Monitoring of performance

Due to the site's age, online continuous composition and differential pressure monitoring were unavailable for every catalyst bed. This data was collected from a mixture of local field gauges and pressure transmitters.

Consequently, the site would complete a catalyst survey for the various catalysts at a monthly frequency, including many local and control system readings. Before each survey, the plant was to be at full rates, with no planned changes to plant rate or other major work occurring.

The field operator completed a detailed field reading sheet and took panel readings. A portable digital transmitter was connected to fittings on the local pressure gauges upstream and downstream of the catalyst bed for differential pressure over the catalyst bed. This would ensure that only the bed differential pressure was being measured, and piping pressure drop was not included if the permanently installed pressure transmitters were not located immediately upstream and downstream of the bed.

The timeframe to complete the monthly survey readings was typically around 2 to 3 hours, with the gas samples for laboratory analysis being collected at a similar time to the differential pressure readings.

The local readings were added to the site database for catalyst performance, and the results were communicated to the site team.

The results were sent to Clariant approximately four times yearly for detailed analysis. The data sent to Clariant was normalized to reflect the flow rate changes between each catalyst performance evaluation.

The data reviewed internally consisted of the following:

- Differential pressure over the catalyst bed
- CO slip after the HTS reactor
- Inlet, outlet and catalyst bed temperatures

Internal graphs of performance were reviewed monthly. Following advice from Clariant and based on downstream carbon monoxide (CO) slip, the inlet reactor temperature increased over time, as shown in Chart 4.

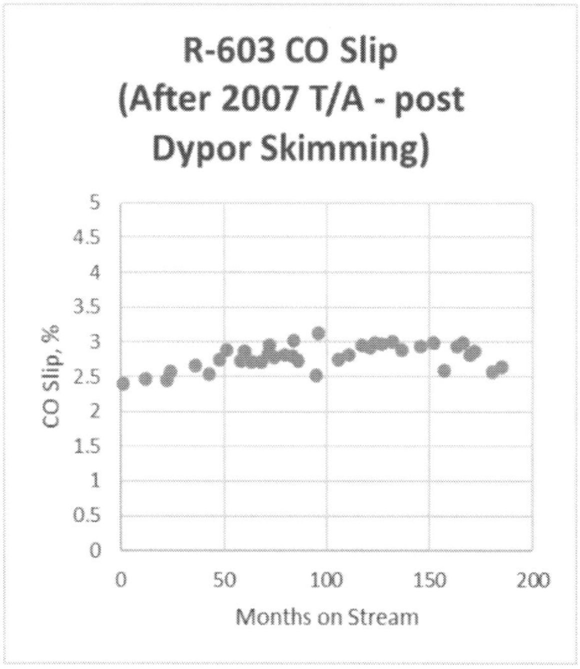

Chart 3. R-603 CO slip over time (After Dypor Skimming)

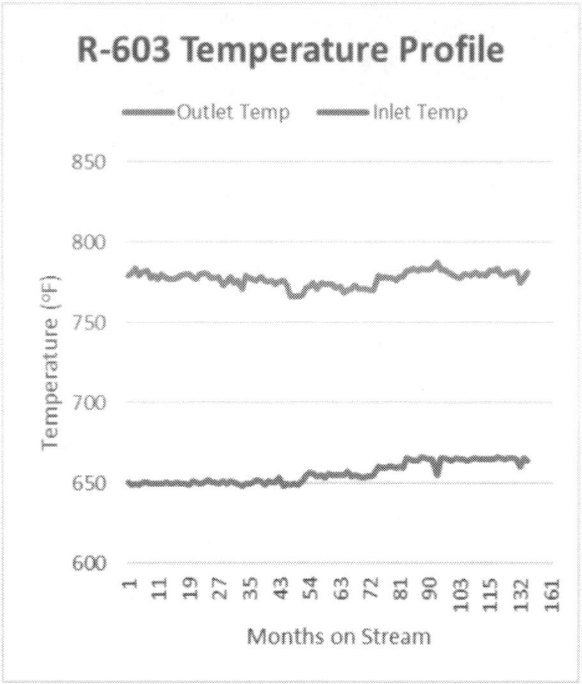

Chart 4. R-603 Inlet reactor temperature over time.

The option for purchasing a "spare charge" in case of emergency change out was considered in the 2020 campaign, as the next turnaround would require a charge of catalyst anyway, the option to hold the catalyst on site for a more extended period to act as a spare charge was considered.

Limits were placed on the trigger for differential pressure over the reactor that would require a spare catalyst charge to be ordered should a rapid change out be required.

However, this outcome did not occur, and no spare catalyst charge was required.

Following the skimming of Dypor, the observed pressure drop (differential pressure) showed minimal escalation, reaching only 6 psi post-skimming (far lower than 20 psi observed in 2007). This marginal increase in differential pressure throughout 180+ months of operation contrasted with the significant pressure drop challenges encountered during the 2007 incident. The subdued rise in differential pressure post-skimming underscored the effectiveness of the remedial measures, including removing accumulated fines and degraded material from the catalyst bed. This outcome reaffirmed the reactor's resilience and stability, demonstrating its ability to maintain optimal operational parameters even after prolonged operation. Post-skimming's controlled differential pressure elevation underscored the reactor's improved performance and highlighted the enduring benefits of proactive maintenance and catalyst management strategies.

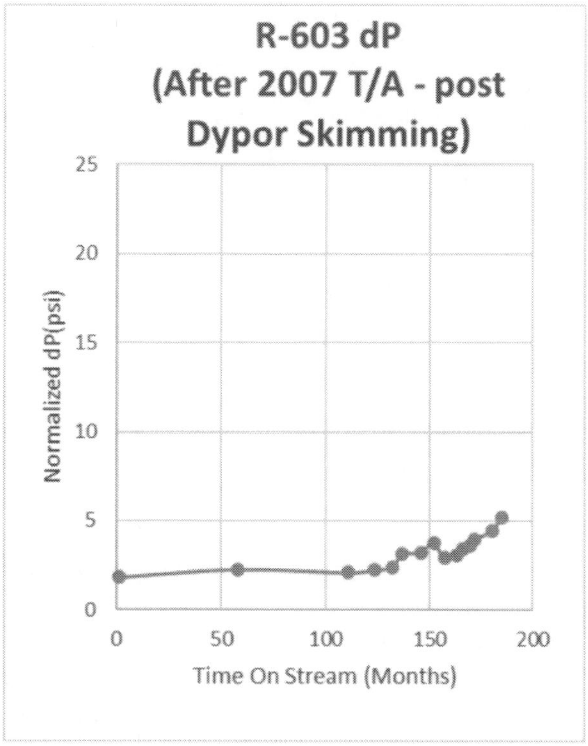

Chart 5. R-603 differential pressure (After Dypor Skimming)

As years elapsed, the HTS reactor continued to operate with ShiftMax 120, steadfastly meeting performance requirements without needing catalyst replacement. The catalyst's longevity and durability became evident as it endured the rigours of continuous operation, maintaining its catalytic activity and reactivity over extended periods.

A comprehensive analysis of exotherm run lengths across the years revealed intriguing insights into catalyst behaviour and performance trends. Plotting exotherm run lengths over time, accompanied by marginal increments in inlet temperature to enhance reaction kinetics, unveiled a compelling narrative of catalyst resilience and stability. Despite incremental changes in operational parameters, including inlet temperature adjustments to optimize reaction kinetics, the catalyst exhibited remarkable stability in temperature profiles and minimal signs of deterioration.

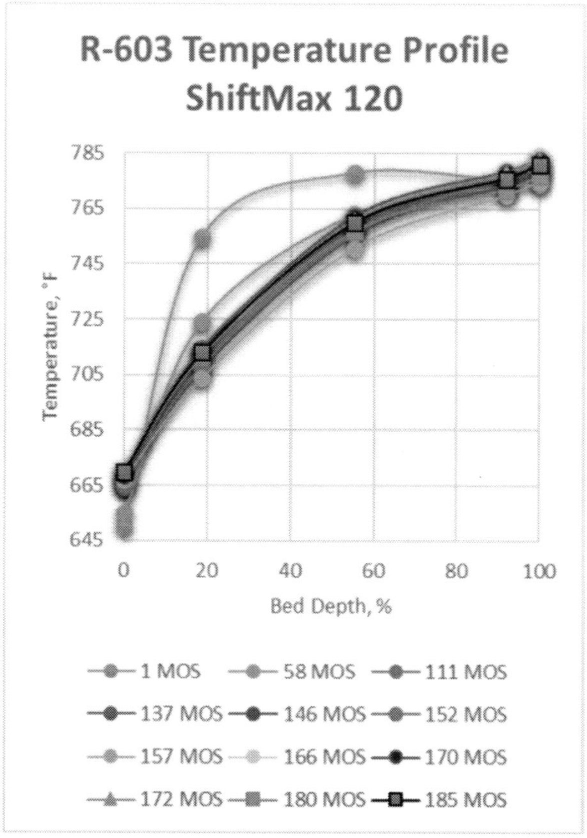

Chart 6. R-603 temperature profile for 2007 charge over time

After 16 years of continuous operation, the CO slip out of the reactor consistently meets the required performance of < 3%, demonstrating the sustained effectiveness of ShiftMax 120. This level of performance ensures that the reactor maintains optimal functionality, meeting the ammonia production requirements necessary for Gibson Island's operations.

Another thing that contributed to the long life was the very high uptime on the plant's front-end catalysts, which Clariant supplied, with few front-end outages, especially in the last few years of the campaign.

Other Operational Challenges for HTS

The catalyst was always kept under nitrogen during any plant upset, with positive pressure continuously monitored during operator rounds to ensure no risk of air ingress.

During startups, the inlet valve to the shift converter would pass slightly, leading to the catalyst bed heating up slowly over time.

Large amounts of condensate were typically drained from the downstream drain valve before putting gas through the reactor. This was evidence that the catalyst was exposed to moisture on every startup.

During major turnarounds, the reactor was blinded, but on routine plant trips or short outages, the vessel was only isolated with valves and put under nitrogen.

Conclusion

Even though a high-pressure drop was experienced in the first six months of operation, it was steady after replacing the hold-down layer (Dypor) with larger size (2") alumina balls.

The catalyst was operated at a lower inlet temperature for better CO conversion, as advised by Jim Richardson, the then Clariant's subject-matter expert on catalysts. Close monitoring and periodic performance evaluations have contributed to one of the longest-known runs of HTS catalysts.

Acknowledgement

We want to thank Jim Richardson for his great tech support during this longest run of HTS until 2015, when he retired.

Re-activation of poisoned ammonia synthesis catalyst at OCI Nitrogen

The startup of an ammonia synthesis converter is an integral part of the overall plant startup and is typically undertaken by operators following well-established plant procedures. After an unexpected shutdown of their ammonia loop, OCI Nitrogen had a suspicion of catalyst poisoning and sought the help of Topsoe.

Topsoe and OCI Nitrogen made a revised startup procedure together and carried it out in just a few days, thereby ensuring the catalyst's continued performance and avoiding a costly premature catalyst replacement.

Christian Rehman Hansen
Topsoe A/S

Ahmad Ghani, Patric de Konink, Luc Murrer, Gerd Decramer
OCI Global

Introduction

A smooth startup of the ammonia synthesis loop is crucial for overall plant startup. If the startup of an ammonia synthesis loop does not progress as expected, it is essential to pause and identify the cause so that the correct action can be taken to avoid damage to the catalyst and equipment. If this simple rule is not followed, it can lead to the untimely replacement of the catalyst and/or equipment, resulting in significant costs and safety hazards.

Normal startup

During a typical (hot) startup of the ammonia converter, the ammonia synthesis catalyst is fully active. This allows for a relatively quick startup without requiring strict monitoring of various operating conditions. The loop pressure can generally be maintained at a higher level (50-80% of normal operating pressure), and the control valves of the converter can be partially open. The startup heater (SUH) is then fired at a high capacity. Once the temperature reaches a range of 200-250°C (392-482°F), the ammonia synthesis reaction often begins, which is observed as an increasing exotherm in bed 1 of the ammonia converter.

The developing exotherm in bed 1 helps initiate the reaction in downstream beds. Once the reaction is achieved in all beds, it becomes a matter of adjusting parameters to achieve normal operation within a few hours.
A normal startup is something that most plant operators and engineers have ample experience.

with, as it typically occurs multiple times in a year.

First startup (activation)

There are some differences between the initial startup of the ammonia synthesis catalyst and a normal startup. The main difference is that the catalyst is inactive and needs to be reduced from its oxidized form (Fe_3O_4) to its active state (Fe) using H_2.

$Fe_3O_4 + 4\ H_2 \rightarrow 3\ Fe + 4\ H_2O$

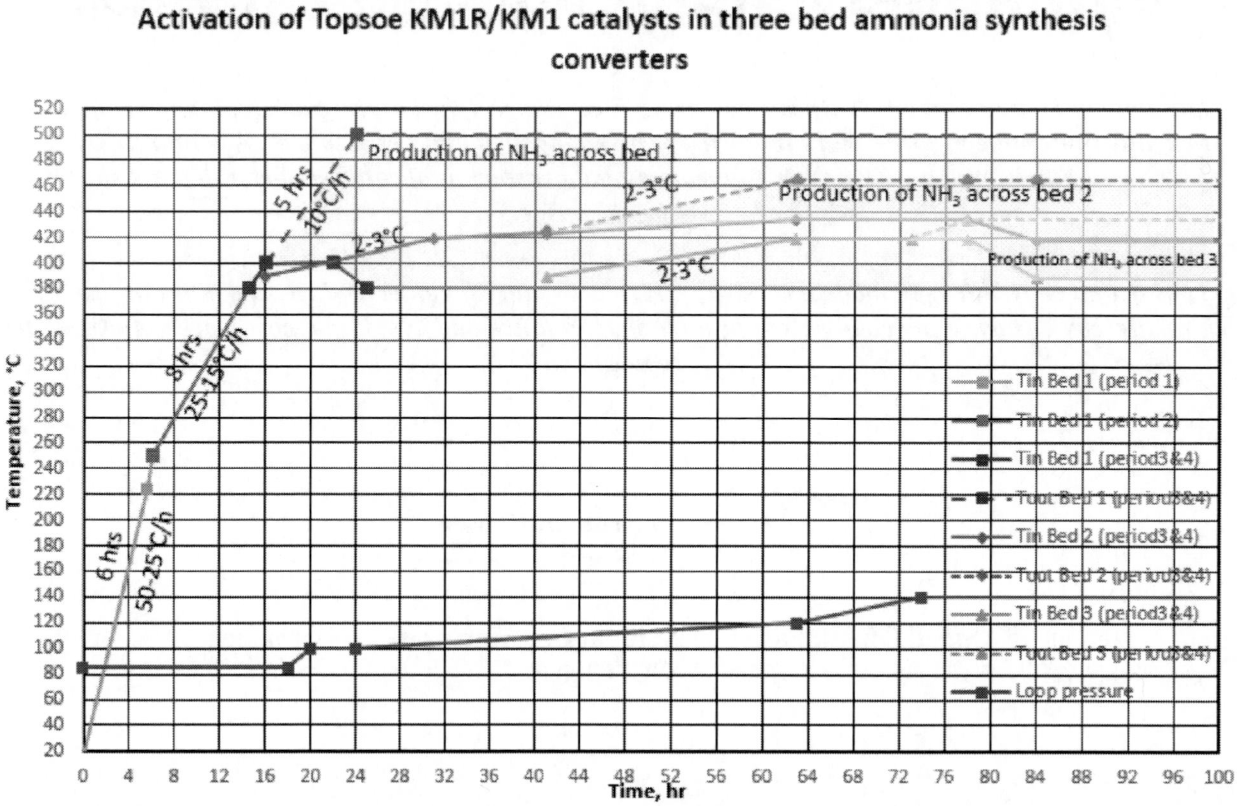

Figure 1. Activation timeline for 3-bed ammonia synthesis converter

Because of this, it is typical to begin with a loop pressure of 80-100 bar g (1,160-1,450 psi g) and with most control valves closed or slightly open. The unreduced catalyst is typically reduced at temperatures around 400°C (752°F), and as it becomes active, the ammonia synthesis reaction will start. This is observed as a developing exotherm in bed 1 of the ammonia converter.

Activation of the catalyst leads to the formation of catalytically active Fe crystallites. Smaller crystallites favor high activity. On the other hand, exposing these crystallites to high water concentrations increases the risk of sintering, thereby lowering activity. For that reason, it is important not to heat the catalyst too rapidly, and thus, the activation process needs to be carefully controlled, not only for bed 1 but for all catalyst beds in the converter.

Activation is not something that plant operators and engineers typically have a lot of experience with, as ammonia synthesis catalysts are usually replaced every 10-20 years. Figure 1 shows the optimized timeline for Topsoe KM catalyst activation in a radial flow 3-bed converter with inter-bed heat exchangers.

Issues at OCI Nitrogen

OCI Nitrogen has a 1,660 MTPD Bechtel-licensed ammonia plant, AFA-2, which utilizes a Topsoe S-300 converter in the ammonia loop. During normal operation, the loop pressure suddenly increased and within 1-2 minutes the exotherms in the converter – starting with bed 1 - started decreasing until the production stopped completely. OCI Nitrogen started investigating the cause of the upset. There were several scenarios to investigate and check up on, and while this was ongoing, they decided after a while to also initiate the startup procedure by starting the SUH (defined as hour 0) and monitor its progress. During this startup, they encountered some trips of the SUH, which were unrelated to the topic of this paper. However, despite reaching an inlet temperature of approximately 290°C (554°F) in bed 1 of their S-300 ammonia converter, with a loop pressure of around 150 bar g (2,176 psi g), there were no signs of the ammonia synthesis reaction taking place. At hour 15, they decided to stop heating the catalyst and await the investigation results to determine the cause of the plant trip and the unexpected lack of reaction.

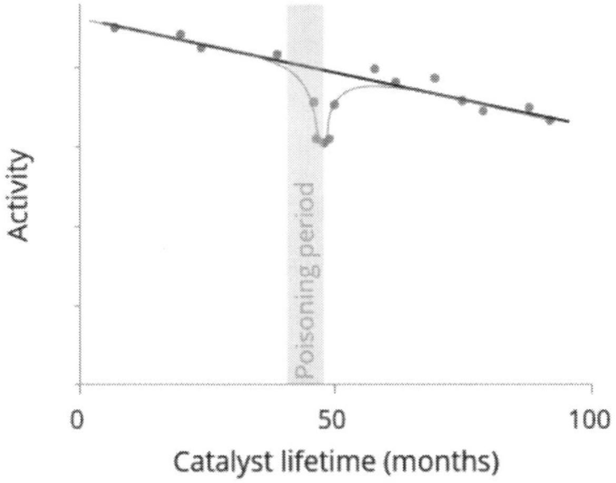

Figure 2. Recovering catalyst activity

It was discovered that a molecular sieve that purified the make-up gas (MUG) to the loop was malfunctioning. This malfunction resulted in an unknown amount of water or steam entering the loop, poisoning the catalyst. The overall loss of reaction in the converter happened in a matter of minutes after lining up the faulty molecular sieve.

When water or other oxygen-containing compounds reach the catalyst, it causes oxidation and renders the catalyst inactive. Fortunately, this process is reversible, especially for KM catalyst, which is reputed for its oxygenate poisoning resistance and recovery due to the way it is designed and manufactured. This is demonstrated in Figure 2 from an industrial case study.

After identifying the issue and confirming that it was not causing any further problems, the team at OCI Nitrogen was prepared to resume operations. However, they wanted to ensure a safe restart to prevent any adverse effects on the catalyst and equipment. To achieve this, they contacted Topsoe A/S, the supplier of both the catalyst basket and catalyst.

The issue and its implications were discussed during a quickly scheduled virtual meeting. It was clear from the startup that the catalyst was inactive, as no reaction took place. The exact oxidation level was uncertain, but it was estimated that the catalyst could be compared to a pre-reduced catalyst. It was likely that the poison effect primarily affected the catalyst in the first bed. If this estimation was correct, it shouldn't take long to activate the catalyst, as the reduction reaction of a pre-reduced catalyst begins at 250°C (482°F). However, as water was released from the first bed, it would poison the catalyst in the lower beds.

Based on previous experience with similar issues, Topsoe suggested following the same procedure used for catalyst activation. Although this approach may be considered conservative and time-consuming, it ensures that the catalyst's activity is re-established, thus avoiding premature catalyst replacement. There was a deviation from the procedure in terms of pressure. Higher pressure than usual was allowed, as the flow in the converter was already reasonably high. The primary reason for initially keeping the pressure low is to ensure proper flow distribution.

Awareness at re-activation

Although the re-activation process should be approached similarly to a normal activation, as mentioned in the previous section, there are some specific considerations to remember. Since it was assumed that the poisons entered the converter from the converter inlet, it was likely that the catalyst in the downstream beds was still active. This meant that the downstream beds could initiate the ammonia synthesis reaction before the catalyst in bed 1. Avoiding this scenario as much as possible by keeping the downstream beds cooler is important. We want to prevent the catalyst from being hot and producing ammonia while being poisoned by the upstream water released from the re-activating catalyst. However, this can pose a challenge as the control valves, which regulate the cooling flow, are typically only slightly open during an activation or re-activation process.

Figure 3. Hopper with catalyst being lifted by crane

Consequences of poor activity

If the re-activation process is not carefully controlled, it can decrease activity for the remaining catalyst lifespan as discussed previously. There have been instances where this has been observed, resulting in significant losses in production. For example, in one case for an African client, the activity decreased to approximately 80% of the expected level, leading to a loss of around 70 metric tons per day (MTPD) in a 2,000 MTPD plant. Over a year with 350 production days, this amounts to a loss of 24,500 metric tons of ammonia. At this point, it becomes worth considering catalyst replacement instead of accepting such a significant loss for 10-20 years. However, there are also safety risks associated with catalyst replacement.

Here are some of the major risks involved in replacing the catalyst in an ammonia converter:

- The plant must be completely shut down, and precautions must be taken to install proper blinds, close valves, and prevent gas leakage during the work.
- Heavy equipment lift requires at least one crane in areas with other ongoing activities and personnel assisting with the lifting. See Figure 3 for example.
- Working in confined spaces.
- Working under a nitrogen atmosphere.
- Working at heights.
- Passivation of catalyst.

It should be evident that these risks alone are sufficient reasons to avoid catalyst replacement unless necessary. With proper care of the catalyst and equipment, replacement should only be required every 10-20 years in a plant.

Re-activation

Once OCI Nitrogen and Topsoe had agreed on the procedure to be followed, the OCI Nitrogen team made slight adjustments to the loop conditions in accordance with the procedure. They proceeded with a heating rate of approximately 5°C (9°F) per hour for the first bed, the typical limit for a pre-reduced catalyst. By hour 30, there were already small indications of a reaction occurring in bed 1, as observed from the exotherm in Figure 4.

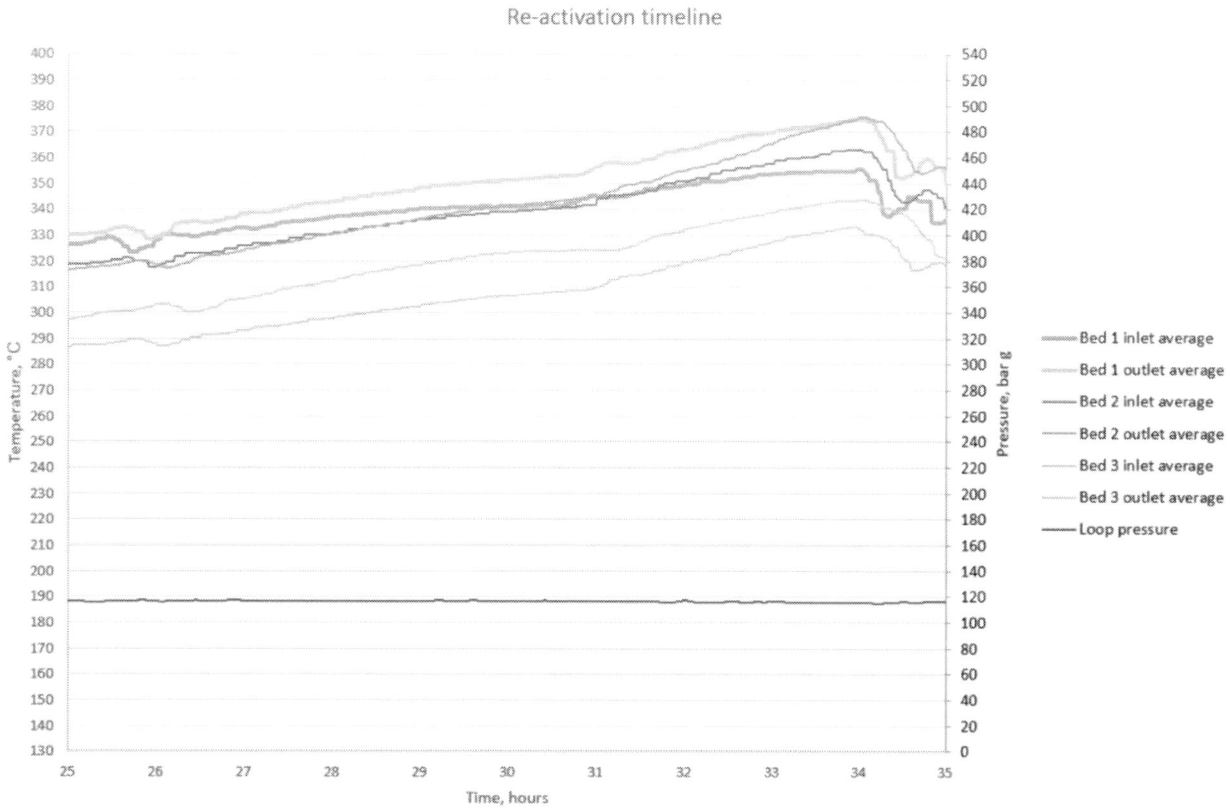

Figure 4. Midway progress in re-activation

A few hours later, at hour 34, temperatures were slightly decreased throughout the converter, as shown in figure 4. The exact cause of this decrease is not well understood. The same situation occurred again 6 hours later, at hour 40, but to a greater extent, as depicted in Figure 5. It is believed that this was due to the control valve being opened too rapidly, resulting in excessive cooling while the gas in the system was still relatively cold. It took approximately 5 hours to regain control of the situation, primarily by closing the control valve again.

In the following hours, there was a noticeable increase in the exotherm, particularly in bed 1. Additionally, an increasing exotherm was observed in bed 2. These combined exotherms indicated that the catalyst was becoming more active. It became progressively challenging to control the heating rate of the bed inlet temperatures, but the temperatures also started to fall within the normal range.

Rapid opening of the control valves were made as the plant approached normal operating conditions. These adjustments resulted in temperature excursions observed 62 hours after the initial heating of the startup heater (SUH), as shown in Figure 6.

Approximately 64 hours after igniting the startup heater (SUH) following the initial trip, the loop pressure was back in the normal range, and the re-activation was considered complete. The only remaining task was to further optimize the parameters of the ammonia loop.

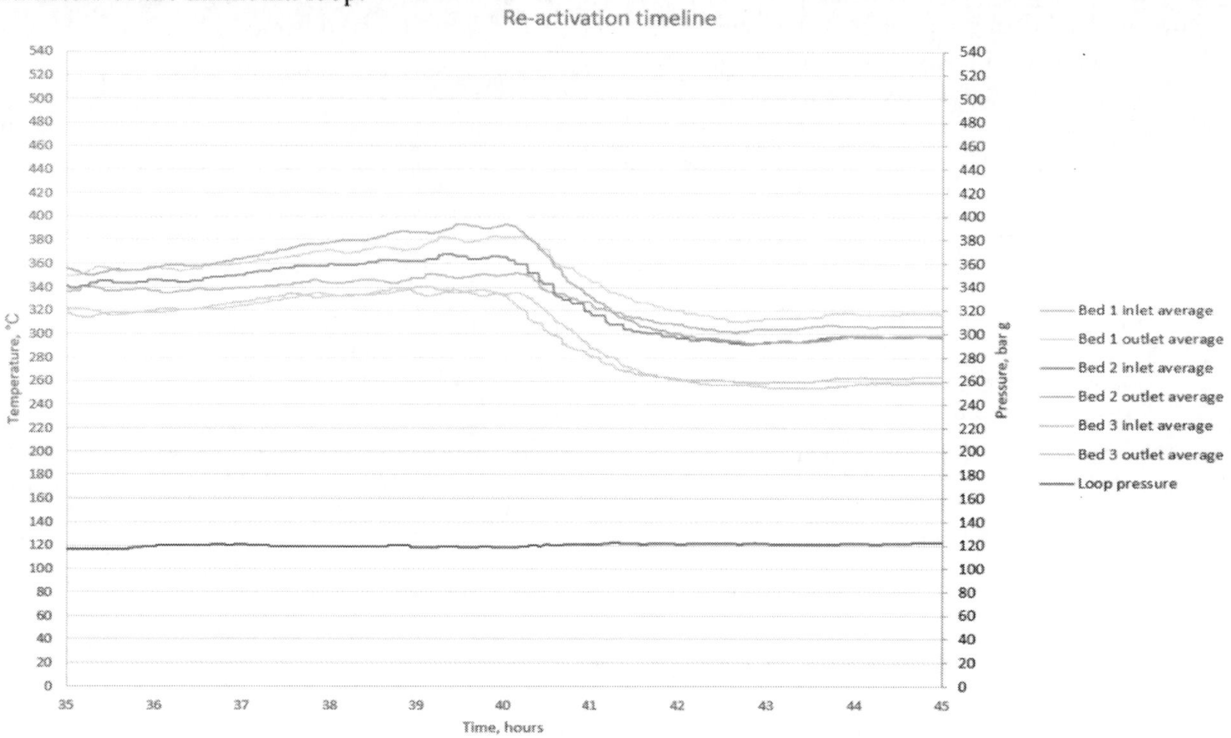

Figure 5. Temperature loss

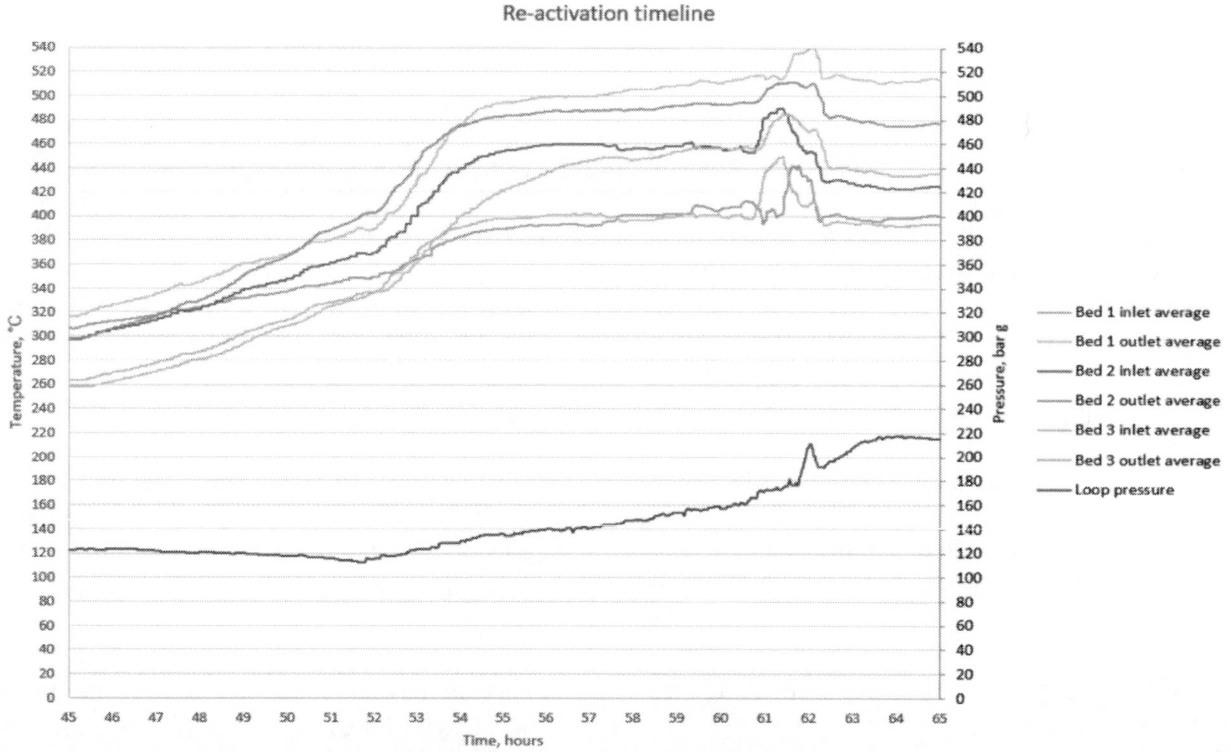

Figure 6. End of re-activation

Success or failure?

A quick way to determine the success of the re-activation is to check if the production level matches similar conditions observed in the past. However, a more detailed approach to assessing the success of the re-activation is to conduct a performance evaluation, which involves conducting mass and heat balance checks. A performance evaluation also provides a measure of the overall catalyst activity, taking into account catalyst age and operating pressure.

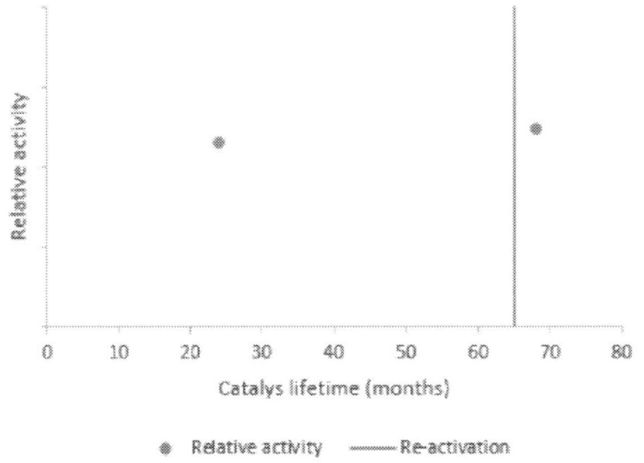

Figure 7. Relative activity before and after re-activation

In Figure 7, the catalyst activity is divided by the expected activity to compare the activity before and after the poisoning incident. As shown, the relative activity of the catalyst remains the same, indicating that it has regained its pre-incident activity level.

Conclusion

Thanks to the actions taken by OCI Nitrogen in collaboration with Topsoe, the catalyst was successfully re-activated in a controlled manner. As a result, the activity of the catalyst was not affected by the poisoning incident. This allows OCI Nitrogen to continue operating the plant under the same conditions and with the same output without the need for a costly and risky premature replacement of the catalyst.

Demolition of Secondary Reformer Air Insert and Transfer Line Repair Without Replacing Incoloy 800H/HT liner

In Turnaround 2018, Yara Pilbara had planned to skim the top layer of catalyst from the secondary reformer (103 MD). To skim the catalyst, access to 103 MD was only possible via air insert removal and it was planned as normal work with 8 hours of duration, but surprisingly, the air insert got stuck into its sleeve opening. After multiple failed efforts of conventional air insert removal, a demolition approach was taken so that access can be made to replace catalyst. Due to this surprise event, unplanned and unconventional demolition methodology, access to the secondary reformer was obtained after 12 days instead of 8 hours. Once access was obtained, the turnaround team encountered another discovery of severe damage of Transfer Line (107MD) Incoloy 800H/HT liner, which could only be observed via secondary reformer when the air insert has been removed.

This unknown and unexpected event led to extra work during turnaround execution and impacted schedule and cost, which was not accounted for in planning.

Mukesh Kumar
Yara Pilbara Fertilizers

Marcel Markus
Yara International

Introduction

Yara Pilbara fertilizers (YPF) operates KBR purifier technology based 2450 TPD ammonia plant since 2006. An air insert is installed in the secondary reformer since the start of operation and was removed in previous turnarounds with less effort than in 2018. During the turnaround in 2018, YPF had planned to skim the catalyst of the secondary reformer, and to get access, demolition of the air insert was done rather than removal. After demolition of the air insert, severe damage was found on transfer line which was repaired without replacing the Incoloy 800H/HT liner.

This paper details YPF experiences as described below:

- Multiple failed efforts to remove air insert with conventional methodology.
- Unconventional demolition methodology.
- Installation of new air insert.
- Repair of damaged section of transfer line without replacing Incoloy 800H/HT liner in

2018 and condition assessment of refractory repair after 5 years of operations in year 2023.

Air Insert and Transfer Line Basic Design and Construction Features

Air insert is installed as a "**Plug type design**" into air insert sleeve opening of Secondary Reformer as shown in *Figure 1*. Following are the key dimensional features of the air insert.

Length: 1925 mm [75.8 in]
Outer diameter: 518 mm [20.4 in]
Air insert sleeve opening I.D : 530 mm [20.86 in]
Air insert I.D : 200 mm [7.87 in]

Material of construction: Incoloy 800H/HT.
Type of Refractory: Dense high alumina.

The key point of existing design is, to do any internal inspection of secondary reformer or replacing catalyst, the only access inside the vessel can be obtained via removing the air insert. It has two lifting lugs which are 180° apart from each other.

Transfer line directs the reformed gas from primary reformer to secondary reformer and is directly welded to its nozzle , just below the air insert elevation as shown in *Figure 1*.
Following are the key dimensional features of the transfer line.

Inner diameter: 562 mm [22.12 in]
Pressure shell material: SA 516 GR70.
Liner material: Incoloy 800H/HT.
Liner thickness: 5 mm [0.19 in]
Refractory material: Bubble Alumina.
Design pressure: 43.5 bar (g) [630 psig].
Design temperature: 205°C [400°F] for CS shell.

Reformed gas flows from primary reformer to secondary reformer and enters at approx. 739°C [1362°F], just below the air insert. On other hand, hot process air enters at approx. 500°C [932 °F], just above the air insert, which passes through air insert and then both streams mix up together in the **combustion zone** where temperature reaches up to **1327°C [2420 °F].**

Air Insert Demolition

During Turnaround 2018, as a **1st effort**, air insert removal process started on **12th July** and 55T crane was attached with a force on each lug up to 4T. This lifting process continued for 2 days in the same manner and approx. 150 mm [6 in] lift was observed only. It was clear that the air insert will not move upwards, when it is only pulled by the lifting lugs with a force of 4 tons.

As a **2nd effort**, it was decided to apply up to 8 T of force on each lifting lug via crane to lift the insert. Due to safety reasons (if lugs sheared off) a larger crane than 55T was required and a 400T crane was used, as it was the only crane available next to secondary reformer being used for some other work. 8T of force was applied on each lifting lug, but air insert did not move more than 150 mm [6 in] as before.

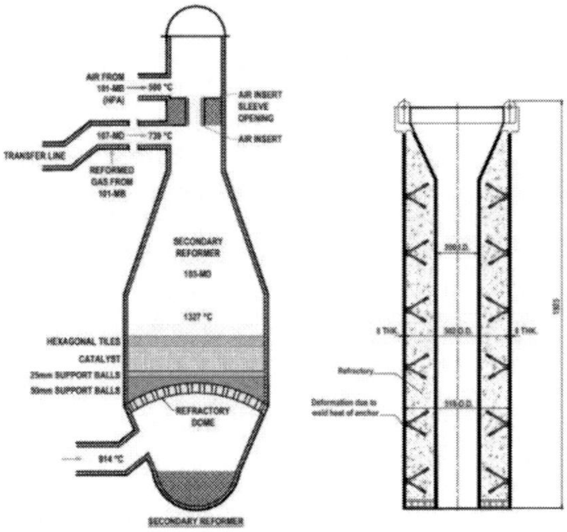

Figure 1: Secondary Reformer and Air Insert simplified schematics.

It was evident that air insert was stuck at the bottom, therefore some force or push from the bottom was required to release.

As a 3rd **effort**, a rope was inserted through the air insert hole of 200 mm [8 in] as shown in *Figure 2* and lift was attempted again with a push from the bottom via crane, but this effort was also unsuccessful.

During multiple failed attempts, it was noticed that the crane was not allowing us to put force fully and accurately in the vertical plane, therefore as a 4th **effort** a hydraulic jack was used instead of a crane, to lift the insert. The lifting process continued for 24 hours, but the results of movement were the same with slight improvement.

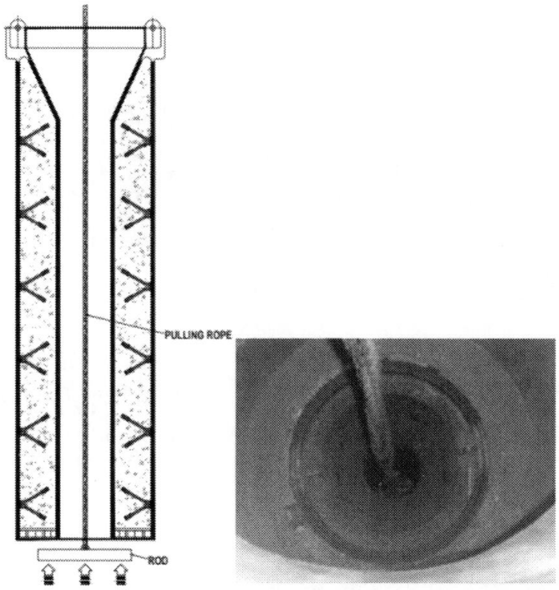

Figure 2: Air insert pull out from bottom.

As a 5th **effort**, it was decided to un-stiffen (by making a cut in the line in a longitudinal direction) the air insert, so that it can lose its integrity. In two shifts (24 hours), a few cuts were made in a longitudinal direction of air insert inner metallic shell side by thermal lance and the process for lifting started again with jacks, but there was no improvement in movement.

As a 6th **effort**, the conical section of air insert was cut off and 750 mm [30 in] refractory was demolished by gun hammer. After making longitudinal cuts in the air insert and considering an adequate amount of refractory has been removed, the lifting operation with the crane and jacks were done again but the air insert did not move any further.

Hence on 23rd July, after discussion with stakeholders (operations /safety / inspection / maintenance) the following options were attempted.

First Option: Weld two additional lifting lugs and try to pull out with a crane or Jack.

Second Option: If the first option does not work any further, cut the top ring and push air insert inside of vessel.

As a 7th **effort**, the first option was tried and was unsuccessful. Hence top ring was removed as shown in *Figure 3* and procedure for pushing down of the air insert inside the vessel was commenced. A critical risk was evaluated to determine if the falling insert could damage the bottom dome.

Much of the refractory and metallic parts were previously removed, which made the air insert much lighter in weight. Additionally, there were hex tiles and catalyst which would act as a cushion for the dome, therefore risk consequences of damaging the dome were low.

Figure 3: Left - Removed top ring and Right – Demolished refractory.

As an 8th **effort**, an I-beam welded with a circular plate was lowered into the insert and an attempt was made to hydraulically ram (110T) the steel beam down. Only 50mm [2 in] of movement was achieved before the weld between circular plate cracked and the steel beam was buckled and become unstable.

As a **9th effort,** a new calculated tool (against buckling and compressive stress) was prepared as shown in *Figure 4*, which consisted of 8-inch schedule 160 pipe and a blind flange which was welded on both ends.

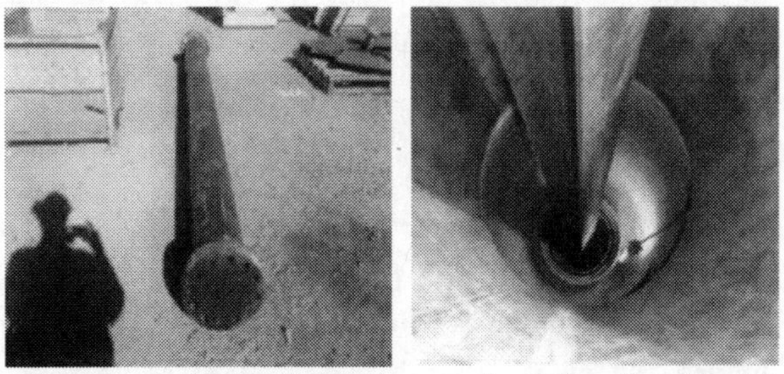

Figure 4: Left -New pipe tool, right: Buckled I-beam tool.

This new tool in combination with an electric jack as shown in *Figure 4* & *Figure 5* moved the air insert approx. 100mm [4 in] more within 2 hours. Movement became much easier, and air insert was eventually and finally pushed into the secondary reformer as shown in *Figure 6* without any damage to the dome on **26th July 2018.** Upon inspection of dropped air insert, it was found that the bottom part was expanded and in a mushroom shape as shown in *Figure 7* (Possibly due to extreme heat) which was preventing movement.

Figure 5: Illustration of jack installation

Figure 6: Air insert fallen into secondary reformer

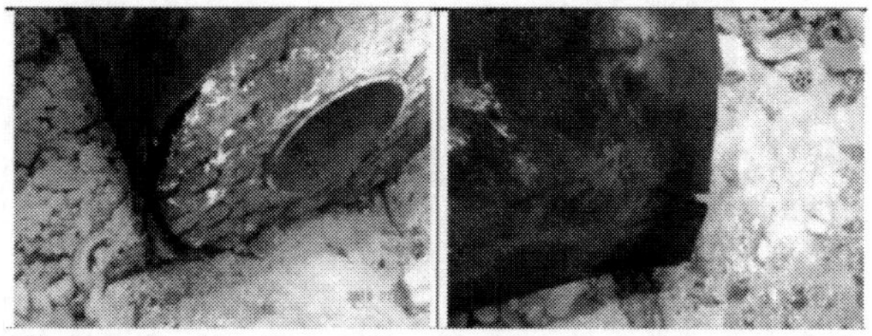

Figure 7: Air insert bottom part mushroomed due to overheating.

The dome was inspected post insert dropping and found to be in good condition as shown in *Figure 8*. As the length of the insert was less than the vessel's diameter, it did not hit the refractory installed on shell during fall.

Figure 8: No damage at bottom dome after insert drop.

Transfer Line Damage and Repair

The transfer line of YARA Pilbara is a water-cooled system with a steel Incoloy 800H/HT liner inside that is in front of one layer of bubbled alumina refractory castable. This refractory is a 100mm [4in] thick layer of Plicast LWI 606 KK.

The 1st inspection of the Transfer line from the manway side, which is at the end of transfer line was done on **9th July 2018** by crawler camera, and it was found in good condition as shown in *Figure 9*. Second inspection from the secondary reformer side could only be done once insert was removed. Upon inspection, another surprise damage was found on the transfer line dog leg section.

Figure 9: No damage on transfer line from manway side

The damaged area was just after the dogleg and at the location of a thermocouple, as shown in

Figure 10.

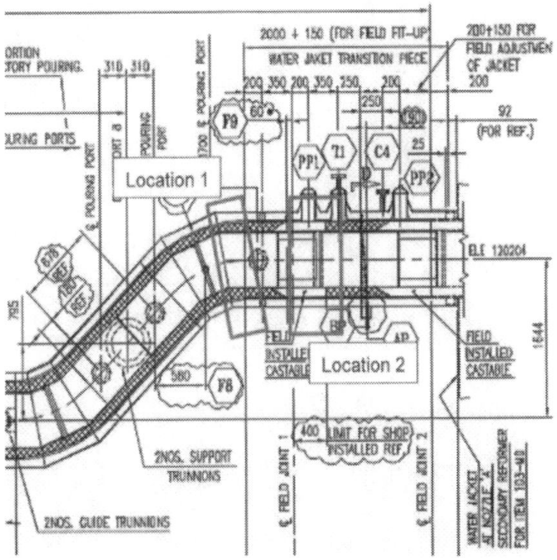

Figure 10: Damaged red highlighted area on transfer line dog leg section.

Location 1: The liner was buckled in that area and approx. 200 mm [8 in] of the refractory (adjacent to support ring) was no longer protected by the liner as shown in *Figure 12.*

Location 2: Bulged liner of 475mm [19 in] near T1 thermocouple nozzle a shown *Figure 11.*

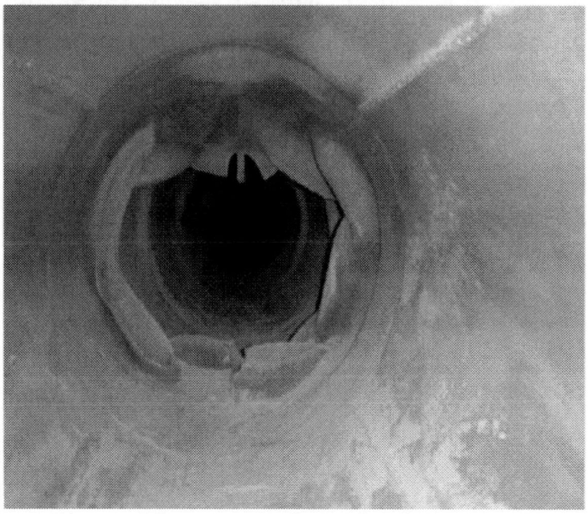

Figure 11: Damaged Incoloy liner near T1

Figure 12: Left – Refractory erosion 60 mm Right – liner moved 200 mm.

Repair of Transfer Line

After careful consideration of operational and personnel safety, the decision was made to do a temporary repair without replacing the inner incoloy 800H/HT liner in that area. For the temporary repair, another refractory castable was selected rather than bubble alumina. The repair plan was as follows:

- Remove the buckled inner liner as far as possible.
- Remove the existing refractory in the area without inner liner.
- Weld new SS anchors on the pressure shell, as shown in *Figure 13*
- Cast the bottom part (between 16:00 and 20:00) with a tabular alumina base refractory castable. This refractory material is normally used as hot face material in transfer line systems with a two-layer system.
- Gunning the top part.
- Install thermocouples on the outside to measure the water temperature in water jacket in case of refractory erosion (as in repair no Incoloy 800H liner was installed to protect refractory).

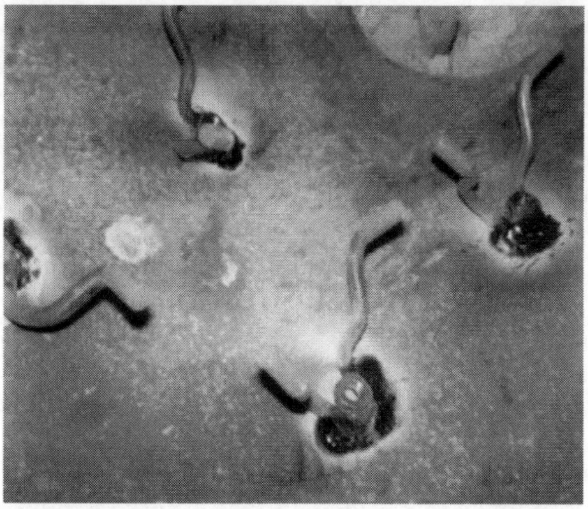

Figure 13: New anchors welded to transfer line pressure shell (Modification from existing design)

Figure 14: Final refractory repair without liner

This temporary repair was originally meant to last for at least **3 years,** and on the positive note, it lasted for approx. **5 years** with safe and reliable operations.

Cause of Liner Buckling

The inner liner is installed with overlaps of 25 mm [1in] to allow the thermal expansion of the Incoloy 800 H/HT liner as shown in *Figure 15.* After removing, the liner plate was inspected, and no metallurgical structure damages were found, and the following two potential causes were concluded.

1. Between the backing strip and liner slip fit was required, which became loose, consequently resulted in liner buckling and sliding from its base location and created an unprotected refractory zone.
2. On the backing plate, liner was welded with a field weld of 5 mm [0.2 in] for location 2 as shown in *Figure 15* & probably this weld also failed.

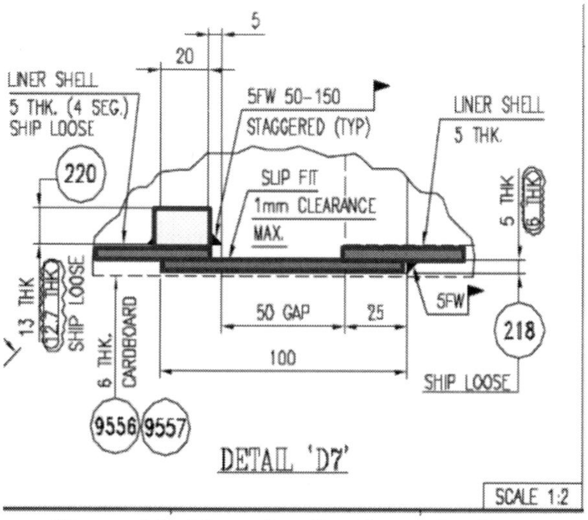

Figure 15: Liner details of transfer line

Air Insert Installation

After transfer line repair and catalyst skimmed in secondary reformer, air insert installation commenced on **8th August 2018**. Before installation of the new air insert, **a dummy wooden frame type air insert** was made and inserted into the sleeve to ensure that the new insert will smoothly fit into the sleeve opening. It passed through due to frame flexibility.

Hence the air insert installation process was started but it got stuck after 900 mm [36 in] of insertion. After 6 hours of effort, it was decided to take the insert out and remove high points that were visible to the naked eye. During the visual inspection of the air insert, the following findings were noted.

1. A very minor local deformation was observed across full length of the air insert wherever anchors are welded (due to anchor weld heat).
2. Because the insert was made of plate, and due to smaller rolling diameter of 518 mm [20 in] during shell forming, circularity was also not maintained.

Both above-mentioned high points were grinded off across full length and installation process was tried again but insert got stuck again at the same elevation after 900 mm [36 in]

Hence, 3D scan was performed on air insert and sleeve opening to know where exactly interference was occurring in between high points (deformation areas) of both components.

Based on outcome of 3D scan, it was found that the air insert sleeve opening was also deformed. It was not possible to do any repair on the air insert sleeve opening, therefore grinding of air insert outer shell and sleeve opening both were done. Precautions were taken not to reduce the thickness less than 4 mm [0.2 in] of air insert outer shell. Most of material removed from the air insert was at the bottom section of 750 mm [30 in] where the maximum interference was noticed based on 3D scan.

Figure 16: Grinded air insert to remove high points and check with dummy template.

Finally with all above efforts, air insert was installed on **11th August 2018** as shown in Figure 17. After installation, it was noticed that the air insert was not fully straight i.e., rested at a small angle, which was finally accepted by technical team for startup.

Figure 17: Installed air insert

Condition Assessment After 5 Years of Operation

The existing transfer line repaired section was replaced in Turnaround 2023. We took the opportunity for a condition assessment of the repair done during Turnaround 2018 and found that the castable refractory area was still in good condition. Refer to *Figure 18* which illustrates refractory condition after 5 years of service.

Installed air insert in 2018 was needed to be removed during turnaround 2023 for internal inspection and catalyst skimming purpose. Unfortunately, this time the air insert got stuck again. Therefore, it was decided not to remove it and catalyst skimming was postponed to turnaround 2026.

Figure 18: Castable area refractory without Incoloy 800H/HT liner in transfer line dog leg section after 5 years of operation

Path Forward

YPF is working on solutions to eliminate the air insert removal issue. The following are two key options, which are under evaluation.

1st option – Cut the approx. top 12 meters [65.6 ft] section of secondary reformer and adopt the modified design with a new manway introduced in the secondary reformer.

2nd option – Cut the approx. top 6.6 meters [21.6 ft] section of secondary reformer and replace with a like to like design including following changes in fabrication.

1. Make Air Insert forged rather than plate material to achieve tight tolerances and circularity.
2. Increase thickness of outer shell of air insert to provide more stiffness and if grinding is required during installation , adequate margin in thickness is available.
3. Install four lifting lugs rather than two.
4. Perform 3D scan of air insert and sleeve opening to ensure dimensional accuracy and fit up.
5. Do trial fit at workshop before dispatch.
6. Explore potential for better heat resistant materials.

Failure Occurred at the Primary Reformer Tubes of Ammonia 1B Plant at PT Pupuk Kujang

PT Pupuk Kujang is a government-owned company that produces ammonia, urea, NPK, and other petrochemical products. The company operates two ammonia plants, each with a total production capacity of 330,000 tons per year. On Saturday, July 21, 2018, at 18:30, the lighting on the east side of the primary reformer suddenly tripped. During on-site observations, operators discovered a flame bursting out of the primary reformer wall. Pupuk Kujang suspected a rupture in primary reformer tube row 1, number 48, and, based on this condition, decided to shut down the Ammonia 1B plant. Upon inspection, it was discovered that there was a rupture on primary reformer tube row 1, number 48. Pupuk Kujang's root cause analysis indicated that the rupture of the primary reformer tube was caused by creep, initiated by a hot spot in the primary reformer tube. This paper describes the events leading to the failure of reformer tubes and the root cause analysis of the failure. It also discusses lessons learned, corrective efforts to solve the problems, and preventive measures so that these incidents do not recur.

Zulkarnaen Arsadduddin, Maryono, Andryan Pradipta Putra, Ronald Andrean Keswantono
PT Pupuk Kujang, Indonesia

Introduction

Pupuk Kujang is one of a government-owned company in Indonesia that produces ammonia, urea, NPK, and other petrochemical products. The company operates two ammonia plants, each with a total production capacity of 330.00 tons per year. The company's plants are built on 510 hectares of land in Dawuan, Cikampek, Karawang Regency. The company built the second plant, Kujang 1B, in 2003 with design capacity of 1,000 metric tons per stream day of ammonia production. The plant uses the Reduced Energy Ammonia Process licensed by Kellogg Brown & Root, Inc. (KBR). The construction was carried out by the main contractor, Toyo Engineering Corporation (TEC) Japan, and supported by two domestic contractors, namely PT Rekayasa Industri and PT Inti Karya Persada Teknik [1].

On Saturday, July 21, 2018, at 18:30, the lighting on the east side of the primary reformer suddenly tripped. During on-site observations, operators discovered a flame bursting out of the primary reformer wall, specifically located at the east bottom side. The flame burned the lighting cable around the wall of the primary reformer. Pupuk Kujang suspected a rupture in primary reformer tube row 1, number 48, and, based on this condition, decided to shut down the ammonia 1B plant [2].

Upon inspection, it was discovered that there was a rupture on primary reformer tube row 1, number 48, located between the catalytic tube and bottom reducer. Out of the total 192 tubes, one tube was ruptured, three tubes had more than 4% creep, and the rest showed no visual indications. These tubes had been in service since October 2004, approximately 14 years, and are HP modified micro-alloy tubes. Pupuk Kujang's root cause analysis indicated that the rupture of the primary reformer tube was caused by creep, initiated by a hot spot in the primary reformer tube.

This paper describes the events leading to the failure of reformer tubes and the root cause analysis of the failure in Pupuk Kujang. It also discusses lessons learned, corrective efforts to solve the problems, and preventive measures so that these incidents do not recur.

Process Description

The feedstock to the Kujang's ammonia plant is natural gas, steam, and air. The ammonia plant process sequence consists of several sections (see Figure 1): Feed Gas Treatment, Primary Reformer, Secondary Reformer, Shift Converter, CO_2 removal, Methanation, Synthesis Loop, Refrigeration, Ammonia Recovery Unit, and Hydrogen Recovery Unit. The ammonia product can be delivered as all warm liquid to an adjacent urea plant, or as cold liquid to storage. The ammonia plant production can be reduced to 80 percent of full capacity. The incident in the paper occurred in the primary reformer section.

Process Description of Primary Reformer

Primary reformer, A-101-B, is a gas fired, processing furnace containing radiant and convection sections (see Figure3). Processing occurs in catalyst packed tubes contained in the radiant section. The desulfurized feed is mixed with medium pressure steam. The process steam is added in a ratio of 3.2 moles of steam per mole of organic carbon in the feed gas. The mixed feed is preheated in the convection section of the primary reformer, A-101-B. The hot mixed feed is distributed to the reformer catalyst tubes, which are suspended in the radiant section of A-101-B. The feed passes down through nickel reforming catalyst and reacts to form hydrogen, carbon monoxide, and carbon dioxide.

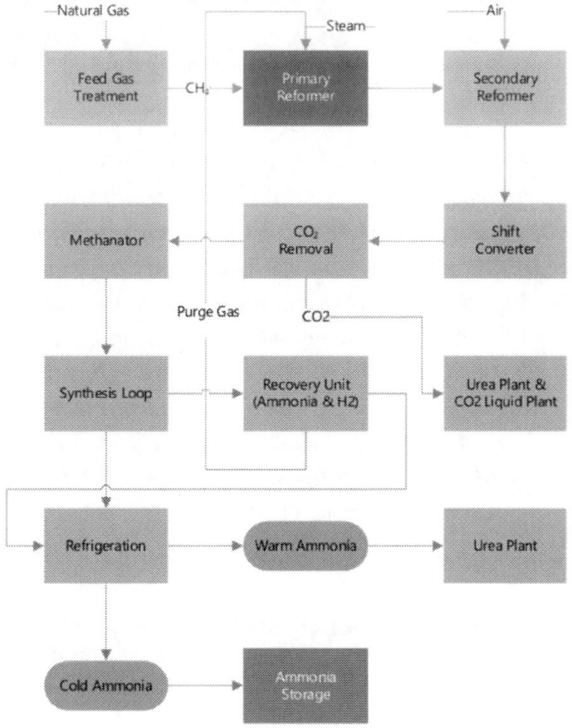

Figure 1. Ammonia 1B block diagram

The reformer tubes catalyst is split into four equal and parallel sub-headers. The 4 sub-headers distribute the flow of hot mixed feed to 192 tubes, with 48 tubes for each sub-header (see Figure 2). Each tube is packed with two types of nickel reforming catalyst, with the lightly alkalized catalyst placed on top and the standard catalyst on the bottom. The tubes extend downward through the radiant section of the reformer to a horizontal collection header. Each tube has a multi-holed catalyst support plate in the bottom of the tube that the reforming catalyst rests upon. These 192 catalyst-packed tubes constitute the primary reforming element of the process. Each catalyst tube holds 0.0948 m^3 [3.35 ft^3] of catalyst for a total of 18.2 m^3 [6,427 ft^3] of catalyst. The catalyst tubes made of HP Modified (25Cr–35Ni+Nb) for the catalytic tube (upper part) and (20Cr-32Ni+Nb) for the pig tail (lower part) with maximal tube wall temperature in normal operation of 922°C [1,692°F]. It is designed to run for approximately 140,000 services hours or about 16 years.

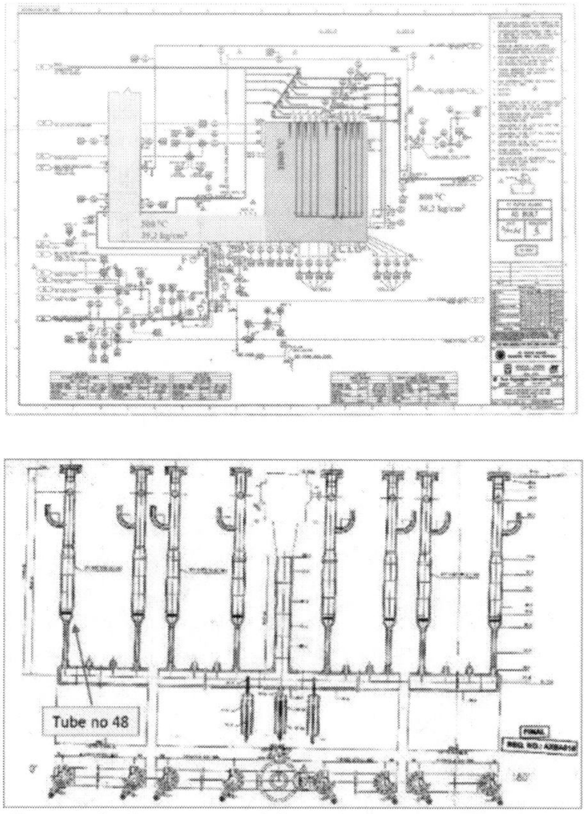

Figure 2. Drawing of flow diagram and catalyst tube configuration of primary reformer

Each of the 4 collection headers contains a centrally located riser tube that directs the flow upward through the radiant fire box to the Primary Reformer Effluent Transfer Line, A-107-D. The transfer line is a refractory lined, water jacketed transition piece connecting the primary reformer with the Secondary Reformer, A-103-D. The use of internal manifolding at the outlet of the catalyst section leading to the riser tubes allows the reformed gas to continue to pick up heat in these risers and collector pipes while exiting the radiant section.

The reactions taking place in the primary reformer consist of steam methane reforming reaction and water gas shift reaction, which are equilibrium-restricted and strongly endothermic. The heat for the endothermic reforming reaction is supplied by 70 fuel gas burners located on either side of the rows of tubes with 14 burners in each of the five rows (see Figure 2). There are two sizes of burners in the reformer arch section. The burners located between the rows of catalyst tube have a normal heat release capability of 6.82 GJ/hr [6.46 MMBtu/hr] each. The burners between the furnace wall and the outer rows of catalyst tubes have a 4.44 GJ/hr [4.21 MMBtu/hr] capability. The outer burners are designed smaller because they only require supplying heat to one row of catalyst tube instead of two, and the catalyst tube located on the sides also receive radiant heat from the furnace walls. The furnace arch burners are down-firing and develop a reformed gas temperature from 500°C [932°F] at the inlet tubes to 799°C [1,470°F] at the outlet of the catalyst tubes. The pressure at the outlet of the catalyst tubes is 36.2 kg/cm^{2}g [5,149 psig] and the outlet from the primary reformer will contain about 66 volume percent hydrogen and 13.4 volume percent methane, on dry basis.

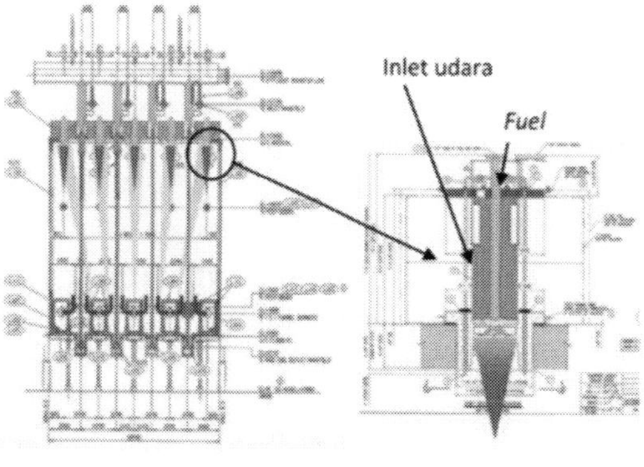

Figure 3. Radiant section of primary reformer.

The primary reformer is designed for maximum thermal efficiency, about 92 percent. The outlet manifold for the catalyst tubes and the outlet "riser" pipes, are located inside the firebox. The convection section includes coils for mixed feed preheating, process air preheating, high pressure steam superheating, feed gas preheating, boiler feed water preheating, and combustion air preheating. A-101-B is also equipped with forced draft fan, A-101-BJ1 and induced draft fan, A-101-BJ2. Reformer flue gas entering the reformer stack (A-101-BI) shall be 140°C [284°F] minimum in normal operation [1].

Chronology of Events

Shutdown Activities

During 2018, Primary reformer in ammonia plant could not be operated more than 95% of its maximum capacity [2].

- In January – March 2018, the tube wall temperature across Row I of A-101-B is measured more than 922°C [1,692°F].
- Pupuk Kujang operated the primary reformer of Ammonia 1B with maximum riser temperature of 780°C [1,436°F] to lower the tube wall temperature of A-101-B.
- From March – June 2018, the tube wall temperature of A-101-B started increasing again (> 922°C [1,692°F]).
- Pupuk Kujang decided to lower the riser temperature to 770°C [1,418°F].
- On Saturday, July 21, 2018, at 18:30 hours, the lighting on the east side of the primary reformer suddenly tripped. During on-site observations, operators discovered a flame bursting out of the primary reformer wall, specifically located at the east bottom side. The flame burned the lighting cable around the wall of the primary reformer.
- Pupuk Kujang suspected a rupture in primary reformer tube row 1, number 48, and based on this condition, decided to shut down the Ammonia 1B plant on July 21, 2018, at 21:00.

Internal Visual Inspection

Figure 4 shows the detailed condition of tube row 1 number 48 after experiencing failure. The catalyst tube is located within insulation, so the damage position can be observed after the insulation is removed. The insulation condition in the bottom header area is fairly good. Leakage is visible in the tube caused by a significant crack. The crack is located at the reducer between catalytic tubing and pigtail.

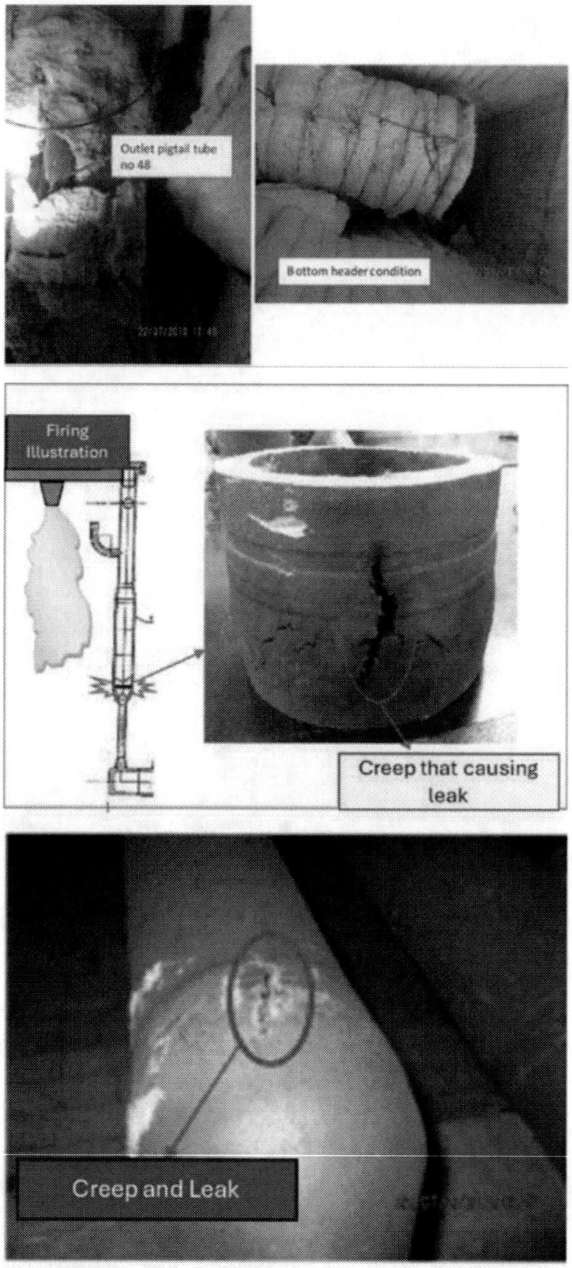

Figure 4. Leakage in tube Row 1 number 48.

Material Inspection

The chemical composition data and microstructure analysis indicated that the two materials of the tube (catalytic tube and pigtail) had different compositions but were still of the same type, namely HP Mod Nb material (see Figure 5). In terms of composition, both were in accordance with the specifications required. There were no anomalies observed in the microstructure, indicating that the materials were in good condition. Therefore, it could be confirmed that it was not due to defects or material-related aspects.

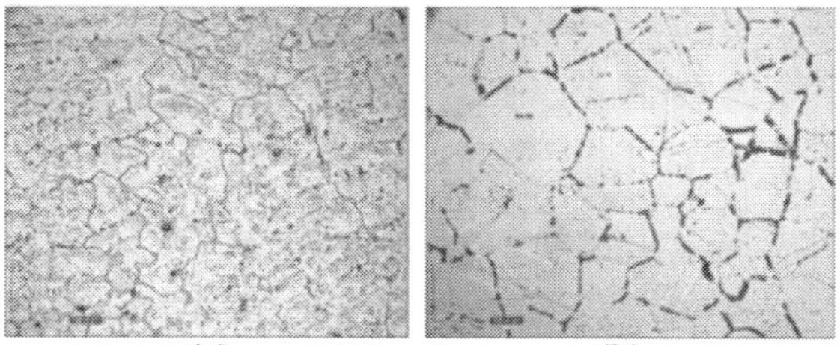

(a) (b)

Figure 5. Metallographic results (microstructure of the reformer tube in aging condition which typically has 2 precipitate phases; Chromium Carbide & G-phase), (a) microstructure of catalyst tube material, (b) microstructure of reducer tube material.

It was clearly visible that the leakage was due to cracks propagating around the reducer as seen in Figure 4. Upon closer examination, cracks propagated in the area where the reducer was post welded. Furthermore, the cracks formed mainly originated from the inner surface of the tubing. The phenomenon, explained in Figures 6.

According to creep test conducted by PT Pupuk Kujang (see Table 1), four tubes needed to be uninstalled. All of these tubes have creep percentage of more than 4%. Therefore, PT Pupuk Kujang decided to uninstall them to prevent the same failure.

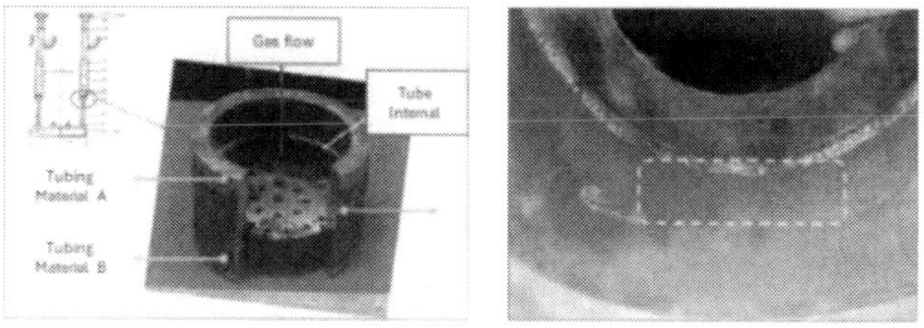

(a) (b)

Figure 6. (a) Construction material on the reducer of tube (b) Crack initiation in the inner surface of reducer

Table 1. Creep test report by PT Pupuk Kujang [2].

Laporan data creep Reformer A-101-B

Row	Tube No	Aktual Creep reducer		Persentase creep	Standar Reducer	Aktual creep tube	Persentase creep	Standar Tube
		Point 1	Point 2					
1	1	391	391	0	391	400	0,63	397,5
	2	392	392	0,26	391	401	0,88	397,5
	24	391	392	0,26	391	401	0,88	397,5
	25	392	392	0,26	391	402	1,13	397,5
	47	392		0,26	391	403	1,38	397,5
	48	432		10,49	391	404	1,64	397,5
2	1	391	392	0,26	391	399	0,38	397,5
	2	391	392	0,26	391	399	0,38	397,5
	24	391	392	0,26	391	401	0,88	397,5
	25	392	392	0,26	391	402	1,13	397,5
	45	391	391	0	391	401	0,88	397,5
	46	391	392	0,26	391	400	0,63	397,5
	47	404	405	3,58	391	402	1,13	397,5
	48	391	391	0	391	399	0,38	397,5
3	1	391	391	0	391	398	0,13	397,5
	2	391	391	0	391	399	0,38	397,5
	24	392	392	0,26	391	401	0,88	397,5
	25	391	391	0	391	403	1,38	397,5
	47	391	391	0	391	399	0,38	397,5
	48	392	392	0,26	391	399	0,38	397,5
4	1	391	391	0	391	398	0,13	397,5
	2	391	391	0	391	399	0,38	397,5
	24	391	391	0	391	402	1,13	397,5
	25	391	391	0	391	403	1,38	397,5
	45	392	392	0,26	391	401	0,88	397,5
	46	391	391	0	391	402	1,13	397,5
	47	407	407	4,09	391	398	0,13	397,5
	48	404	409	4,60	391	402	1,13	397,5

Macro Observation (Cross-Sectional Welding Area)

Figure 7 displays a cross-sectional view of the weld where leakage occurred. Micro-cracks were mostly occurred in the metal of the reducer's inner surface, while the metal in the catalyst tube was comparatively better. This difference was due to the catalyst tube being higher specifications than the reducer tube. The concentrated pattern of micro-cracks in the inner area suggests that damage started from the inner surface. The identified crack shapes and dimensional changes in the tubing indicated the occurrence of creep. Creep mainly happened around the reducer, which is the catalyst's end area.

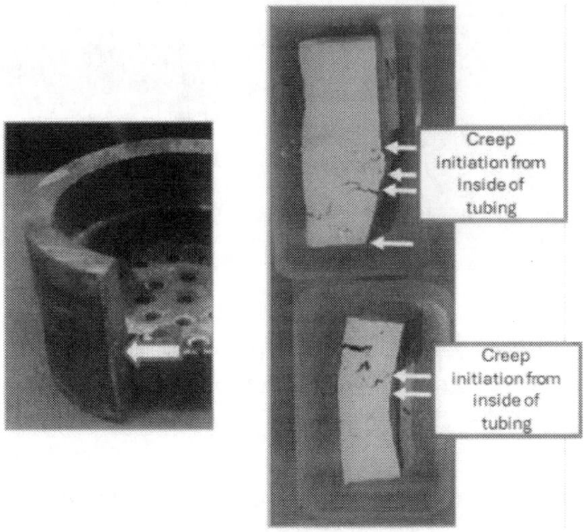

Figure 7. Cross sectional surface on tube Row 1, number 48 [2]

Catalyst & Flame Condition

Of all the catalyst present in the tube, on average, it experienced breakage (see Figure 8 and 9). Additionally, carbon deposits were observed. This condition indicated that the side reaction of steam methane reforming occurred inside the tube. Based on the condition of the catalyst and the location of the damage that occurred, it is possible that a hot spot occurred around the reducer area. Temperature measurement data also supports this condition.

Figure 8. Catalyst Condition Inside the Tube [2].

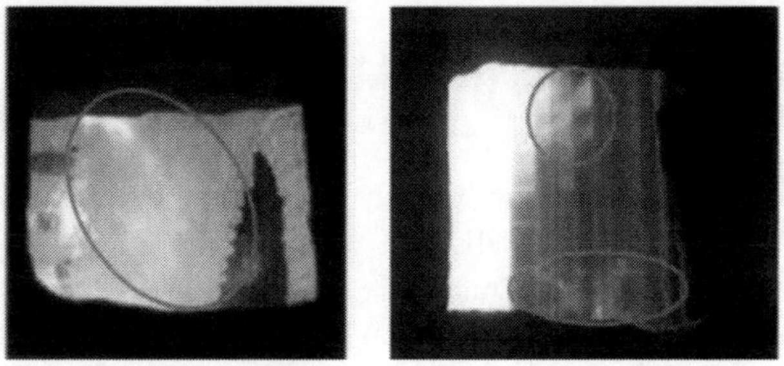

Figure 9. The radian section condition. (left) erratic flame from burner; (right) hot band on catalyst tube [2]

Pupuk Kujang did some catalyst sampling across the primary reformer tube and found:

- The catalysts on Row I, numbers 47-48 and Row IV, numbers 47-48, are covered with carbon deposits, especially those located 3-4 meters from the top flange (top).
- The bluish appearance of the catalyst on the upper part of the unloaded tube indicated the formation of spinel (nickel aluminate). Spinel is inactive species in nickel reforming catalyst and forms when there is high local temperature on the reformer tube. The inactive side on catalyst will lead to tube hotspots and carbon deposits on catalyst.
- The fire at the edge rows (Row I and Row IV) is seen spreading to the adjacent rows.

Natural Gas Composition During 2018

The composition of natural gas entering the Ammonia 1B plant (see Figure 10) experiences fluctuations. The CH_4 content varies from 81.5% to 86.7%, the total C_{2+} content fluctuates from 6.5% to 9.5%, and the inert gas content $CO_2 + N_2$ fluctuates from 6.5% to 10.5%. The natural gas composition for the previous two years was quite stable, with CH_4 content at 80% to 83%, C_{2+} content at 7% to 9%, and inert gas content at 9.5 to 11.5%. Composition fluctuations affect the flame pattern generated in the burner. Additionally, this composition also affects the S/C ratio for the reaction requirements in the primary reformer to prevent carbon formation reaction.

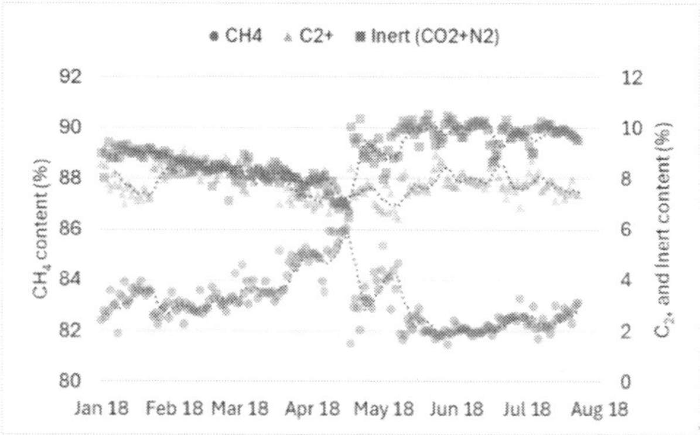

Figure 10. Natural Gas Composition in 2018

Root Cause Analysis

Based on several facts revealed from the failure incident of Primary Reformer tube (A-101-B) in Row I number 48, creep phenomenon occurred along the hot tube. Reviewing the conditions of the tube, creep is a condition of material property degradation due to exposure to high temperatures. There are several factors that possibly could cause the hotspot in catalyst tube including natural gas composition fluctuation, sulfur poisoning, and low steam to carbon (S/C) ratio. The root cause analysis diagram is illustrated in Figure 11.

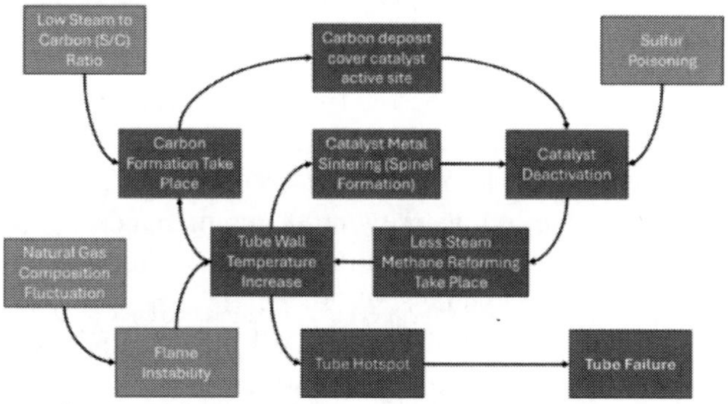

Figure 11. Root Cause Analysis Diagram

As shown in the diagram, tube hot spot in primary reformer is caused by increase in tube wall temperature. Meanwhile, the increase in tube wall temperature either can be caused by flame instability or catalyst deactivation. Furthermore, catalyst deactivation is part of cause-and-effect loop which can be triggered by several factors as shown in Figure 11.

Flame instability can be caused by natural gas composition fluctuation. Natural gas composition can be seen in Figure 10. The fluctuation of natural gas composition can be classified into two categories consisting of heavy hydrocarbons (C_{2+}) fluctuation and inert gases (CO_2 and N_2) fluctuation. The heavy hydrocarbons fluctuation led to more heating value of natural gas when C_{2+} content increased. In the other hand, inert gases fluctuation led to less heating value of natural gas. The fluctuation of natural gas heating value can cause instability of natural gas velocity and pressure in radian section of primary reformer and create flame instability. Erratic flame in the primary reformer can lead to local hot spot in catalyst tube caused by local tube wall temperature increase. Furthermore, the flame instability also causes carbon formation and catalyst metal sintering to take place on catalyst. That will make the catalyst deactivate and worsen the tube wall temperature of catalyst.

Another possibility that could cause tube hotspot is a low steam to carbon (S/C) ratio and sulfur poisoning. A low S/C ratio will lead to carbon formation on catalyst, which will cover the active sites of catalyst, eventually causing catalyst deactivation. Sulfur poisoning can also cause catalyst deactivation because it reacts with the active component of the catalyst.

The primary reformer catalyst was installed in April 2014. Since then, the S/C ratio had always been operated more than design basis (3.2 mol/mol) as shown in Figure 12. The S/C ratio is also greater than minimum ratio of 1.0-1.2 mol/mol, which is required to prevent carbon formation on steam reforming catalyst at 700-800°C [1,292-1,472°F] and 40 bar [580 psi] [3]. The catalyst performance regarding methane (CH_4) leak was also satisfactory, with a value less than design basis of 13.4%-vol. Based on the data, that is unlikely that low S/C ratio was causing the catalyst tube hotspot. Also, the failure occurred locally at catalyst tubes number 47-48 in Row I and

IV. If the failure was caused by a low S/C ratio, all of the primary reformer catalyst would be deactivated, and the methane slip would increase significantly.

The sulfur poisoning was also unlikely to cause catalyst tube hotspot because the H_2S content in the primary reformer feed gas had been managed to be lower than 0.1 ppm-vol. This is the allowable sulfur content that ensures the primary reformer runs smoothly [3]. The sulfur content can be seen in Figure 13.

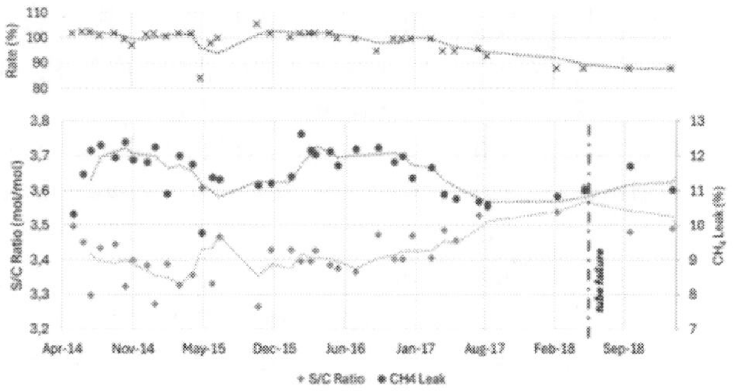

Figure 12. S/C ratio profile [2]

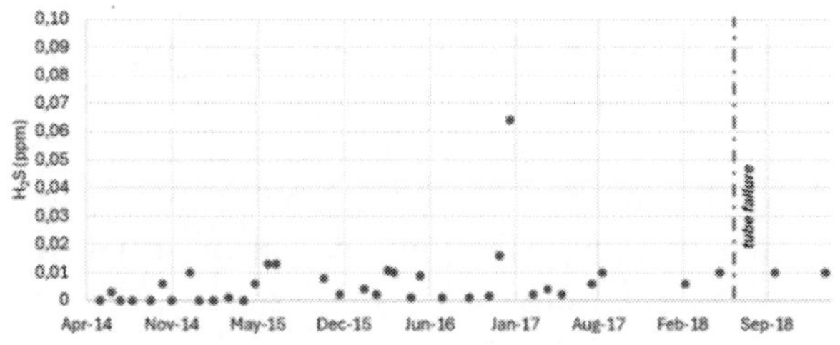

Figure 13. Sulfur content in primary reformer feed gas [2]

Failure Mode

Based on all the data gathered, the failure mode that occurred in primary reformer tube Row I, number 48, was due to creep. The creep was a result of hotspot conditions originating from the inside of the tubing, especially in the reducer area (catalyst end). The significant increase in tube wall temperature was indicated by field measurement results reaching a peak point of more than 922°C [1,692]. The fluctuation of natural composition caused erratic flame patterns on the radian section of the primary reformer. This led to the local increase of tube wall temperature at catalyst tube number 47-48 in Row I and IV. High temperature on the tube catalyst induced carbon formation on the active sites of catalyst and causing metal sintering of the catalyst. This led to further catalyst deactivation, which worsened the tube wall temperature.

Improvement Effort

Several corrective and preventive improvement actions have been taken after shutdown until now to prevent damage in the primary reformer such as:

- Replacing catalysts in several tubes prone to hotspots during 2018 such as: Row II number 15, Row III numbers 17, 18, 19, 20, 21, 25, 40, 41, and 42.
- Replacing catalysts in Row I number 46 as a precautionary measure against hotspot.
- Replacing catalysts in Row I number 26 to ensure accurate tube catalyst pressure drop measurements.
- Removing tubes Row 1 numbers 47 and 48 and replacing the caps at the bottom header of these rows.
- Removing tubes Row 4 numbers 47 and 48 and installing plugs at the outlet pigtail numbers 47 and 48.
- Adjusting the alarm values of TI 1315 A-D and TI 1316 A-D (reformed gas exit catalyst tube temperature) from the original 840°C to 815°C [1,544°F to 1,499°F].
- Continuing preventive maintenance actions such as maintaining TI-1314 (temperature riser) values at a maximum of 770°C [1,418°F] until the 2019 turn around, regularly measuring the temperature of the reformer tube skin, and adjusting burners regularly if there are disruptions in the tube skin temperature or gas composition.
- Replacing primary reformer catalysts during the 2019 turnaround.
- Conducting a computational fluid dynamic (CFD) analysis for arch burner regarding changes in natural gas composition.

- Conducting a Lifetime Remaining Assessment during the 2022 turnaround.

Present Primary Reformer Tube Condition

A remaining lifetime assessment (see Figure 14) for the reformer tubes of the Primary Reformer Unit (A-101 -B) at PT. Pupuk Kujang has been performed with the software Tube Life from MP (Merged with mother company Institut Dr. Foerster GmbH & Co. KG) in April 2022. It delivers a complete analysis according to the Tube Life methodology. This report summarizes the results of the remaining lifetime assessment study undertaken for Ammonia 1B plant primary reformer.

The remaining life assessment is based on the modified Dyson Model which primarily works on the principle of creep with respect to specific loading, temperature, and time exposure of the reformer tube to various designs and operating history. It will provide remaining life calculations based on experience and expert judgment for fee-for-service (FFS) decisions, replacement strategy, and purchasing support of overall tube life management. The model includes estimates of an equivalent metal temperature to cause strain at a given time (or age of the tube). The equivalent temperature approach allows for estimating future damage based upon damage accumulated in the past. Based on the assessment, all the tubes are operating well within "normal creep" after 16 years of service with additional life expected ranging from 8 years to greater than 15 years.

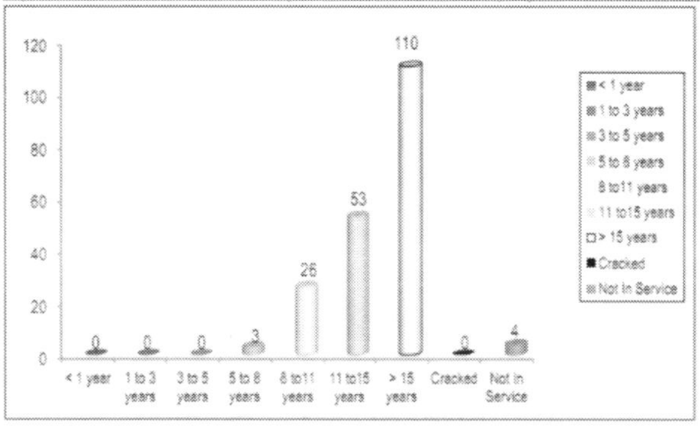

Figure 14. Distribution of tube remaining lifetime in primary reformer of Ammonia 1B Plant PT Pupuk Kujang [2].

Lesson Learned

To prevent similar damage from occurring in the reformer tube, several lesson that can be learned and implemented, including:

- A lot of effort was required to keep the primary reformer running smoothly. The most effective method to prevent primary reformer tube failure is by thoroughly examine the composition of natural gas entering primary reformer and adjust the suitable operating parameter based on the data. Moreover, routinely check the tube wall temperature of primary reformer tube to verify the burner adjustment.
- Good teamwork between the operations team, process engineers, and maintenance team was a key to successfully achieving optimum operating parameter and its smooth operation.
- Since primary reformers in Ammonia 1B plant mostly run at full loads and above design production rates, it is important to still operate within the design envelope without exceeding the design parameters.

Summary

The damage to the primary reformer tube was caused by the phenomenon of creep. The hot spot condition in the damaged area led to a critical condition exceeding the creep resistance of the tube material in that area resulting in failure. Hot spot occurred due to an increase in temperature caused by deactivation of catalyst in the tube. Deactivation of catalyst caused by fluctuation in natural gas composition.

References

[1] KBR, Pupuk Kujang, Toyo Engineering Company: "1,000 MTPD Ammonia Plant operating Instruction Manual", August 2004

[2] Operation records (process historical data, log book, log sheet, etc.)

[3] Twigg, Martyn V.: "Catalyst Handbook Second Edition", 1989

Challenging Repair Solutions for Recurring Leaks in a High-Pressure Carbamate Condenser

The High-Pressure Carbamate Condenser (HPCC) is a critical heat exchanger in the high-pressure loop of any Urea Plant. The carbamate condenser in Yara Brunsbüttel has been in operation since 2007 and faced various problems throughout its life. This exchanger is a kettle type exchanger, where the carbamate solution flows in the tubes and exchanges heat with condensate on the shell side, generating steam. In the first years of operation, several tube leaks have occurred due to flow distribution issues in the outlet channel of the exchanger due to flow stagnation. Following process and design improvements, the plant was returned to operation with several tubes plugged. The equipment operated from 2015 until 2021 without further problems, but inspection findings from the 2021 turnaround revealed further signs of degradation. In 2023, the urea plant in Brunsbüttel was shut down due to leak of the HPCC. The urea plant was out of operation for thirteen days to repair the HPCC. This paper will discuss the identified Root Cause Analysis (RCA) and the lessons learned from the last event, that was recognized as three different causes. Although, the equipment was fully inspected during the last turnaround, it was difficult to identify the location of the leaks that occurred in 2023. The solution to understand the capabilities of the ammonia leak test in a potentially contaminated and corroded leak path was interesting and was accomplished by testing a laboratory scale reproduction of the exchanger.

Giulio Paci
Yara International ASA

Hauke Hauschild, Mathias Schmahl
Yara Brunsbüttel GmbH

Introduction

The Yara Brunsbüttel urea plant was designed by Snamprogetti which was revamped by Casale in 2010. The synthesis section uses a self-stripping technology.

The High-Pressure Carbamate Condenser (HPCC) was not modified during the urea revamp and has been installed in 2007, following Snamprogetti's recommendation.

A simplified sketch of the synthesis section is presented in Figure 1.

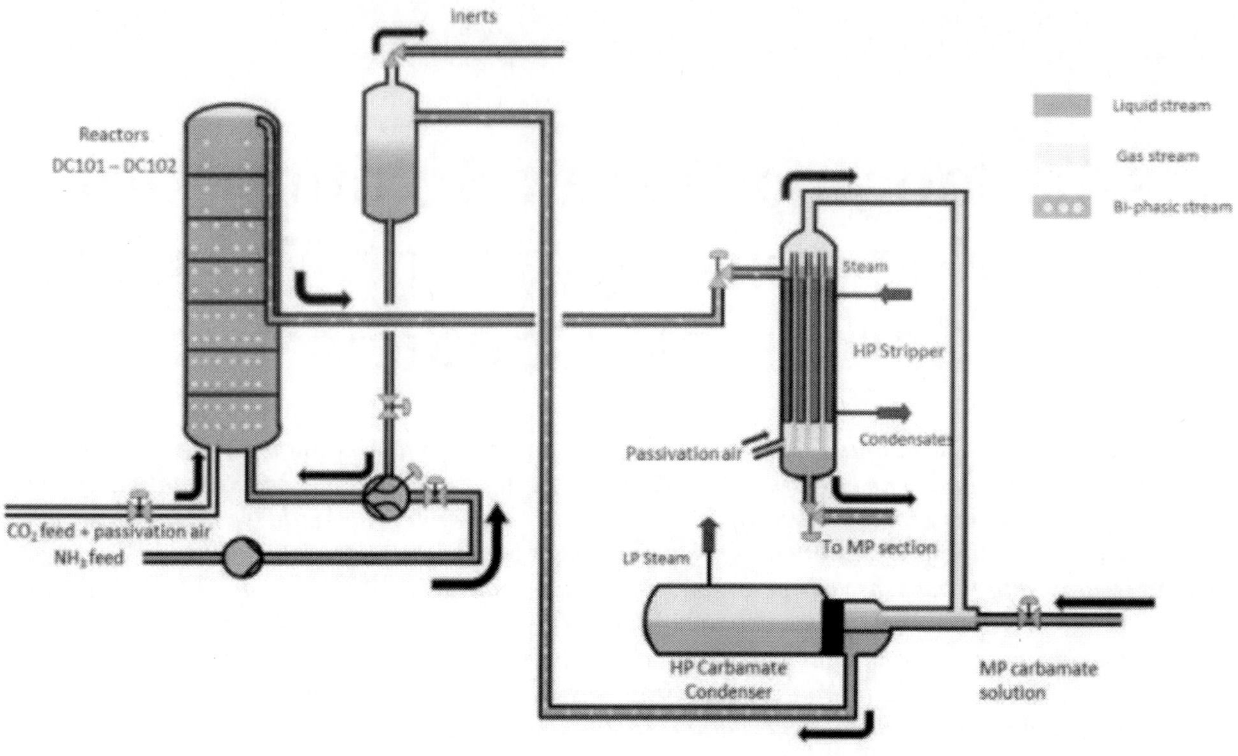

Figure 1. Snamprogetti HP Urea synthesis sketch.

Background

The original HPCC was made of 25Cr-22Ni-2Mo material, and first failed soon after initial startup (four years). The leaks were located on the outer tubes. Failure mode at that time was attributed to an improper distribution.

In 1990, after eleven years of operation, the flow was reversed by modifying the internals of the HPCC channel. The tubesheet behind the internal pipe started to corrode rapidly due to condensation of gas - the pipe prevented carbamate solution with adequate passivation oxygen from reaching the tubesheet – as shown in Figure 2.

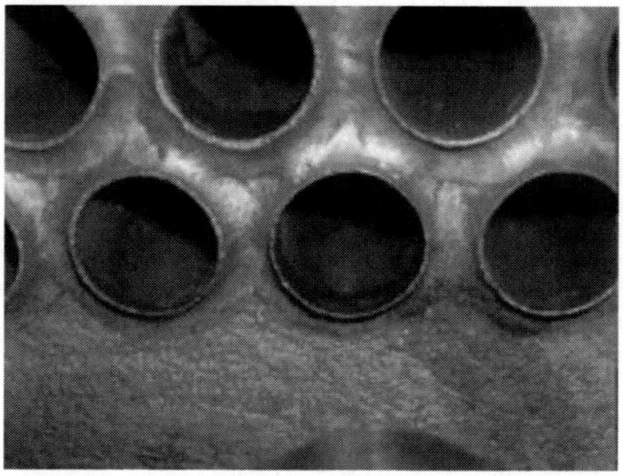

Figure 2. Condensation corrosion in the inlet tubesheet.

After fifteen years of operation with this circulation scheme, the former configuration was resumed until 2007 when this equipment was replaced by a new one that is the subject of this paper. These flow schemes are shown in Figure 3.

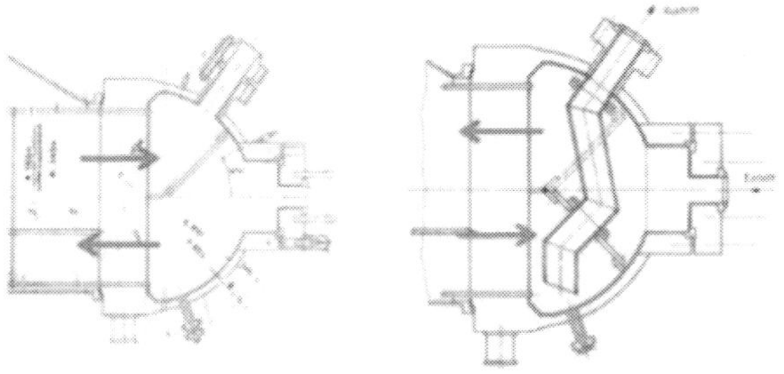

Figure 3. Different flow configurations left 1979-1990 & 2004-2007; right 1990 – 2004.

The new HPCC was commissioned in 2007. It is a horizontal, kettle-type condenser with process flow on tube side and steam generation on the shell side. The tubesheet consists of carbon steel forging protected by 12 mm (0.5 in) of 1.4466 (UNS S31050) corrosion resistant weld-overlay.

The tube bundle contains 1500 "U" tubes with outer diameter of 19.05 mm (0.75 in) and 2.11 mm (0.083 in) wall thickness and made of X1CrNiMoN25-22-2 (UNS S31050) austenitic stainless steel. In the channel there is an internal partition plate designed to split the process flow into two passes. The process fluid (urea carbamate solution) is fed to the top half while the outlet is located at the bottom half.

Modification for the new HPCC

Several modifications were made, compared to the original HPCC, following the suggestion of the licensor. The main modifications were:

- Process flow direction.
- Orientation of inlet and outlet process nozzles.
- Removal of partition plate in the shell side.

The flow direction was changed to comply with the new Snamprogetti design. The other reason for the modification was to reach the best heat transfer with the top to bottom flow configuration.

The orientation of process nozzles was modified compared to the original exchanger (Figure 4 and 5).

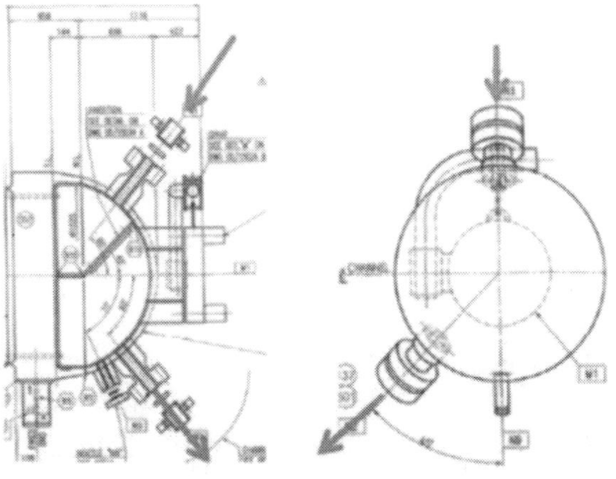

Figure 4. Orientation and flow configuration of the new HPCC.

Figure 5. Installed HPCC. Note the orientation of outlet nozzle.

To increase the tube length, the plate used to keep the bundle flooded on the shell side was removed. Thus, the bundle might not be completely submerged with steam condensate (Figure 6).

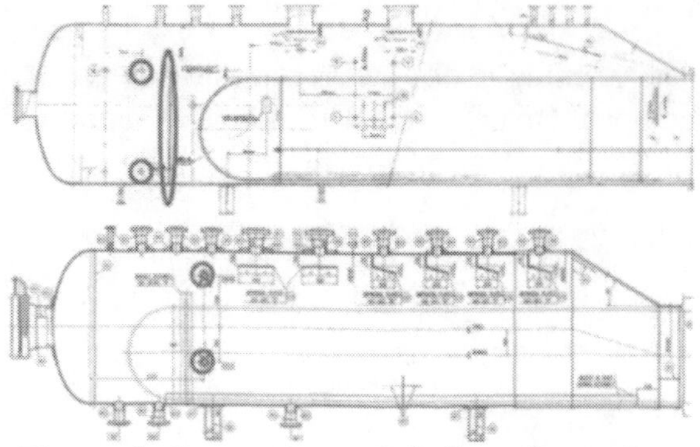

Figure 6. Comparison of shell configuration of old HPCC (top) and new HPCC (Bottom).

Unexpected HPCC failure in 2014

In May 2014 the Brunsbüttel urea plant was shut down due to an increase of conductivity in the boiler feed water (BFW) of the HPCC. This increase of conductivity was linked to a urea carbamate leak into the steam. After inspection, the leak was localized to tubes in the bottom section (process exit) of the channel.

Inspection for material loss using eddy current testing found that the tubes located in the bottom left tubesheet were corroded. The corrosion began in the tube just past the tubesheet and only affected the bottom tubes. (Figure 7)

The immediate action was to plug eighty tubes in this location. The partition plate between the inlet and outlet chambers had been found to be loose, meaning that some bypass was occurring during operation.

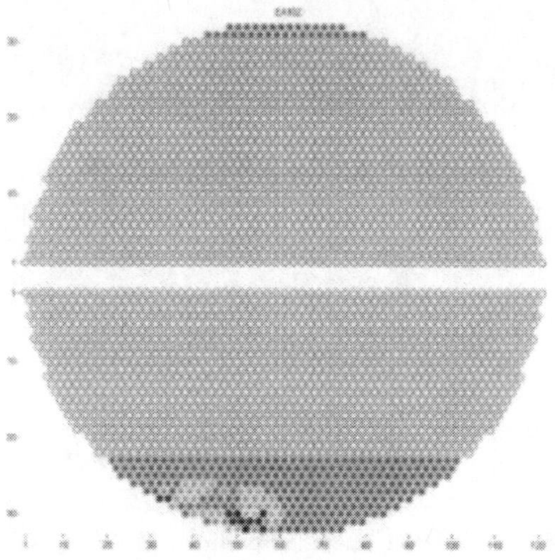

Figure 7. Extent and location of damage of HPCC.

This failure was totally unexpected, as it happened after only seven years of service and because the equipment was thoroughly inspected in the 2010 turnaround and was assessed to be "like new"..

Root cause analysis of the 2014 failure

Based on the early failure of the new HPCC, a technical team was established to conduct a Root Cause Analysis (RCA) of this premature failure.

Several considerations and hypotheses were considered, including:

- Lack of passivation agent (oxygen),
- Poor circulation/carbamate stagnation possibly caused by longer tubes,
- Liquid level build-up in the outlet channel,
- Effect of loosened partition plate between inlet and outlet channel,
- The effect of tubes not totally submerged, and
- Crystallization inside the tubes during block-in periods.

At the end of the investigation the most likely root cause was due to liquid build up in the outlet channel caused by the new position of the carbamate solution outlet nozzle. This root cause was also proved by calculation of fluid velocity and pressure drop (Figure 8 and 9).

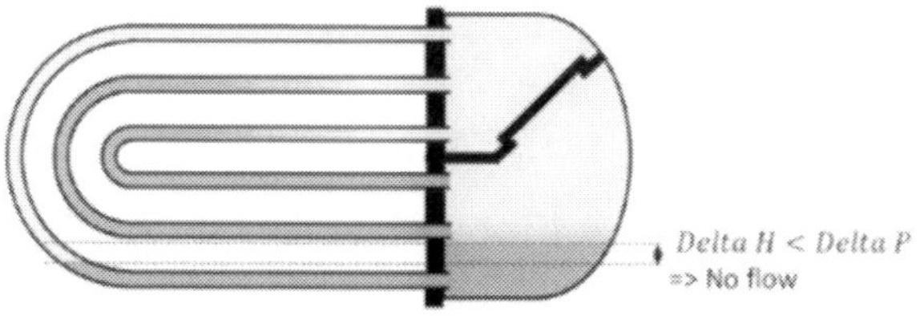

Figure 8. Consequence of liquid build-up in the bundle.

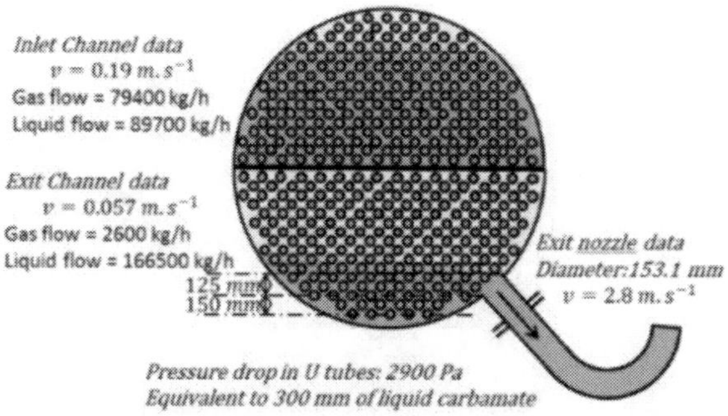

Figure 9. Cut view of HPCC channel with some physical data.

Proposed modification and follow-up

The plugging of eighty tubes (5% of surface area) has increased the pressure drop across the heat exchanger. In addition, the remaining tubes are located near the exit nozzle, so it has been considered that no design modification was required. Nevertheless, it has been decided to measure the thickness of all tubes, using eddy current testing, during the next turnaround or any other suitable opportunity to validate this theory.

These recommendations were successfully completed, and the urea plant was returned to operation. The plant, and HPCC, operated without any problem until 2023.

Reliability issues and 2023 Urea plant outage

In March 2023 the urea plant in Brunsbüttel was shut down due to a leak of HPCC, which was identified again by increased conductivity on the shell side of the exchanger.

In hindsight, some indications of the failure could have been identified during the previous inspection during a turnaround in 2021, but the peculiarities of the failure discovered in 2023 made the prediction difficult. It is worth noting that when high-pressure urea equipment is inspected, the best technique and the first to do remains visual inspection. In this case the HPCC channel was not easy to access due to the presence of a welded partition plate and very tight space inside. This made the visual inspection of the tubesheet weld-overlay and tube to tubesheet welds very challenging.

However, there were observations during the 2021 turnaround that revealed poor workmanship from the manufacturer during the construction of the equipment. For example, intergranular corrosion was observed in the overlap strip weld-overlay in the tubesheet. Based on the characteristics of this observation, it was decided not to take any actions during the 2021 turnaround. At that time, it was thought that repairs to these areas would have created more problems in terms of material sensitization and metallurgical degradation in the material that was already corroded.

It is well recognized that intergranular attack is one of the main forms of corrosion in urea high-pressure synthesis especially in the welds and heat affected zones due to improper weld procedures or poor welding quality.

The urea plant was out of operation for thirteen days to repair the HPCC. At least three leaks were identified after a dedicated inspection plan. These three leaks were found to be of a different nature compared to what was observed in 2014.

Others potential leakages were also discovered during the inspection of the condenser, and since different tubes had already been plugged, it was decided to replace the HPCC in the next planned shutdown, to eliminate the risks of these failures (lower efficiency) and the need to achieve more urea production.

Leak number 1

The first leak was identified in the weld on one tube plugged during the 2014 failure. It is likely the corrosion started from a weld defect from a start and stop arc welding, as can be observed by a magnified picture (Figure 10). These features can happen during extensive welding job and especially in the moment in which the welding arc stopped and the shielding gas protection is lost.

These defects present possible points of corrosion when the plant is in-service, because the defects are characterized by small porosities in the affected area.

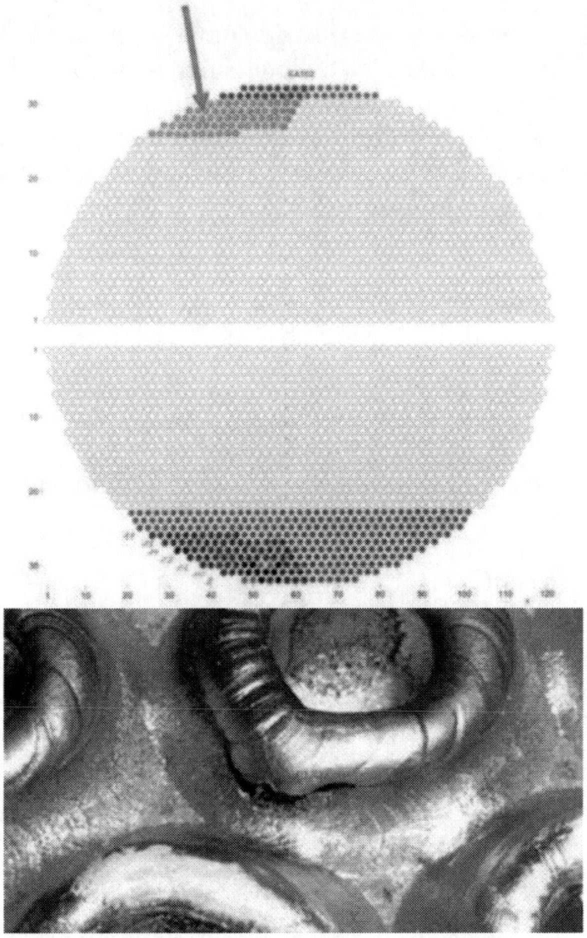

Figure 10. identification of the leak number 1 in the tube map and detail of defect.

It is important to mention that the leak was discovered by a simple soap test, pressurizing the shell side by air (Figure 11). However, the usually more sensitive ammonia leak test normally used to find leak in high-pressure equipment was conducted and surprisingly did not indicate a leak.

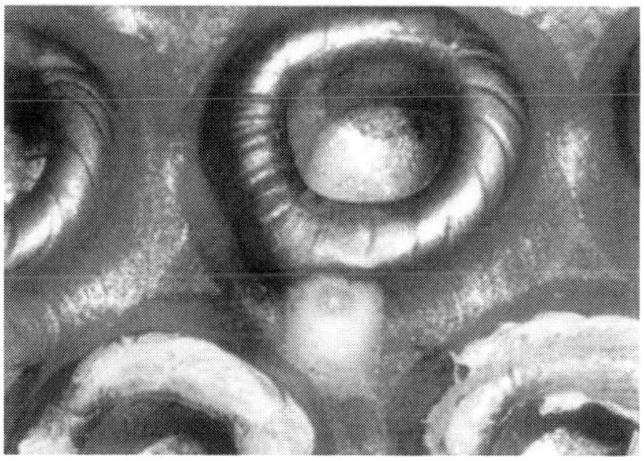

Figure 11. Positive bubble air test with shell under pressure.

Based on this discovery, it was decided to inspect for similar defects and make repairs needed to reduce the risk of future leaks.

It was discussed and agreed that similar defects would be ground out and weld repaired. In order to reduce the heat input, it was agreed to adopt smaller diameter filler rod for the repair and use only stringer bead TIG technique. In addition, it was also taken into consideration to reduce the interpass temperature as much as possible (maximum 110°C (230°F)), compared to 150°C (302°F) which was used in the original welding procedure.

The repair was accomplished by weld build-up, as a plug interfered with other repair options..

After discussing ways to improve the visual inspection and it was found that doing using magnified macro-pictures of the tubesheet could help identify defects missed by the naked eye.

Following this strategy, others potential tube-to-tubesheet welds leak were discovered (Figure 12).

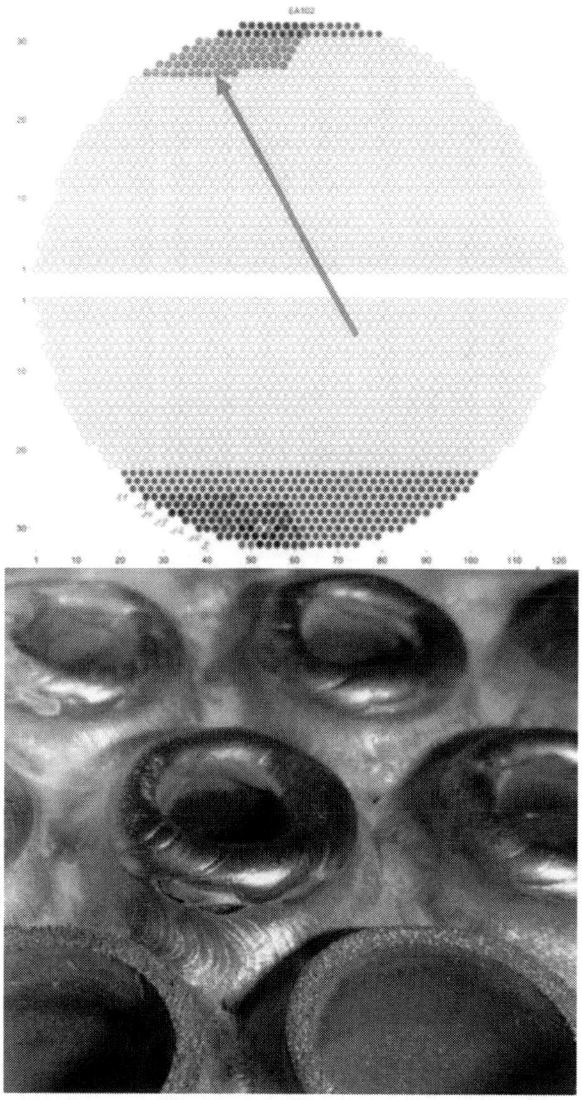

Figure 12. Potential future leak point.

Leak number 2

The position and characteristics of the second leak show correlation with the intergranular corrosion of the tubesheet weld-overlay observed in previous turnarounds, for which no actions were taken.

It seems that leak started from a pinhole in the tube-to-tubesheet welds, but also the leak position suggested that there would have been a trigger effect of the poor welding procedure during tubesheet overlay. In fact, this zone has undergone at least two heat cycles, during weld-overlay and during tube-to-tubesheet welding (Figure 13).

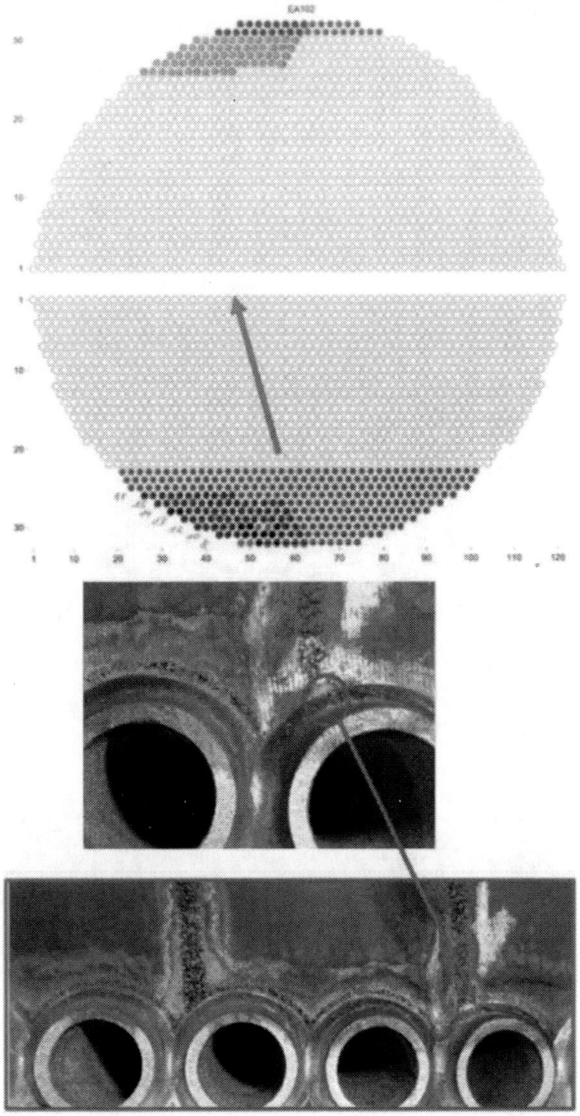

Figure 13. Leak point number 2.

This failure pointed to the fact that this was somehow a systematic defect on the tubesheet that meant there was the potential of numerous defects existing and likely future leaks (Figure 14).

Figure 14. Future potential leak point for failure number 2.

It was then recommended to focus future inspection in the areas of "double" heat affected zones, or better said, in the area where there is the overlap area among tubesheet weld-overlay and tube-to-tubesheet welds.

Corrosion was also seen between the neighboring tube-to-tubesheet weld which is also partly around manufactured tubesheet effects. Proper repair by grinding was nearly impossible, time consuming, and difficult work.

To repair this defect the tube was trimmed and plugged.

At the end some recommendations were given:

- It was recommended to only grind the areas of "double" heat effected zones in the top row of the outlet section and bottom row of inlet section, in order to remove the intergranular corroded area; after grinding dye penetrant testing was required for confirmation.
- When manufacturing new equipment, electroslag welding should be the preferred overlay technique for the tubesheet, and improvements in welding procedures are required, along with proper manufacturing follow-up on workshop with specific witness points during production.

Leak number 3

Leak number 3 was identified as a typical type of defect in the tube-to-tubesheet welds of high-pressure equipment in a urea plant. The character of defect was a pinhole. The most likely RCA was an embedded defect that corroded and opened-up during plant operation. (Figure 15)

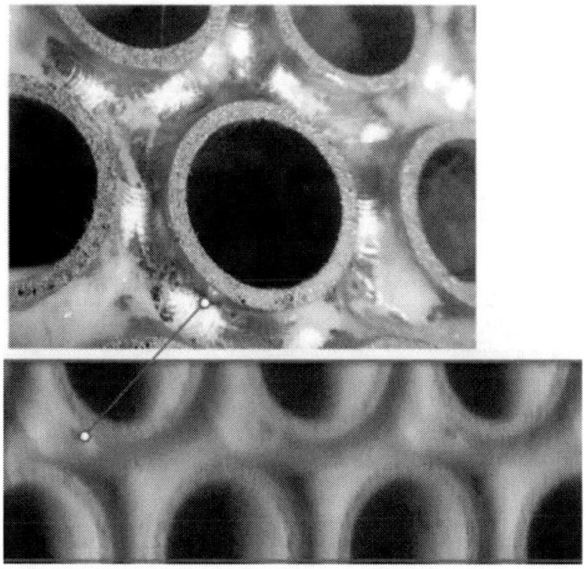

Figure 15. Tube leak number 3 and further potential location near the leaking tube.

With the aim to preventively identify these types of defects, the possibility of using P-scan ultrasonic testing was studied. Unfortunately, due to the roughness of the internal surface of the tubes, due to slight corrosion/etching, this technique was not able to detect these kinds of indications.

Since the ammonia colorimetric test also failed to detect this leak, it was also decided to investigate the reason why a method with such high sensitivity was unsuccessful.

It is important to add that this leak was discovered together with the leak number two with air/soap bubble test, but after the repair of leak number 1 was accomplished.

By grinding the pinhole, it was discovered that the urea-carbamate solution came into contact with the carbon steel pressure bearing resistant part. Figure 16 shows a cavity behind the trimmed tube.

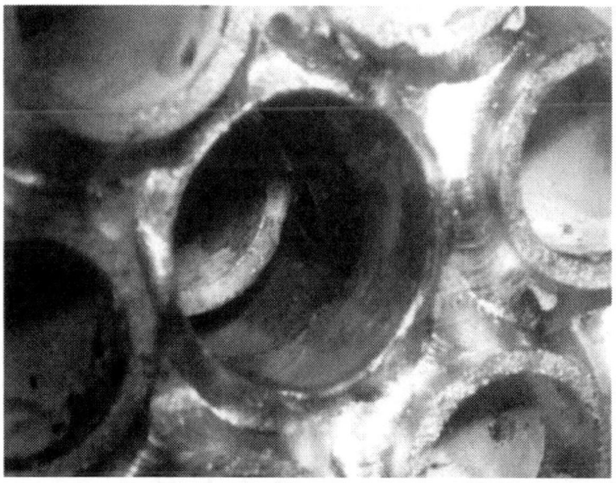

Figure 16. Cavity in the carbon steel tubesheet.

With the help of an endoscope, the corroded area was thoroughly identified and assessed (Figure 17). Based on the strength calculation, it was decided to plug the tube affected by the pinhole using the normal technique, without reinforcing the hole in the tubesheet.

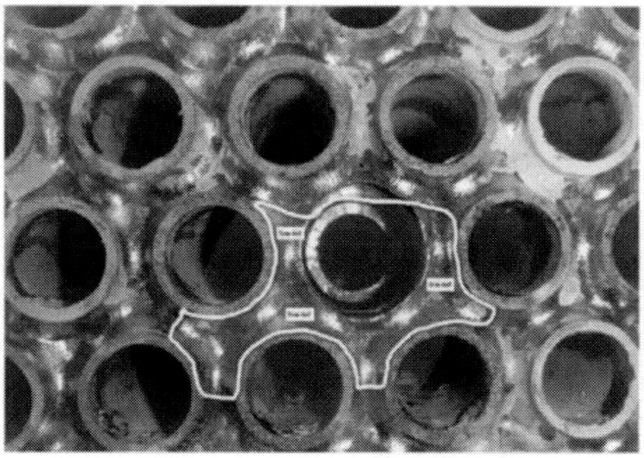

Figure 17. Area affected by corrosion.

Small scale NH_3 leak test in laboratory

Due to the difficulties found during the ammonia colorimetric leak testing and the fact that it has been used a sensitive paste that during the test would have changed the color from yellow to red/pink in case of leakage, a small-scale laboratory test was performed to understand the reason this test failed.

It is important to highlight that the ammonia sensitive paste did not change color during the various leak tests, but finally soap bubbles were observed as sign of the leaks.

It was suspected that the ammonia test was unsuccessful due to the presence of the moisture/water in the leak path, especially for leak number 3. However, it has not been excluded that other possible contamination would have stopped the detection or that the paste density affected the sensitivity of the test.

Reproduced NH_3 leak test

A simplified layout was prepared to reproduce in scale, simulating the HPPC leak with glass laboratory materials.

The glass container was filled with 50 ml (0.011 gal) of ammoniacal water (5% concentration). 1.5 ml (0.0033 gal) of water was introduced in the glass container and allowed to flow into only one tube (Figure 18). Both outlet tubes were sealed with a plate where the NH_3 sensitive paste has been applied.

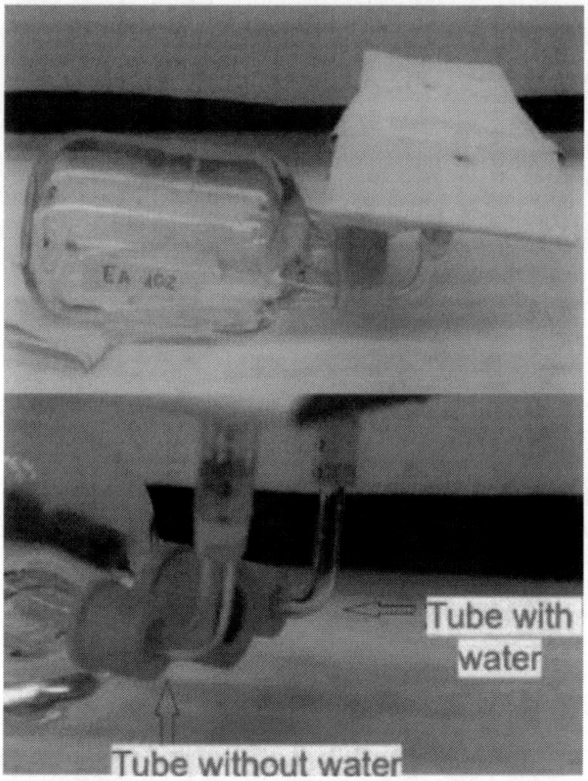

Figure 18. Layout of laboratory NH_3 test.

With this method, it has been possible to prove that water contamination or in general presence of humidity/moisture may influence the result and the detection sensitivity of ammonia leak test. The dry tube (no water presence) immediately showed the paste color change (Figure 19), while the tube with droplets of water started to change the color after 15 minutes of ammoniacal water exposure.

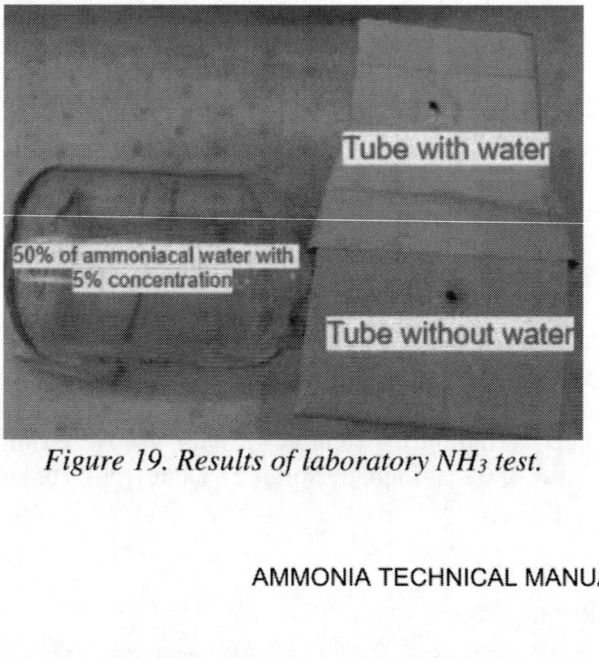

Figure 19. Results of laboratory NH_3 test.

Conclusion

Even though the failure of 2014 was mainly due to a design concern, the different leakages found in March 2023 have been characterized mainly by poor workmanship during construction [1], especially during the tubesheet weld-overlay and tube-to-tubesheet welding.

There was also the opportunity to understand more about some advanced non-destructive examination like (P-scan ultrasonic examination and ammonia colorimetric leak testing).

Regarding the P-scan technique, it has been proven that this is a good method to detect defects in the tube-to tubesheet weld of new equipment but can be difficult to use after equipment has been in high pressure urea synthesis service. The reason is that to obtain a good quality of the result the internal surface must be polished with no roughness.

Finally, it was also proved by small-scale laboratory test that even the ammonia leak test can be affected by the presence of moisture/ water in the leak path. This was a good demonstration that confirmed the high confidence of this technique to detect leak in high-pressure equipment but also to tune it based on the contamination in which the equipment could have been subjected.

References

[1] Roberto Gorza, "Prevention of a Potential Catastrophic Failure of a High Pressure Condenser in a Urea Plant", AIChE 2014

Safety Aspects of Green Ammonia Production

Over the years, the ammonia industry has maintained excellent safety standards, but accidents still happen too frequently, as each accident is one too many. All changes to known operation processes and modes introduce new safety challenges that must be addressed appropriately. Production of ammonia from hydrogen and nitrogen is not a novel technology concept. However, having the hydrogen produced from electrolysis is a reasonably new development, as is the concept of using hydrogen (and possibly oxygen) as co-feed to ammonia plants feeding on natural gas. The presentation maps out safety aspects and considerations for such cases. It provides an overview of what an electrolysis provider can and should do to secure operational safety for the electrolysis unit, which produces the hydrogen for the ammonia plant.

Esben Sorensen, Glenn Rexwinkel
Plug Power

Venkat Pattabathula
SVP Chemical Plant Services

Introduction

There are several safety aspects that must be considered when building and operating an ammonia plant. Many of these relate to personnel safety in connection with the commissioning and pre-commissioning of the plant and safety aspects related to the storage and transportation of ammonia. The focus of this paper is to address changes in the safety landscape related to adopting green hydrogen supplementation to existing ammonia production by electrolysis or external supply as a novel concept. The most common hazard associated with hydrogen is its propensity to ignite, as illustrated in Figure 1, reproduced by Eck et al. (2021)[1].

Introducing hydrogen produced by electrolysis as an element to the ammonia process configuration requires not only that the electrolysis process itself should be completely safe but also that sections in the ammonia process where process conditions change because of this novel concept are analyzed to understand how such change can be made without increasing the overall process risk compared to traditional process configurations.

Feeding pure hydrogen to a specific part of the ammonia process naturally means that the risks associated with higher hydrogen partial pressure in that part of the process increases compared to what it was in the original configuration. From a safety point of view, two main aspects of increasing the partial pressure of hydrogen in a given process section should be given focus:

1. Increased risk of explosion (or detonation) as the rate of flame propagation for a combustible gas generally increases with the partial pressure of hydrogen in the gas

2. Increased risk of HTHA (High-Temperature Hydrogen Attack) as mapped by the Nelson curve

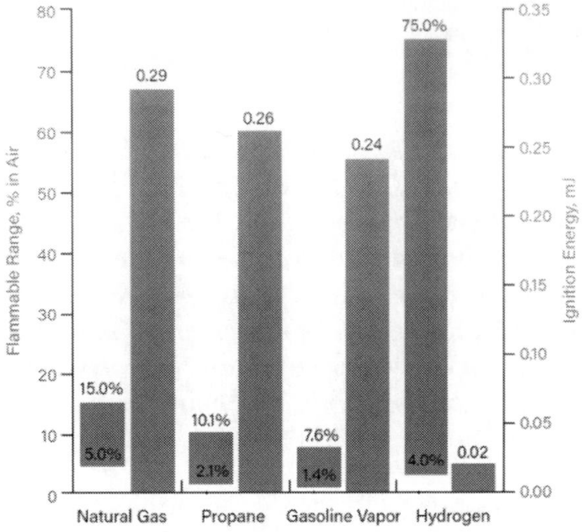

Figure 1 Flammability range and ignition energy of hydrogen compared to other fuels.

A third safety aspect that should also be borne in mind is the risk of attaining unexpected, explosive conditions during abnormal operation due to the targeted increase of hydrogen partial pressure during normal operation. Incomplete purging in connection with equipment inspection or venting of hydrogen-containing gas at an unsafe location can create conditions where explosive mixtures of oxygen and hydrogen could be formed. The risk of attaining explosive conditions is naturally higher in the process sections with high hydrogen concentrations.

There are, of course, other safety aspects to be considered as well however, not all of them are negative. In cases where the increased hydrogen partial pressure has been attained at the expense of a reduced CO partial pressure, there may in fact, be certain safety aspects that are improved – e.g., lower risk of personnel poisoning by CO or carbonyls.

Hazards related to Green Ammonia Processes.

Production of ammonia by reacting hydrogen with nitrogen is not a new concept and hence does not require any specific analysis to identify hazards related to green ammonia production. On the contrary, there are certain operational and safety benefits to introducing pure hydrogen rather than producing it through the normal reforming process. There are, however, certain new safety aspects associated with the hydrogen production by electrolysis and novel aspects also result from the aim to fully utilize a variable power profile as such dynamic green ammonia processes typically are controlled in a different manner than traditional processes[2]. The safety aspects related to such novelties should of course also be considered.

Within the context of analyzing novel risks related to green ammonia production, the most pertinent safety aspects to be discussed in the following sections can hence be divided into:

- Safety aspects related to revised design result from requirements of enhanced process dynamics or other new design requirements.
- Safety aspects related to green ammonia revamps – i.e. sending hydrogen as supplementary feedstock to an already existing ammonia plant and
- Safety aspects related to the production of hydrogen by electrolysis.

Safety Aspects Related to Process Dynamics
Efforts to optimize plant performance towards purely green operation typically lead to changes in process control, as discussed by Kakoti et al. (2023)[3] and/or reactor design, as discussed instance by Rizzi et al. (2016)[4] and Rizzi et al. (2019)[5]. Implications of such novel concepts on the plant safety should, of course, be carefully considered and addressed. These considerations will typically be made on a case-to-case basis and will, therefore, not be discussed further here.

Safety Aspects Related to Green Ammonia Revamps

Most commercial-scale ammonia processes where a green revamp would be an option have an overall process configuration as discussed by Dybkjær (2013)[6] and shown in Figure 2.

Various flavors of this generic process configuration exist (Dybkjær, 1995)[7].

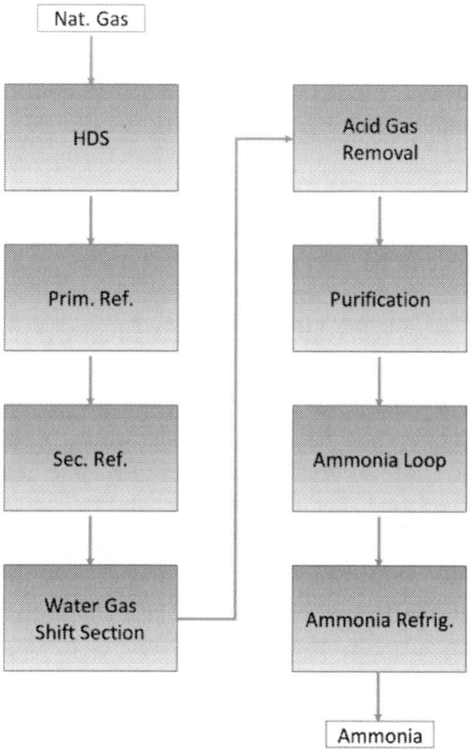

Figure 2 Overall process configuration for commercial-scale ammonia plants

KBR developed a concept where the syngas purification takes place before introduction to the synloop in a so-called Purifier™. This adjusts the N_2/H_2-ratio in the syngas and thereby enables the reduction of the steam reformer size (Pattabathula et al., 2017)[8]. Other authors discussed in some detail the implications of adding hydrogen to a conventional process based on two-step reforming. Examples are Kakoti (2022)[9], who discusses options for boosting production capacities of processes built under Topsoe license, and Mielke (2023)[10], who discusses partial replacement of feedstock with green hydrogen in plants designed by Uhde. The implications discussed by these authors seem relevant for most ammonia processes, and are hence discussed below.

Implications of adding hydrogen to a traditional Two-step Ammonia Process:

Adding hydrogen to an existing ammonia plant naturally means that the ratio between reactants (H_2/N_2) is increased above the optimal value, and some other design or operational change is required to restore that balance. The obvious way to achieve that in a traditional two-step ammonia process is to shift the load from the primary reformer to the secondary reformer. Increasing the load in the secondary reformer means that a more significant fraction of the hydrogen and CO in the syngas is burned, and at the same time, the amount of nitrogen added to the syngas is increased. For small disturbances (up to say 5-7% excess hydrogen) there hence exists a certain shift in load, which will restore the balance between H_2 and N_2 to the desired value.

There are certain intrinsic safety considerations that must be made in connection with such a change in plant operation. In the first place, the temperature rise across the secondary reformer increases. This is a safety concern, but it will typically also require that the inlet temperature to the secondary reformer is lowered so that the exit temperature from the secondary reformer does not exceed the mechanical design temperature of its outlet system. Excessive reduction in the primary reformer load, however, reduces the overall steam production from the unit. Hence, an appropriate balance has to be found, which both secures sufficient steam generation and avoids excessive temperatures exiting the secondary reformer. Addressing these issues simultaneously typically defines the limit for how much (green) hydrogen can be added to the plant and, hence, how much the plant capacity can be boosted if there is capacity available in downstream units.

An important implication of the above-mentioned modifications is that the Process Air Compressor (PAC) will operate at an increased load. For plants where this is impossible, an alternative to increasing capacity is to add oxygen co-produced in the electrolyzer unit, which produces green hydrogen for the unit and sends it to the PAC. Thereby, the load can be shifted from the primary reformer to the secondary reformer without increasing the load on the PAC, but such a change requires verification that the increased oxygen partial pressure in the PAC is acceptable from a design and safety perspective. Adding oxygen to the PAC could be attractive for plants where the primary reformer is the bottleneck or for feedstock-constrained plants. The overall result of such a modification is not an increase in production but merely a partial shift from grey to green production. At current green ammonia prices of 700-1400 $/ton[11], such modification could be quite attractive for many plants. Importing nitrogen to the process is another way to reestablish the stoichiometric ratio and will, according to Kakoti (2022) in some cases, enable > 10% capacity increase by green hydrogen import without other major modifications.

Implications of adding hydrogen to a Purifier™ based Ammonia Process:

On the surface, adding hydrogen to an ammonia plant with a purifier seems to entail less plant modifications than what was presented above. Maintaining process conditions (P,T) in the purifier will, in principle, ensure that the desired H_2/N_2 ratio is maintained even if the inlet stream to the purifier has an increased hydrogen content. This mode of operation requires that hydrogen is added upstream of the purifier and hence entails the safety implications related to an increase in hydrogen partial pressure as described previously. An alternative modification is to operate the purifier at a slightly elevated temperature, evaporating more nitrogen into the process stream and then adding the hydrogen downstream of the purifier. Both cases have to account for potential bottlenecks on liquid nitrogen supply and evaporation capacity in the purifier. A detailed analysis must be conducted on a case-to-case basis to determine the best configuration. Overall, the safety precautions related to adding green hydrogen to a purifier-based ammonia process seem relatively minor.

Hazards related to Hydrogen Production by Electrolysis.

The previous section of this paper addressed the implications of feeding pure hydrogen to the ammonia process. This section focuses on the safety aspects of the electrolysis unit itself.

There are several types of electrolysis units commercially available, but the dominating technologies can be grouped as

- PEM (Proton Exchange Membrane)
- AWE (Alkaline Water Electrolysis)
- SOEC (Solid Oxide Electrolysis Cells)

Of these, only PEM and AWE are available on a large scale to enable commercial-scale ammonia production.

While certain of the safety aspects are different for the different technologies, there are also many safety aspects that are similar, as discussed in some detail by ISPT (2023)[12]. To facilitate an in-depth analysis of these, a thorough analysis will be made on PEM technology as offered by Plug Power – simply because this is the electrolysis technology for which the authors have the best insight. Most, if not all, of the safety aspects discussed below are equally relevant for all types of electrolysis but there are, on the other hand, certain risks associated with the SOEC and AWE technologies which must be taken into consideration and are not covered here as such risks don't exist in the case of PEM. For SOEC, such considerations relate particularly to the safety aspects of producing pure hydrogen in partially ceramic structures being heated up to 750-800 °C

(1400-1500°F) while in the case of AWE, the high concentrations and amounts of KOH in AWE systems demand additional safety measures.

The predominant risk associated with electrolysis is that a partition separating the anode side (with a high concentration of oxygen) from the cathode side (with a high concentration of hydrogen) ruptures and creates an explosive mixture of hydrogen and oxygen. This risk is intrinsic in all types of electrolysis technologies. There is a huge pressure on the development of electrolysis technology to reduce power consumption and make the technology more compact. The way to increase efficiency is to reduce the electrical and molecular resistance of the membrane or diaphragm between the Anode and Cathode sides by applying thinner membranes or diaphragms, but this naturally also leads to an increased risk of uncontrolled migration of hydrogen to the oxygen side or vice versa. Similarly, the push towards more compact layouts leads to geometries where oxygen and hydrogen in high concentrations are present in locations very close to each other.

This scenario and how to rectify its associated dangers will be the main focus of the subsequent part of this section. The analysis is relevant for all electrolysis manufacturers, irrespective of the type of technology.

There are essentially three types of pressure configurations available on the market:

- Atmospheric electrolysis
- Pressurized alkaline electrolysis
- PEM configurations with a pressure gradient across the membrane.

Plug's configuration is of the latter type, where the anode side operates close to atmospheric conditions while the cathode side operates at 40 bar g (580 psig).

Safe design of PEM units.

We start by a short introduction to Plug Power's PEM technology to facilitate the understanding of the following analysis.

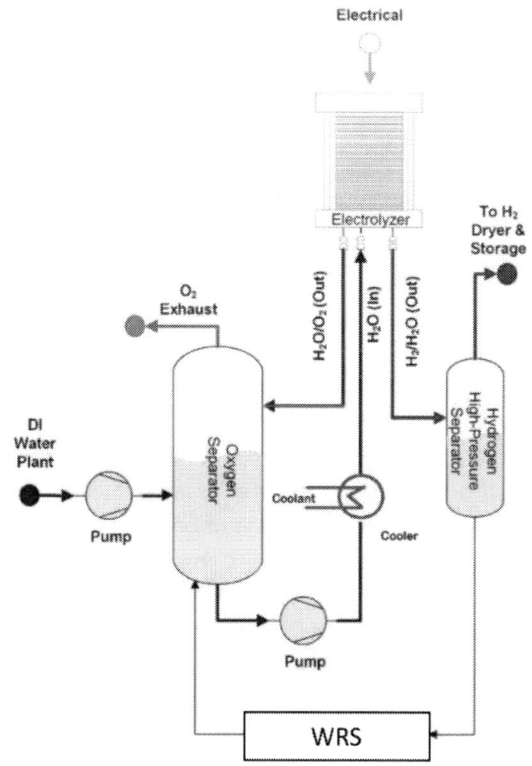

Figure 3 Plug Power's PEM Process Configuration

The process used by Plug Power to produce hydrogen by PEM is shown conceptually in Figure 3. The stack (marked "Electrolyzer" on the sketch) is the core of the process where water is split into hydrogen and oxygen. Protons formed within the stack permeate to the downstream "Cathode side" of the stack, where they recombine with electrons to form a mixture of hydrogen and water, which is then separated into its constituents in the downstream "Hydrogen High-Pressure Separator". Hydroxyl ions, conversely, are constrained to the upstream "Anode side" where they donate electrons to the stack and recombine to oxygen and water separated in the

"Oxygen Separator". The Anode side operates at close to atmospheric conditions, whereas the Cathode side operates at a pressure exceeding 40 bar g (580 psig). Water is recovered from the cathode side and recycled to the anode side via the water recycle system (WRS), which reduces the amount of hydrogen transferred to the anode side. The process configuration indicated in Figure 3 has many benefits from a safety point of view, as described below.

In the first place, the fact that the Cathode side is operating at a much higher pressure than the Anode side eliminates the risk that oxygen in relevant quantities should flow to the cathode side. This eliminates ~50% of the risk scenarios during normal operation (during which only the anode system can attain a flammable composition) whereas both the Anode and Cathode systems could achieve explosive conditions if both were operating at the same pressure. The high pressure on the Cathode side is generated electrochemically - i.e. without any use of compression, so any safety aspects related to compression of hydrogen up to the supply pressure are also avoided.

Moreover, the worst-case explosive scenario is now governed by the max operating pressure of the low-pressure side – i.e. the slight overpressure at the Anode side - instead of the hydrogen supply pressure, which determines the max pressure of an explosive mixture in a system with a balanced pressurized configuration.

Hydrogen, in small concentrations, will always be present in the oxygen section. There are two main mechanisms for hydrogen to end up in the oxygen section:

1. Hydrogen carry-under and
2. Hydrogen crossover.

Carry-under is the transport of hydrogen dissolved in return water from the cathode separator to the anode separator. Crossover is hydrogen diffusion from the high-pressure cathode side through the proton exchange membrane to the low-pressure oxygen side. These mechanisms are well understood and accepted.

Plug ELX systems are equipped with safeguards that shut down the system to a safe state when elevated hydrogen concentrations are measured in oxygen or when elevated oxygen concentrations are measured in the hydrogen. These safeguards are adequate if the concentration changes are slow and take 5 minutes or more. However, if concentration changes are faster than 5 minutes, a potentially dangerous gas composition can be created before the safeguards can detect the condition. We describe in the following how this potential safety issue has been addressed in our design.

The most common failure mode for a PEM electrolyzer is the development of a pinhole leak between the Cathode and Anode compartments. These leaks do not pose a risk since they can be detected in time, and the system can be shut down safely. In very rare cases, a membrane can rupture, forming a much larger hole. Though very rare, it must be considered.

Since the hole size of a pinhole or even a larger rupture is impossible to quantify, we assume for the design of the safety system the worst-case situation that a rare crossover event can cause a hydrogen-oxygen mixture to occur in the oxygen-water separator that will explode.

For hydrogen and oxygen gas mixtures, the explosive limit range is very wide and gets wider as the temperature increases. The lower flammability limit (LEL) of hydrogen in oxygen is 4 vol% at 20 °C (45°F) and 1 bara (14.5 psia), while the upper flammability limit (UEL) is 95.2 vol%. Schröder et al. (2004)[13] measured LEL and UEL values for hydrogen-oxygen mixtures as a function of temperature and pressure. The results are shown in Figure 4

Between the LEL and the UEL the gas mixture is flammable, and an ignition may result in a deflagration or detonation. Deflagration propagates with the velocity below the speed of sound (sub-sonic) in the unburned mixture. A detonation propagates with the velocity above the speed of sound (super-sonic).

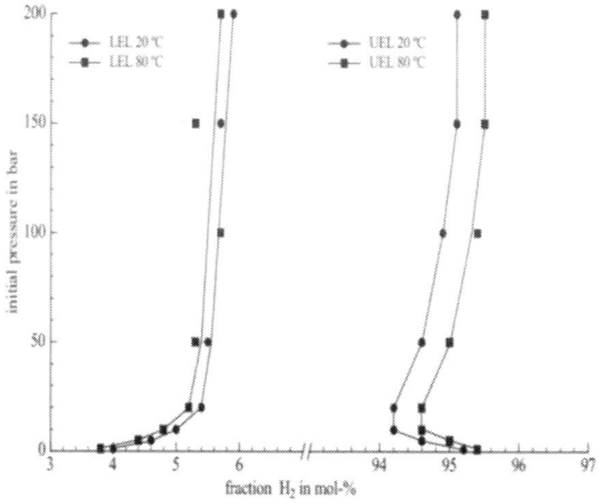

Figure 4 Pressure dependence of the explosion limits of H_2/O_2-gas mixture

Deflagration events in enclosures or confined spaces lead to significant overpressures. The pressure peak is nearly 10 times higher than the initial pressure in oxygen-hydrogen mixtures. During deflagration, the pressure grows almost uniformly within the enclosed space. Deflagration in an enclosure can be mitigated by venting. This is the most cost-effective and widespread explosion mitigation technique.

Detonation is the worst-case scenario for a hydrogen incident. Detonation propagates 2-3 orders of magnitudes faster than deflagration, resulting in pressures at the detonation front more than 15-20 times higher than the initial pressure. The venting technique used to mitigate deflagration events does not apply to detonation events, as there is insufficient time to release the pressure. Where deflagration events produce a uniform increase in pressure inside a closed vessel, detonations tend to show an oscillatory pressure response where very high but short pressure peaks are produced.

For hydrogen-air and hydrogen-oxygen mixtures, the highest deflagration pressure peak is obtained at a stoichiometric composition - see Figure 5. The pressure behavior of detonations looks similar but with much higher and shorter pressure peaks. Consequently, the worst-case scenario for a hydrogen-oxygen explosion is at a stoichiometric composition of 66.6 vol% hydrogen and 33.3 vol% oxygen.

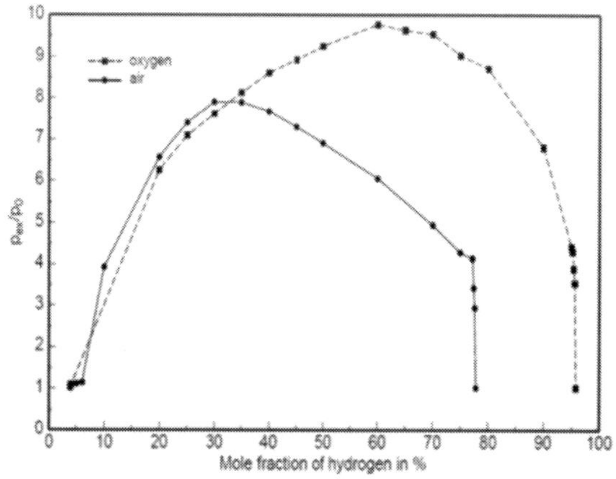

Figure 5 Deflagration pressure of hydrogen-air and hydrogen-oxygen mixtures in a closed vessel at room temperature

The safe operating limits were determined experimentally to be sure that our oxygen-water separators can handle a stoichiometric detonation. Currently, Plug Power supplies 2 different size electrolyzer systems: a 5 and a 10 MW system. These systems use oxygen-water separators of various sizes, and consequently, these need to be explosion tested separately. All systems use the Allagash stack depicted on Figure 6 for the electrolysis process. The pressure vessels and the Allagash stack have been subjected to explosion testing based on European norm NEN-EN 14460:2018 for explosion pressure shock-resistant equipment where permanent deformation is allowed.

Figure 6 The Allegash stack utilized in Plug Power products and tested for safe operation limits.

NEN-EN 14460 specifies requirements for explosion-resistant equipment that will withstand an internal explosion without rupturing and not give rise to hazardous effects on the surroundings. It applies to equipment (vessels and systems) where explosions are an exceptional load case. There are two types of explosion-resistant equipment: explosion pressure resistant and explosion pressure shock resistant. Explosion pressure-resistant equipment is designed to withstand the explosion pressure without permanent deformation and will not harm the surroundings. For explosion pressure shock-resistant equipment, permanent deformation is allowed, provided the equipment will not give rise to hazardous impacts on the surroundings.

IBExU conducted the explosion tests in Freiberg, Germany. The IBExU Institut für Sicherheitstechnik GmbH, IBExU for short, is a technical engineering service company in the field of explosion protection with a tradition going back to 1928. IBExU is an accredited test laboratory and accredited certification center for the testing/certification of equipment, protective systems and components intended for use in potentially explosive areas in addition to safety, control and control equipment for use inside of potentially explosive regions in accordance with Directive 2014/34/EU.

To assess the vessels in terms of explosion pressure shock resistance, static ignition tests with stoichiometric H_2/O_2 mixtures were performed at various initial pressures. The gas volume inside each vessel was flushed at least three times with the stoichiometric gas mixture and then pressurized to the desired test pressure. This gas volume in the vessels was then ignited.

The tests started at a low ignition pressure, and the pressure gradually increased after every successful test. A test was successful when only ductile failure was observed, and no significant pressure peaks appeared 1 m (3ft) from the vessel. The fact that the vessel could be pressurized for the next test is proof that the previous test did not compromise the vessel integrity. The highest initial vessel pressure where no failure was observed is called the safe explosion pressure (SEP).

In total, 21 experiments were performed. where the conditions were gradually worsened until point of rupture. Each test was initiated at ambient temperature – approximately 20 deg.C. The measured maximum pressure peaks were at least 7.3 times the vessel's design pressure and peaks up to 90 times the design pressure were observed. Because the duration of these high-pressure peaks is extremely short (in the order of several microseconds), the vessels can withstand these detonation events without brittle failure. When the test pressure was less than 52% of the design pressure of the vessel (absolute pressures), the vessel passed the test. Only when the test pressures exceeded 52% of the vessel's design pressure, did the detonation cause not only permanent deformation but also the ejection of the full content of the vessel.

Explosion experiments with the electrolyzer stack showed it has a better explosion shock resistance than the separator vessels. Based on these results, it was decided to operate at 65% of the safe explosion pressure of the vessels, with the pressure safety valves set at 75% of the safe explosion pressure. The Plug ELX systems are equipped with SIL safeguards to ensure operation within a safe window.

Conclusion.

A review of state-of-the-art ammonia production and a high-level analysis of the safety aspects of integrating hydrogen production by electrolysis into such process configurations were presented. It was advocated that the novel safety risks associated with such changes are surmountable and that the safety gains related to the possible elimination of the traditional reforming systems are one of the benefits of green ammonia technology. As a result, the main focus in green ammonia production considerations should be on the safety of the electrolysis unit itself.

Subsequently a thorough review of what Plug Power has done to address this issue was presented and it was shown that even the worst-case accidents have been thoroughly considered by Plug Power – both in the way that the technology has been designed and how the underlying assumptions have been tested experimentally.

Overall, the analysis shows that green ammonia production can be achieved in a manner which is no more hazardous than traditional ammonia production. It is believed that the standards to which Plug Power has designed their system could leverage further development of good industrial practice within the space of green ammonia technology. They have certainly proved their worth in the dozens of plants Plug have delivered electrolysis equipment to until now.

[1] S. Eck, M. Snyder (2021); "Hydrogen Safety Fundamentals"; aiche-org/CEP December 2011

[2] G. Genova, S. Panza (2022); "Casale flexible green ammonia plant, the economically viable green production"; Nitrogen and Syngas 2022 in Berlin

[3] A. Kakoti, D. Shrivastava (2023), "Green Ammonia - Safe Approach to Design and Operate a Dynamic Ammonia Plant"; 67th AIChE Ammonia Safety Symposium, 2022 in Munich

[4] M. Rizzi, X. Junhong (2016); "2000 MTD Inert-free Synthesis Loops: A China First in Operation at Henan Jinkai Ammonia Plant." Casale 4th Customer Symposium, Venice 2016

[5] M. Rizzi, L. Redaelli (2019); "Safety and Reliability in Ammonia Synthesis Converters"; 63rd AIChE Ammonia Safety Symposium, 2019 in San Francisco

[6] I Dybkjær (2013) "100 Years of Ammonia Synthesis Technology"; Ammonia Technical Manual 2013 p.101

[7] I. Dybkjær (1995); "Ammonia Production Processes"; Ammonia – Catalysis and Manufacture; Ed: A. Nielsen, Springer Verlag, 1995

[8] V. Pattabathula, C. Morgan, M Hofman, G Colman, A Malhotra (2017); "Commissioning of first world scale KBR Purifier ™ ammonia Plant in North America"; AIChE Technical Manual 2017, p.1

[9] A. Kakoti; "Hybrid Revamp of Existing Ammonia Plants"; Nitrogen + Syngas 2022

[10] B. Mielke; "Reducing the Carbon Footprint with Blue and Hybrid Ammonia Plants"; Ammonia Technical Manual 2023, p.119

[11] https://www.futurebridge.com/industry/perspectives-energy/green-ammonia-an-alternative-fuel/#:~:text=Cost%20of%20green%20ammonia&text=The%20current%20Price%20of%20green,%2C%20to%20%24%20310%20per%20tonne.

[12] Institute for Sustainable Process Technology (2023); "Safety Aspects of Green Hydrogen Production on Industrial Scale". Hydrohub Innovation Program – Public Report

[13] Schröder, V., Emonts, B., Janßen, H., Schulze, H.-P, (2004), "Explosion Limits of Hydrogen/Oxygen Mixtures at initial Pressures up to 200 bar", Chemical Engineering & Technology, 27(8), 847–851

Learnings During Engineering for Autothermal Reformer-Based Ammonia Plant

Ammonia production using autothermal reforming (ATR) of natural gas is a process which allows large reduction of CO_2 emission by concentrating most of the CO_2 production in one single stream which can be sent to sequestration.

In the ammonia industry, there are some reservations against autothermal reforming with its use of oxygen in the process which is one of the main differences to the conventional steam reformer process. This oxygen stream is bringing a potential additional hazard into the ammonia plant. However, in the methanol industry, use of oxygen is a well-accepted practice.

While engineering an ATR-based ammonia plant, Uhde engineers were confronted with these hazards and the proposed means for their mitigation. At some examples the safe handling of oxygen and the know-how-transfer from methanol plant design is illustrated. The hazards can safely be handled by proper mechanical design of the plant (especially piping) and by the safety systems around the ATR.

Another topic is determining the boundaries on how carbon intensity is calculated, which can be either the plant alone (Scope 1) or include utility supply (Scope 2), and the impact of this on plant design.

Dr. Bernd Mielke
thyssenkrupp Uhde

Introduction

Mitigation of man-made contributions to climate change is one of the biggest challenges of our times. Current production of energy, goods, and transportation is strongly linked to the release of carbon dioxide into the atmosphere, where it acts as a greenhouse gas. In order to avoid such emissions, the production processes shall be changed to utilize carbon-free feedstocks and fuels, or unavoidable carbonaceous emissions must be captured before they reach the atmosphere. The transportation sector shall also switch to carbon-free fuels. Truly the best option would be to substitute usage of fossil fuels by hydrogen derived from renewable energy. However, green hydrogen production cannot ramp up as fast as demand will grow. Therefore, a transition period will exist, where low carbon energy and hydrogen carrier, like blue ammonia, will play an important role. [1]

Blue Ammonia Production

In principle, there are currently two alternative processes for industrial scale production of hydrogen from natural gas: The steam methane reforming process (SMR) and the autothermal reforming process (ATR).

In both concepts there are two points where CO_2 emission can occur:

- The reforming section, consisting of the flue gas from the reformer or the fired heater.
- The CO_2 removal unit, where the CO_2 is separated from the process gas.

The amount of CO_2 emitted to the atmosphere can be reduced by capturing the CO_2 and sending it to a sink. If the CO_2 emission to the atmosphere is largely reduced, the ammonia product is called "blue ammonia", often also referred to as "low-carbon ammonia". There is no uniform definition of this term yet, but many projects call for a carbon capture rate of 90 to 95 %. [2]

When comparing the process versions of SMR and ATR one can see that the overall emission for both is similar (around 1.7 t CO_2 per t NH_3 according to Scope 1 (direct CO_2 emissions at site; refer also to definition in the second chapter of this paper) for a typical modern natural gas-based plant), but the split of the CO_2 emissions for the ATR variant is more to the process side rather than to the flue gas side. By using a part of the hydrogen from the product gas as heater fuel (see Figure 1), CO_2 emissions can be further reduced. With return of recycle streams to the process almost all carbon introduced to the plant can be extracted using the CO_2 removal unit. This makes additional installations like flue gas washing units unnecessary. In addition, the ATR process offers advantages in terms of CAPEX and construction effort.

An ATR ammonia process with improved CO_2 capture rate is illustrated in Figure 1.

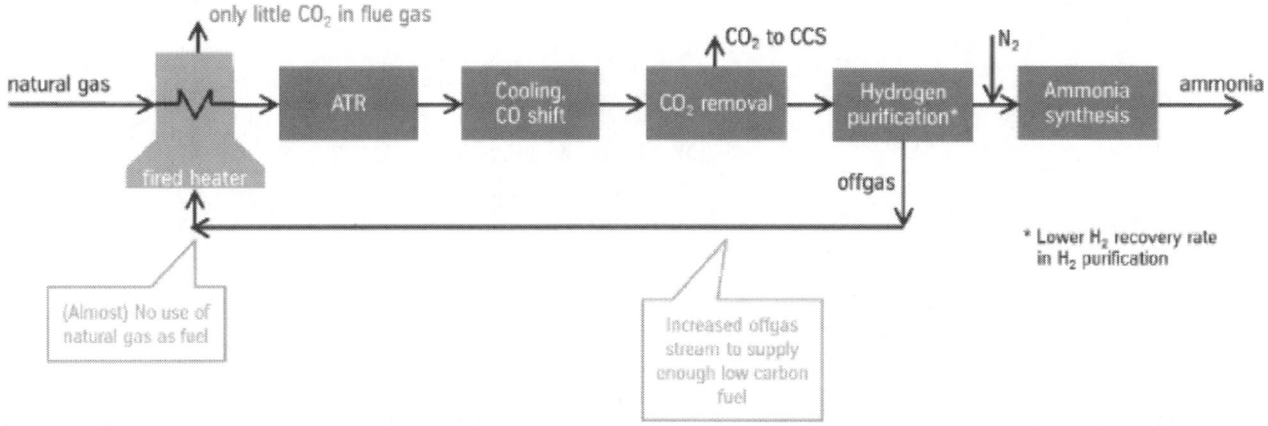

Figure 1. Improved autothermal reformer based ammonia production with reduction of CO_2 emission by 90%.

Engineering for Autothermal Reformer-based Ammonia Plant

In 2022 Nutrien, the world's third-largest nitrogen producer, developed plans to build the world's largest clean ammonia facility in Geismar, Louisiana [3]. thyssenkrupp Uhde was selected as technology provider and partner for this project. The new plant was intended to serve growing demand in agriculture, industrial and emerging energy markets.

The plant for which thyssenkrupp Uhde has conducted the engineering activities, is a 3,500 mtpd single train ammonia plant. Carbon capture rate is in excess of 90%. The captured CO_2 will be sent to a pipeline network and finally stored underground through CCS in combination with enhanced oil recovery. The process involves injecting CO_2 into partially depleted oilfields to force out additional volumes of oil, with CO_2 being trapped and permanently stored.

The plant uses the ATR process concept as described above and depicted in Figure 1.

Oxygen Use in ATR Process

In the autothermal reformer process the energy for the endothermic reforming reaction comes solely from the combustion of pre-reformed feedstock withing the process. Firing the ATR reactor with pure oxygen, rather than with air as in a conventional secondary reformer, significantly reduces the volume of process gas by eliminating nitrogen, which is inert in the reforming section. This is beneficial in terms of equilibrium and equipment size. Table 1 shows a comparison of the heat duties in the ATR and in a conventional air blown secondary reformer.

Reactor	Unit	Combustion Heat Duty
ATR	MW	340
	MMBTU/h	1161
Secondary Reformer	MW	103
	MMBTU/h	352

Table 1. Heat duty liberated by internal combustion for a 3,500 mtpd ammonia plant.

The ATR is a substantial consumer of oxygen. A typical figure is 0.77 t O_2/t NH_3, which is equal to approx. 113 t O_2/h (125 ST O_2/h) for a 3,500 mtpd ammonia plant. Such quantities with the required purity can only be provided by a cryogenic air separation unit (ASU).

Cryogenic ASUs employ the following steps:

- Filtration and compression of ambient air
- Cooling of air and adsorptive removal of water, carbon dioxide, and the higher boiling vapor-phase hydrocarbons
- Deep cooling of the air in the heat exchanger core close to liquefaction
- Distillation of the different constituents
- Removing heat by the evaporation of liquid nitrogen and subsequent depressurization and heat transfer through the heat exchanger core

The liquid oxygen produced from the ASU is stored in cryogenic insulated storage tanks.

Oxygen Properties and Hazards

Oxygen is a colorless, odorless gas with a normal boiling point of -183 °C (-297 °F). Elemental oxygen is slightly heavier than air, and is used in the steel industry, for cutting and welding, water treatment, rocket propulsion and in medical services.

Oxygen itself is not flammable, but most substances can burn (being oxidized) in an oxygen enriched atmosphere, even metals like titanium. At oxygen concentrations as low as 24 vol % the oxidation reaction intensity doubles that under normal air.

To start a fire three elements must be present simultaneously:

- Oxygen atmosphere
- Oxidizable material (fuel)
- Ignition energy

Ignition energy can be provided by a flame, sparks, static electricity, particulate impingement, friction or heat of adiabatic compression. Oxygen fires can take place inside vessels or piping without notice from the outside. [4]

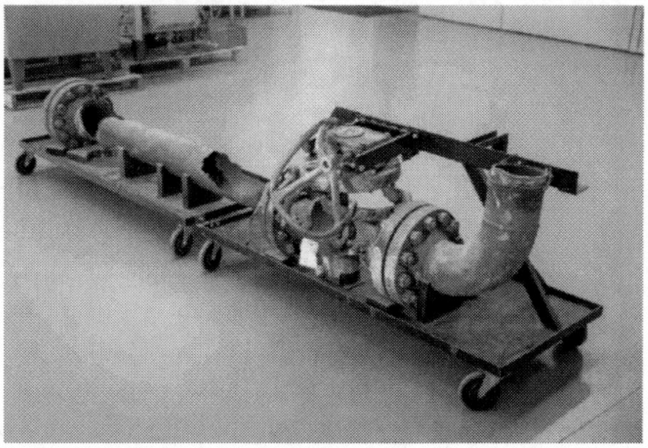

Figure 2. Burnout 6 inch ball valve in industrial oxygen pipeline. [5]

Gaseous oxygen, especially when it's cold, can accumulate in pits, channels or enclosed structures, where the oxygen enriched atmosphere will favor starting a fire.

Liquid or boiling oxygen can cause severe frostbite and burns due to its very low temperature. Also, materials will become brittle and will lose their ductility. Oxygen spills can soak into the ground or into asphalt.

Severe incidents in ASUs have been reported with explosions and multiple fatalities, such as one that occurred July 19, 2019 in Yima City, Henan province in China. According to root cause analysis conducted on several incidents, the main cause is often accumulation of hydrocarbons in liquid oxygen in the main condenser. Traces of hydrocarbons from neighboring petrochemical processing operations can be drawn in with the air, concentrate over time, and form an explosive mixture in the ASU. Vendors have taken steps to mitigate the risk of explosion. These include: pre-purification of the air prior to liquefaction, maintaining the condenser unit in a fully submerged condition, use of adsorbers in the liquid oxygen phase to remove hydrocarbons, and purging liquid oxygen from the main condenser to prevent hydrocarbon accumulation. Beyond mitigation, of equal importance, is the need to monitor the effectiveness of these measures by suitable devices. [6]

Oxygen Related Engineering Observations

The use of pure oxygen in an autothermal reformer inside an ammonia plant is a relatively new field of application, but in principle the same has been done for decades in methanol plants. Oxygen has also been used extensively for gasification of heavier hydrocarbons, including solid materials like coal and petroleum coke. thyssenkrupp Uhde used high pressure oxygen in the gasifier of the Puertollano plant in Spain. This can serve as an example that with the right process design and technical solutions in place, together with the right mindset, the safe handling of oxygen is not a secret.

During the engineering phase for the 3,500 mtpd ATR-based ammonia plant thyssenkrupp Uhde encountered the following learnings:

Material Selection

For oxygen containing piping, stainless steel material ASTM A312, Gr. TP316L was considered. Although copper and copper alloys have been reported as suitable materials for oxygen service, they are not permitted in ammonia plant piping and process equipment.

Piping Design and Plant Layout

- Gas velocities in high pressure oxygen pipeline shall be kept low in the range of 4-5 m/s (13-16 ft/s).
- Elbows shall have a large radius. Pipe routing shall be straight and smooth and without pockets.
- Valves shall be explicitly suitable for oxygen service.
- Welded connections are preferred over flanged connections.
- Oxygen pipelines shall be located at the outer end of a pipe rack, side by side with cooling water piping or other inert fluids as a barrier.
- Cable trays above or below the oxygen line shall be avoided as to maintain its visibility.
- The oxygen valve at battery limit shall be located behind a fire wall, so that it can be closed even in case of fire.

Process Design

Oxygen preheating shall not exceed 200 °C (392 °F).

In case of emergency or process trip the oxygen supply to the burner of the ATR is stopped, and the oxygen is routed to a dedicated vent stack (see Figure 3). It may be necessary to control the closure time (set a minimum and maximum closing time) of cut-off valves to avoid adiabatic compression. Supportive steam is added to the oxygen line to ensure flow in the direction of the ATR, and prevent reverse flow of process gases into the oxygen piping.

Figure 3. Oxygen vent structure in 3D model.

ATR Burner Design

The proper burner design is essential for safe operation, high performance, and long equipment life. The geometry of the burner can be optimized using CFD methods. Referenced burner designs have proven their reliability in industrial references, including in many methanol plants.

Commissioning

Before commissioning, piping and equipment must be cleaned according to ASTM G93 requirements. The focus is on removing organic contaminants like grease or oil as well as particles to the maximum possible extent. Organics and particle impact have been identified to be responsible for many oxygen fires.

Opening a valve rapidly against a closed downstream valve must be avoided, as it can result in high heat being generated through adiabatic compression of the oxygen.

Assessing the CO_2 Footprint

The comparison of different plant concepts is often based on their specific energy consumption (energy consumption per ton of ammonia produced), also referred to as energy intensity. It is clear that for a fair comparison the boundary line, defining what is part of the evaluation and what not, has to be identical for all variants investigated.

Furthermore, a way must be found for handling different forms of energy crossing the boundary line. This is usually done by converting all of the energy sources (e.g. natural gas, electricity, steam) into an equivalent primary energy source (e.g. natural gas), which is the amount of primary energy that would be needed to produce them.

For blue ammonia production, another important parameter is the specific CO_2 emission per mass of ammonia produced, also called product CO_2 footprint. Obviously, the energy consumption of a blue ammonia process is higher than for a conventional one, caused by the additional effort for the following two needs:

First, CO_2 otherwise emitted must be separated from the flue gas and / or the process. In case of a conventional ammonia plant, this is achieved by scrubbing the reformer flue gas. This process consumes power for pumping the scrubbing solvent and steam for its regeneration. Advanced concepts with ATR (autothermal reformer), which reduce or avoid the complication of a flue gas scrubbing unit, lead to additional energy consumption because the process air compressor is replaced by a more energy-intense air separation unit (ASU).

The second contribution is the compression of the CO_2 for export to the receiving facility (for sequestration or utilization). For example, there are projects, where compression to the maximum operating pressure of the pipeline of 17.2 to 19.3 MPa g (2,500 to 2,800 psi g) [7] is required. Without mentioning the export pressure as a site or project-specific parameter, a statement on the energy intensity or carbon footprint is worthless.

The Right Boundary Line

Often, projects and plant concepts are advertised by stating some percentage reduction of CO_2 emissions. The same comparison is made for the energy intensity of processes, also here the boundary line must be clear, otherwise such declaration is meaningless. For example, two plants, both recovering 95 % of their CO_2 emission can have a large difference in the actual CO_2 emission associated to one ton of product. This is illustrated by the following example:

Plant A: The plant is designed to provide all energy needed for ammonia production and for the ancillary processes of CO_2 capture, compression and air separation (if any). This is achieved by means of sufficient steam production and use of hydrogen-rich gas as heater fuel as illustrated in Figure 1. Consequently, there is no significant energy import except of natural gas.

Plant B: The plant is minimized in natural gas consumption and there is only a small amount of waste heat available which is fully used for steam production, sufficient to cater the needs of ammonia production. All other needs, such as for additional CO_2 separation (flue gas scrubbing and / or ASU) and CO_2 compression are satisfied by imported steam or electricity.

It is obvious that plant A has a higher natural gas consumption than B in order to produce the required energy. Thus, its CO_2 emission is also higher, even after reduction by 95 %. However, the smaller emission of plant B can be a good figure on paper, but for a fair comparison one has to add the CO_2 emissions of steam and electricity generation outside the ammonia plant. By doing so, this can more than offset the apparent advantage if it is generated in a coal fired power plant or in a gas-fired boiler with no measures taken to recover their CO_2 emissions.

This is also reflected in the commonly used definitions of Scope 1 and Scope 2 for greenhouse gas emissions [8]. Scope 1 emissions are the direct emissions from the production site, whereas Scope 2 emissions include indirect emissions associated with the purchase of energy as illustrated in Figure 4.

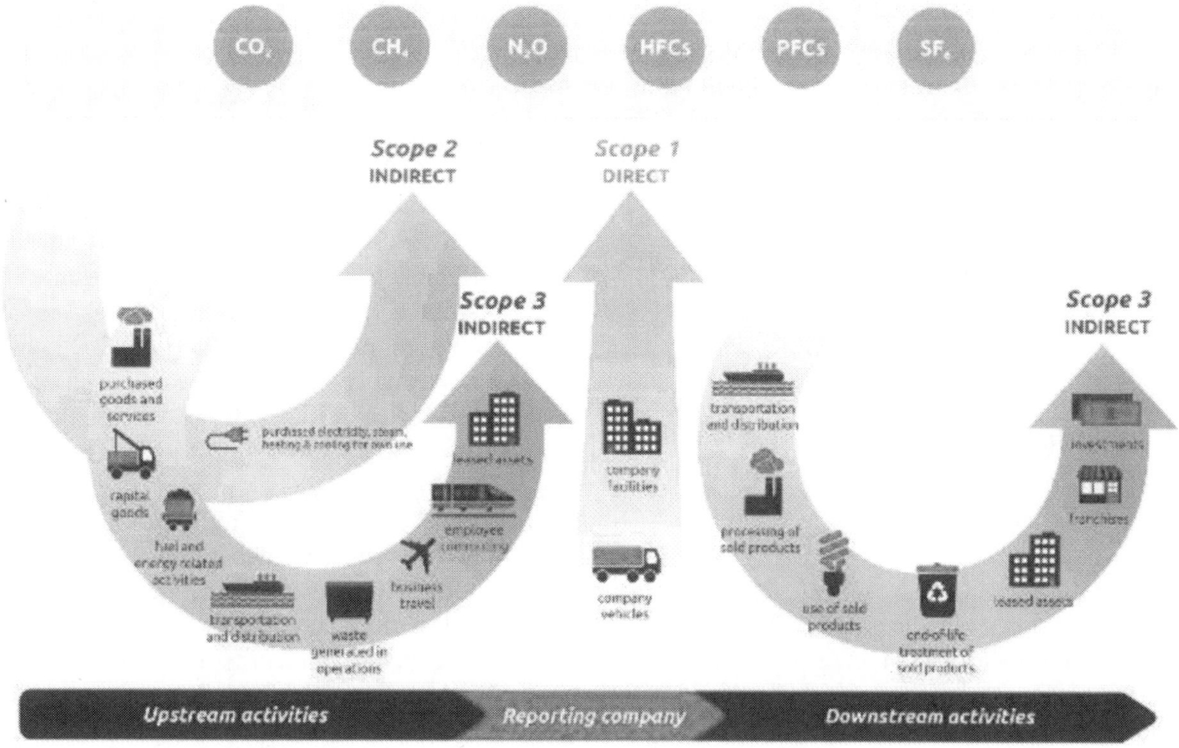

Figure 4. Scope definition for greenhouse gas emissions: Visualization by US EPA [8].

It is worth mentioning that even for identical plants at different locations the Scope 2 CO_2 emissions can differ because the emissions associated with imported electricity depend on the generation structure of the local power supply network.

Carbon Intensity

While the percentage of CO_2 emission reduction is a quite common and understandable definition for the degree of avoidance of greenhouse gas emissions, the better measure for comparing different processes or their products is the carbon intensity (ton CO_2 emission per ton of product).

Advantages of this reference figure are:

- It allows a comparison with other production concepts, like green ammonia.
- It allows inclusion of other emissions which occur over the full value chain of the product, like transport to consumer, cracking to hydrogen, etc.

To support production of products with lower CO_2 footprint, different definitions have been created under different legislations, for example:

US Infrastructure Bill 2021: "Clean hydrogen" is defined as an emission of less than 2 kg of CO_2e at the site of production for each kg of hydrogen produced (2 lb CO_2e / lb H_2) [9]. Applied to ammonia (production of 1 kg (2.2 lb)

NH_3 requires 0.178 kg (0.392 lb) H_2), this defines clean ammonia as associated with an emission of 0.36 kg CO_2 / kg NH_3 (0.36 lb CO_2 / lb NH_3).

European Renewable Energy Directive: To be counted as "sustainable" as per EU taxonomy, the life cycle emission has to be less than 3 kg of CO_2e for each kg of hydrogen produced (3 lb CO_2e / lb H_2) [10]. Applied to ammonia, this defines sustainable ammonia as associated with an emission of 0.53 kg CO_2 / kg NH_3 (0.53 lb CO_2 / lb NH_3)..

While the US Infrastructure Bill mentions the emissions "at the site of production", the EU directive is about "life cycle emission". This shows how important it is to be sure about the proper boundary line.

Standards for certification of a product's carbon footprint have been set by DIN EN ISO 14067.

Summary

Decarbonization of the industrial and the transportation sectors will create new markets for low-carbon ammonia. The projected demand will be so high that only large ammonia plants make sense, preferably with high CO_2 capture rates and only one place where the CO_2 is emitted from the process. Here the ATR process will be superior to the SMR process. thyssenkrupp Uhde has engineered a 3,500 mtpd ATR plant, where an ASU is part of the complex to supply the ATR with oxygen. For established ammonia producers experience in oxygen handling is limited, and it has its peculiarities, but each of them can be managed by transferring well established know how from similar applications to the new field of ATR based ammonia production.

The oxygen supply requires an air separation unit as an additional element of the ammonia plant. The ASU and the effort for CO_2 capture and export lead to higher specific energy consumption (energy intensity) and lower CO_2 emissions (carbon footprint). For its determination it is important to be certain about the proper boundary lines and not to forget CO_2 emissions outside the fence line of the facility originating from processes for its energy supply. The well-known definitions of Scope 1 and Scope 2 can serve as a framework for this.

References

[1] IRENA and AEA, Innovation Outlook: Renewable Ammonia, 2022

[2] B. Mielke, Reducing the Carbon Footprint with Blue and Hybrid Ammonia Plants; AIChE Safety in Ammonia Plants and Related Facilities Symposium, Munich, 2023

[3] Nutrien, press release, 26 Oct. 2022, www.nutrien.com/what-we-do/stories/nutrien-partners-thyssenkrupp-uhde-clean-ammonia-project, accessed on 27 March 2024

[4] EIGA, Doc 04/18, Fire hazards of oxygen and oxygen-enriched atmospheres, 2018

[5] WHA International Inc., https://wha-international.com/real-world-oxygen-fire-evidence-underscores-the-need-for-safety/, accessed on 27 March 2024

[6] ABB, Fidas24 – Preventing explosions in liquid oxygen, 2019, https://new.abb.com/products/measurement-products/measurement-analytics/industrial-gas/preventing-air-separation-unit-explosions-n, accessed on 27 March 2024

[7] Energy Equipment & Infrastructure Alliance, "Meeting the Dual Challenge, Chapter 6, CO_2 transport", 2020, https://www.eeia.org/post/CCUS-Pipeline-Transport-Meeting-the-Dual-Challenge.pdf, accessed on 27 March 2024

[8] United States Environmental Protection Agency, EPA Center for Corporate Climate Leadership, "Scope 1 and Scope 2 Inventory Guidance", https://www.epa.gov/climateleadership/scope-1-and-scope-2-inventory-guidance, accessed on 27 March 2024

[9] US Department of Energy, Alternative Fuels Data Center, https://afdc.energy.gov/laws/13033, accessed on 27 March 2024
[10] European Commission, https://energy.ec.europa.eu/topics/energy-systems-integration/hydrogen/renewable-hydrogen_en, accessed on 27 March 2024

Reliability and Safety of Oxygen blown ATRs to enable Low Carbon Ammonia

In the 1950s and 60s considerable industry learning and sharing took place regarding oxygen handling, predominately through the AIChE ammonia seminars. Today we see an increase of oxygen blown ATR (Auto-Thermal Reforming) deployment within flowsheets designed for low carbon hydrogen and ammonia.

Methanol production has led the way with ATR deployment, harnessing their efficiency and flexibility, providing safety and reliability learning that has value to the ammonia industry. This paper will review the risks, referencing industry learning regarding ATR and secondary reformer operations that have challenged the Integrity Operating Window of these reformers.

Matt J Cousins

Johnson Matthey Low Carbon Hydrogen

Introduction

Flowsheets utilising an oxygen blown ATR (Auto Thermal Reformer) offer a pathway to the lowest carbon intensity products when utilising a natural gas or fuel gas feedstock. This paper reviews the flow sheet options, their benefits and impact on safe reliable operations.

"Safety first", a phrase many of us hear in our day-to-day jobs. Indeed, oxygen plant safety was one of the first themes to be highlighted by the ammonia safety symposium. So, we can return to this data to help enable the safety of our future ATR based flowsheets.

In the 1950s the most available and effective route to ammonia was using POx (Partial Oxidation) of heavy feedstocks. The use of gasification necessitated the use of oxygen. An ASU (air separation unit) often referred to as the "oxygen plant" provided the oxygen. Even within the 1950's, there were already many cumulative years of learning to share, with Linde sharing insight from having over 1200 ASUs in the field in 1957 [(1)].

Over the intervening years natural gas became increasingly available. With this the ammonia process transitioned to use a steam methane reforming (SMR) in a primary reformer, to generate hydrogen, whose product gas passes to an air blown secondary reformer [(2)]. These processes have tended to use air, as this also imports nitrogen into the flowsheet.

The ammonia industry now faces its next transition. To marry the simultaneous demands of increasing production with the need to reduce its carbon intensity.

The long-term answer to this, is to deploy "green hydrogen" technologies that when integrated with an ASU, provide the H_2 and N_2 required to feed an ammonia loop.

The Hydrogen Council report green hydrogen availability is too low and its costs too high to deploy it at scale today. A track record for delivery of green projects is lacking, with 6 GW cited in 2021 to be operational by 2023, yet the delivered operational capacity was 1.1 GW. A headwind problem for green hydrogen is the increased LCOH (levelized cost of hydrogen)-"up 30-65% to 4.5 to 6.5 USD/kg" from 2021. The same report contrasts that blue hydrogen is a cost-effective solution in the short to medium term[3]. This promotes ATR based low carbon (blue) ammonia as the industrial answer to the quandary of increasing production while lowering carbon intensity today.

ATRs can effectively replace the SMR flowsheets used today. An ATR will transform a methane feedstock to H_2 with high efficiency and enable far lower CO_2 emissions.

ATR reforming technologies intensify the reforming process, without externally fired combustion thus keeping the CO_2 at pressure ready for capture.

The ATR process is oxygen blown to:

- Deliver a higher temperature and so a lower methane equilibrium position.
- Drive the kinetics of the reforming reaction, so reducing the ATRs size.
- Reduces the gas volume, with no N_2 flow present in the H_2 production.
- Reduces flue gas emissions, so by-products, from fired equipment.

The attractiveness of working with oxygen-based ATR flowsheets is therefore seeing a resurgence.

The industry's collective operating experience shows that with the correct management of hazards, the risks of handling oxygen can be managed. Part of the process to manage risks is applying learning from the past. So, we should be encouraged to return to the AIChE papers from the 1950s to refresh our understanding. While considering how this is applied to ATR flowsheets for the next generation of ammonia plants to meet our current needs.

Learning from the past.

Consideration of the points listed below are highlighted in the minutes of two AIChE round table events, chaired by Walton in 1958 & 1959 [4][5].

1. air intake location
2. impurities in the air intake
3. materials of construction
4. permissible levels of hydrocarbons
5. lubrication, maintenance of equipment
6. analysis and instrumentation

They are all requirements to engineering the interaction between

a) a flammable material (natural gas),

b) an oxygen source and

c) an ignition source

All are as relevant today as they were 60 years ago. Considering two specific events clearly shows the fundamental messages have not changed over the decades, and the experience and important learning captured by the papers is as valuable in respect of plant safety today.

This is shown by contrasting the 1962 ICI paper [6] regarding a serious explosion in 1959 of an oxygen plant in Billingham, UK with the National Fertilizers Ltd (NFL), India 2011 paper [7] describing a 2010 incident at the start-up of O_2 compressor on their gasifier plant.

The 1962 paper discloses an investigation by ICI and Air Products, after an incident that caused two fatalities. It highlights some "general principles", as well as the likely cause of having non-delineated combustible material in the process, resulting from process modifications.

1. Delineation of an area in which an abnormal O_2 concentration could occur.
2. Exclusion of combustibles from the O_2 area.
3. Minimising the possibility of O_2 leakage from the plant
4. Provide a safe means of disposal for high O_2 concentrations (gas and liquid).
5. Control of concentration of contaminates within air separation systems.

The later NFL report concludes the cause of their incident to be linked to contamination from an overhaul and leakage of O_2 causing the larger fire as the incident took hold.

While the NFL event has related causes and initiated equipment damage, thankfully no injuries were reported. This indicates the benefit of understanding the risk so the impact of them is minimised.

The credibility to state we will not hear reports of similar events may be challenged. Saying "all injuries and occupational illnesses are preventable" is credible. This is the leading principle in JM's group EHS policy and something we embrace in all aspects of our work at JM.

A trend one might conclude from the history with oxygen-based systems, indeed of continuously operating chemical processes, is that under normal operation we encounter far fewer problems. However, during turnarounds and modifications there is increased likelihood of an event. Therefore, ATR designs should consider,

A. Minimising the number of turnarounds, high reliability design.
B. Using proven designs that elevate the basis of safety, employing proven control philosophies and SIS (safety instrumented system) architectures.

ATR Process

The ATR process injects oxygen into the process gas using a burner, situated above a bed of catalyst. The now well mixed hot gas stream passes through the catalyst bed, which converts the hydrocarbon rich gas stream to hydrogen rich gas, using the heat from combustion to drive the endothermic reforming reaction.

An ATR design considers three reactions:

The reforming reaction:

$$CH_4 + H_2O \Leftrightarrow CO + 3H_2 \ (1)$$

The reforming reaction including combustion:

$$CH_4 + \frac{1}{2}O_2 + H_2O \Leftrightarrow CO_2 + 3H_2 \ (2)$$

The reforming reaction including WGS.

$$CH_4 + 2H_2O \Leftrightarrow CO_2 + 4H_2 \ (3)$$

An ATR is an adiabatic reactor and within it the energy balance consists of energy from combustion supplying energy consumed by the endothermic reforming reaction and energy loss from the vessel.

The net difference is the sensible heat energy. The mass of process gases carrying that total heat exit the ATR then pass through the reformed gas boiler where a portion of sensible heat can be used to raise steam. This energy balance is illustrated in figure 1.

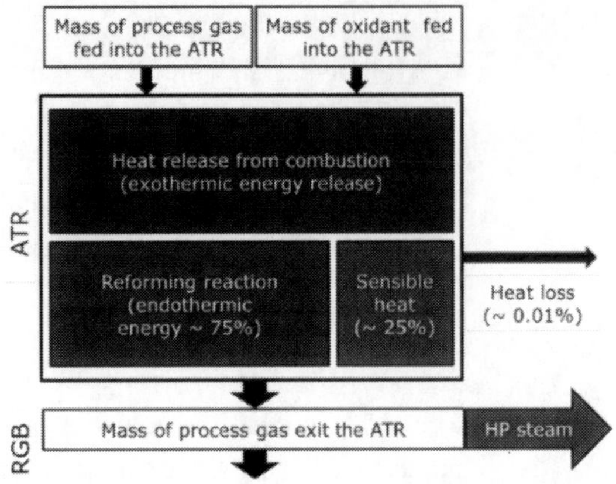

Figure 1. Energies involved within the ATR & RGB, where the ATR is water jacketed and operating at a local S:C of ca. 1.0.

While steam has an energy value, it is not the primary function of the flow sheet, this is to make H_2 that can be reacted to produce NH_3.

Therefore, involuntary steam raising from the ATR, necessary for its operation to ensure a low and stable methane slip, plus that raised from other exothermic reactions, (e.g., WGS and ammonia synthesis), has a positive effect on the energy balance of the overall flow sheet.

However, raising additional steam through the ATR, beyond the involuntary amount needed for its operation, is a result of burning more feedstock in O_2.

- Increasing potential stress in the ATR,
- Increasing the CO_2: product ratio
- Increasing the O_2 demand,
- Decreasing the feedstock : product ratio

It is affecting how much feedstock is consumed by reaction 2, instead of focusing on transforming produced CO, from reforming, with steam to produce H_2, by reaction 3.

Historically burning feedstock or 'waste' gases to raise steam was normal. As we move through the energy transition this is no longer the only option. Importing electrical power is an alternative, and should be compared to using more gas because of the:

- Scope 2 & 3 carbon footprint from gas.
- Shadow costs of dealing with the CO_2.
- Trends for more renewable power to be provided to the grid.
- Costs, today and in the future, of gas vs. renewable electrical power.

Where scope 1 emissions are directly produced from the process. Scope 2 are the indirect emissions from the materials required by the process e.g., from natural gas the process imports. While scope 3 encompasses the remaining emissions up and down the supply chain, and includes all the activities that enable the supplier, e.g., of the natural gas to operate and make their product available.

A - Minimising Turnarounds

Historic AIChE reports have shown us most safety related events occur during a transient or change to the system. Therefore, maintaining stable operations for as long as possible is important to delivering a robust and proven basis of ATR / O_2 safety.

Our observation is that if combustion occurs in proximity to the reforming catalyst, it leads to an ATR system operating less stably and with reduced run lengths. The unique JM longneck ATR design spatially separates the processes of mixing and combustion from the bulk reforming reaction through the catalyst bed. The gases mix in the JM long neck design, due to the action of the burner, where the oxidant entrains both the hot combustion product gas and fresh feed gas in a fuel rich combustion process. Illustrated by Figure 2.

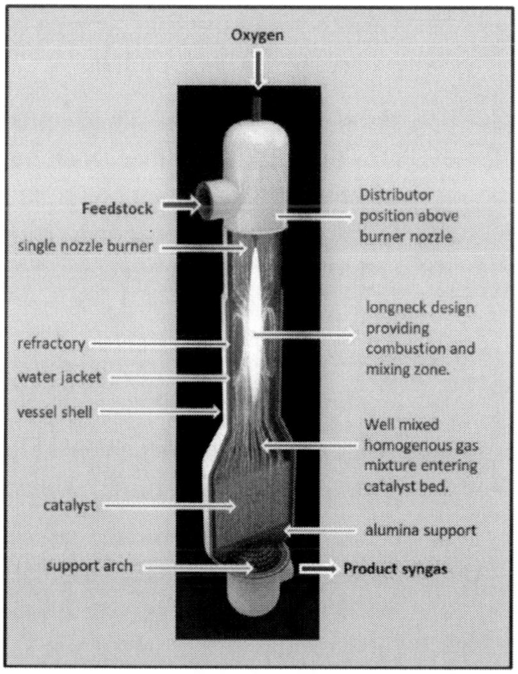

Figure 2: Schematic of JM long neck ATR.

The longneck facilitates space for:

- The oxidant to entrain both the hot combustion product gas and fresh feed gas in a fuel rich combustion process.
- High velocity turbulent gas mixing zone, generating a well-mixed and defined process gas flow, upstream of the catalyst.
- The high velocity provides significant convective cooling of the burner tip.

JM's ATR is a result of detailed engineering design, process simulation with CFD plus many years of industrial operation validating our work.

Figure 3 shows CFD results showing gas velocities and temperatures of the high velocity O_2 burner which is used to give excellent mixing in the long neck of JM's ATR, meaning uniform gas enters the main body of the vessel.

The high velocity O_2 jet from the burner nozzle, enables high mixing rates of O_2 with feed gas. The combustion products become entrained in the high velocity gas flow, that hydraulically prevents bypassing of the mixing and combustion zone by the methane rich feed gas, due to the jet attachment being within the neck of the vessel. This is shown by the velocity plot, the lefthand image within Figure 3.

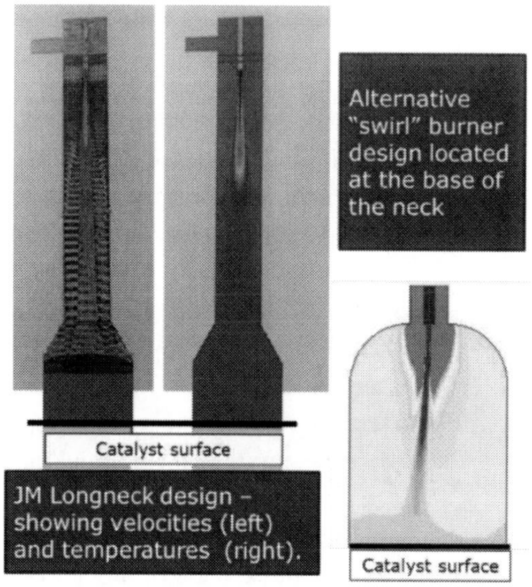

Figure 3: CFD images comparing high velocity gas mixing (used in JM ATR, left) to a swirl burner design (right) that is situated at the base of the vessel neck.

JM's design means:

- Mixing is complete within the neck volume.
- The burner nozzle is cooled by the high velocity gases passing through it, this helps to extend its operating life.

The combination of the long-neck and single nozzle burner design results in long burner life with near zero maintenance requirements. It also supports start-up, shutdown, and steady state conditions, with a comfortable turn down to at least 40% of the design rate. This provides a simple design with lifetimes that are multiples of typical 5-year plant shut down cycles. This has been clearly demonstrated by the record for the longest life of this nozzle type in an air driven ICI designed ammonia plant in the UK. This shows the design concept of a single nozzle to mix oxidant and fuel is robust and valued having been deployed initially in the 1970s with only one intervention. The current burner achieved 27 years life, showing this burner design can provide many years of maintenance free operation.

The image to the right of Figure 3 contrasts the JM single nozzle design, to an alternative swirl burner design located at the base of a shorter neck ATR . This image is a result of JM work to understand the problems encountered by a methanol operator using non-JM technology that employed a multiple nozzle swirl burner to achieve mixing of the O_2 and process gas. The CFD illustrates the differences in the main body of the vessel. Where gas mixing and combustion are taking place closer to the catalyst volume in the main body of the vessel. In this case the closeness of the gas mixing and combustion process to the catalyst bed was shown to be a problem, resulting in catalyst movement and damage from heat where the gas jet impacted the bed.

The robustness of JM's ATR design means target tiles are not required on top of the catalyst volume.

Contrasting this to swirl gas mixing burner designs, where target tiles are commonly used, we observe a difference. With mixing and combustion taking place in the main body of the vessel, closer to the catalyst bed, there is an increased risk of damage to the top layers of catalyst, and because these designs utilise multiple lower energy jets the heat is released from combustion closer to the burner nozzles. This will lead to increased rates of metal oxidation and reduced burner lives. In addition, because this happening closer to the top of the catalyst bed there is a higher propensity to observe catalyst volatilisation, that has required specialist proprietary catalyst solutions [8].

JM's robust design limits pressure-drop growth and is another contributor to enabling long operating cycles without need for maintenance intervention. This sustained high activity and low-pressure drop has been reported in an earlier paper [8].

The catalyst science and optimised operating conditions also contribute to the stability and longevity of the ATR operating cycle.

Higher catalyst temperatures favour lower methane slip, however the risk of catalyst damage also increases, as this initiates reactions that increase volatilization of the catalyst support.

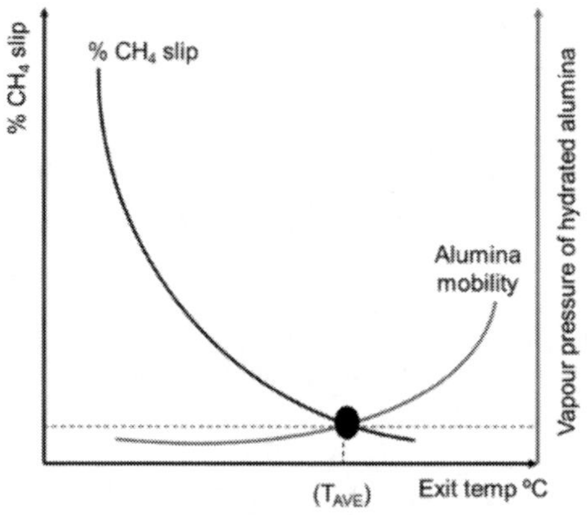

Figure 4 Plot of methane slip reducing and increasing rate of alumina volatilisation with increasing temperature.

The catalyst is designed to be highly uniform in pellet size and shape. This provides a uniform packing arrangement, and so uniform voidage and pressure drop across the catalyst bed. Maintaining this arrangement is critical to maintaining a long and stable operating ATR cycles. The reliability of a JM ATR was reviewed over a 5-year operating window. This paper showed the JM longneck ATR provided:

-methane slip below the design value

-while using oxygen below the design value

-with pressure drop below the design value

Achieving this is a combination of the stable process conditions and the robustness of the catalyst [9].

Stepping back, the ATR must be considered in the context of the ammonia flow sheet. The stochiometric equation reported by Max Appl,[10] is defined below for today's SMR air-based process, equation 4.

$$CH_4 + \underbrace{0.3035\ O2 + 1.131\ N2}_{1.4345\ air} + 1.393\ H2O \rightarrow CO2 + 2.262\ NH3\ (4)$$

This has set the air requirement to supply the required nitrogen to produce ammonia.

We can adapt this for an oxygen-based flow sheet to make low carbon hydrogen. The O_2 level, selected by defining x in the equation below, equation 5, means a targeted exit temperature is achieved due to combustion.

$$CH_4 + xO_2 + 2(1-x)H_2O \leftrightarrow CO_2 + (4-2x)H_2 (5)$$

The other variable is the steam addition, this is required to drive the reforming reaction, reaction 1. Then later to transform all the CO to CO_2, via the WGS, as set out in equation 3.

The global steam addition can be adjusted downstream of the ATR, with steam addition into the WGS section.

Allowing the S:C inlet the ATR to be adjusted independently, effecting the harshness of the conditions and the performance of the ATR. Lowering the S:C effects the performance as follows :

- Increases the methane slip for the same exit temperature, which effects the tail gas composition.
- Makes the operating conditions harsher within the ATR.
- Reduces steam raising capability.

Although a higher exit temperature on day one provides a favourable equilibrium position with respect to equation 1. However, on subsequent days Figure 4 shows the risk of increasing the exit temperature. High temperatures instigate a process that negatively impacts the catalyst.

The desire for higher exit temperatures, or high recycle, is exemplified if the S:C inlet the ATR is lowered. For the same feed: O_2, at the same exit temperature Figure 5 compares the equilibrium CH_4 slip for a S:C of 0.6 and 1.3 inlet the ATR.

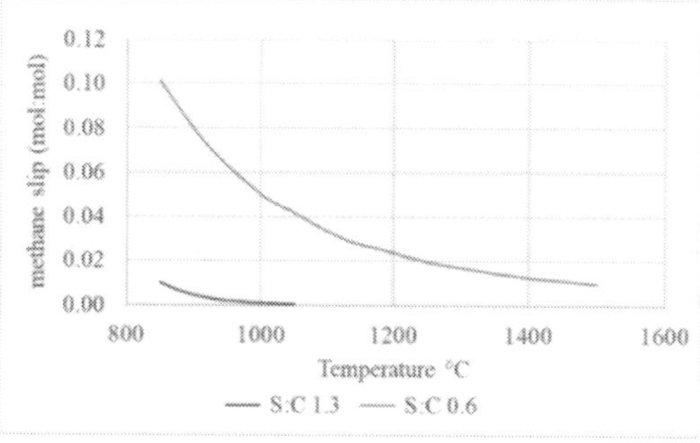

Figure 5. Plot of equilibrium methane slip at S:C 0.6 and 1.3 inlet the ATR.

A way to reduce the CH_4 slip is by addition of more O_2, to increase the temperature, necessitating more combustion, providing a less efficient CH_4:NH_3 (feed : product) and increases the size of the ASU as more O_2 is required.

This would increase the catalysts temperature and while this would facilitate more reforming, the risk is negatively impacting the lifetime of the ATR cycle we are seeking to extend.

Therefore, to retain an efficient loop with low inert (CH_4) levels, while not overheating the catalyst would mean either; more tail-gas is recycled, or a higher purge of tail gas is needed (which is burnt).

So, when thinking of the S:C at the ATR to make blue hydrogen/ammonia we should consider conditions that:

- Minimises CH_4 slip, while not unduly stressing the ATR.
- Long and highly efficient operating cycles between plant shutdowns.

In doing this the ATR operation will

- Minimise CH_4 (feed) : NH_3
- Minimise Carbon Intensity : NH_3

B – Proven Design and SIS Architecture

The main body of the ATR vessel is a refractory lined design, able to safely operate with gas temperatures in the vessel within the range of 800 – 2000°C.

The ATR Refractory design is critical to the system operation. The refractory separates the hot process gas from the pressure shell of the vessel. This protects the vessel shell from high temperatures and reduces heat loss which aids energy efficiency.

Damage to the refractory can be caused by flame impingement. This would allow hot gases to contact the vessel shell and cause hot spots and potentially lead to rupture. Other impacts of flame impingement are potential catalyst damage that might adversely impact conversion and/or pressure-drop.

Two features of the proven JM design are the single nozzle burner and the water-jacketed vessel.

The single nozzle design is simply aligned by the inlet flange and the process gas distributor plates around the burner feed pipe. This gives robust positioning meaning the burner tip is less likely to move from its intended position which could cause flame impingement and damage the refractory.

JM's water- jacketed design maintains the shell at constant and relatively low temperature. This reduces the expansion of the shell that can cause cracking of the refractory and allow gas tracking into the refractory resulting in hot spots. This can cause further expansion which can lead to an amplification of the cracking, leading to more gas tracking through the refractory. The constant shell temperature helps hold the refractory in compression limiting its propensity to cracking. The water level in the jacket is maintained automatically and water supply rate is monitored. An increase in water supply to the jacket indicates excessive heat loss from the ATR that can be investigated further.

The SIS is arranged to trip the plant to a safe condition. On a plant trip, the feed and oxidant will be isolated, and the plant purged to flare, to remove flammables. The ATR burner is purged with steam to ensure no flammables reverse flow into the oxidant system.

Steam is purged through the main process lines as the pressure is reduced via a controlled depressurisation system. The SIS system considers the following scenarios:

Loss of Ignition in the ATR

Combustion with oxygen supplies energy within the ATR. If combustion is lost, so unburnt oxygen can pass downstream, meaning there is potential for a flammable mixture to be created. To protect against this:

- The ATR feed gas temperatures are set significantly above the auto-ignition temperature, so the ATR burner will stay lit.

- Upon start-up, a low inlet temperature permissive is utilised which must be met before oxidant can be added to the process.
- In the event of a low ATR inlet temperature post start-up, a trip isolates the oxidant supply and begins a normal plant trip to remove flammables from the process.

Flammables into Oxidant System

If process gas flows backwards into the oxidant supply line and mixes with oxidant it is likely to ignite due to the temperature of the gases. To protect against this:

- The oxidant supply line contains two high integrity non-return valves. These are specified, to be of different types to avoid common failures.
- Start-up requires a successful burner purge (with steam) and ensuring the differential pressure between the oxidant and the process is sufficient for oxidant to flow into the process before the oxidant is added to the process.
- During normal operation a small continuous flow of steam is passed to the burner. This is to prevent reverse flow of hot process gas into the oxidant system in the event of an oxidant flow failure.
- A proven trip system protects against the steam or oxidant supply pressure being lower than the process pressure. The trip isolates the oxidant line from the process in the event of a trip.

Oxidant Supply

The ATR design must also supply oxidant to the burner, while not causing premature ignition in the supply system. The prevention of premature ignition involves selecting the correct metallurgy, cleaning the equipment using the correct methods and limiting the temperature of the oxygen prior to addition into the process by heating with saturated steam, rather than superheated steam.

Comparative Assessment of Reforming Technologies

It is desirable to a project developer to minimise risks, which we have shown is aided by reducing the number of plant outages. Avoiding frequent planned maintenance, during which the ATR process will deliver no value, and only cost. We have discussed this is primarily caused through an overly stressed ATR, which may have:

- Combustion occurring close to, or at the catalyst surface.
- Short burner life.
- Poor mixing of combusted gases.
- Low catalyst stability, causing high pressure drop, and high methane slip.

These unwanted features add planned or unplanned maintenance to the operation of the ATR flowsheet, illustrated in figure 6.

Additional maintenance routines expose organisations to the hazards that are isolated during normal operation. These topics were detailed in respect of a paper, describing safety lessons learned in nickel catalyst handling in methanol plants [14].

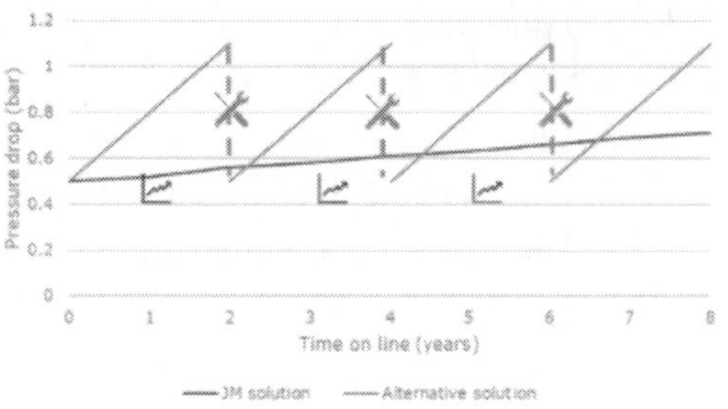

Fig. 6: Comparative assessment of short and long maintenance routines.

As well as the transient start / stop risks we have discussed relating to the ASU, there are also obvious risks of maintenance. When ATR operation has suffered catalyst degradation, removing which has been likened to "mining"[15].

It is our responsibility to design and operate processes that avoid taking free flowing nickel catalyst particles, that can be vacuumed from a vessel, to make fused rock that results in the need to place people at the ATR catalyst bed, manually mining catalysts, because this elevates the risk, mining is recognised as one of the most hazardous occupations [16] plus there is added Ni-exposure risk.

Conclusions

Historically, the largest cause of incidents with O_2 process units has been unstable operation, or periods of operation following plant outages.

Key learnings from the past include:

1. Delineation of areas where abnormal oxygen concentration could occur.
2. Exclusion of combustibles from the oxygen area.
3. Minimising the possibility of oxygen leakage from the plant
4. Provide a safe means of disposal for high oxygen concentration streams (gas and liquid).
5. Control of concentration of contaminants within oxygen systems.

To minimise interventions, the operating conditions of the ATR should be considered as part of the whole flow sheet, whose purpose is to make ammonia in the safest manner, with the highest feedstock efficiency, and moving forwards with the lowest overall carbon intensity. ATR designs should be reviewed to ensure they…

A. Minimise the number of turnarounds.
B. Use proven designs employing proven control philosophies and SIS.

This paper highlights that unnecessarily stressed designs decrease reliability, preventing long operating cycles, and increasing the likelihood of unstable operation and the need to make interventions.

Figure 7 illustrates the motivations behind the JM ATR design.

Factor	'Alternative' ATR	JM longneck ATR
ATR runtime		
Reliability		
Maintenance		
Ni-exposure		
Feed consumption	Highest	Lower
Overall, Carbon Intensity	Highest	Lower

Fig. 7: Comparative assessment of JM longneck ATR design to alternatives.

As the ammonia industry faces its next transition, ATR flow sheets can provide an answer to marry the simultaneous demands of increasing production with the need to reduce its carbon intensity.

References

1) Oxygen Plant Safety. E. Karwat Linde. AIChE. Chemical Engineering Progress 1957. Vol 55. No 4.
2) Solutions to problems with Secondary and Auto-Thermal Reformers. M Humphrys, J R Brightling, R Ebadi. Asia Nitrogen and Syngas 2015.
3) Hydrogen Insights 2023. Hydrogen Council McKinsey & Company. December 2023.
4) The operators report on safety in air and ammonia plants Chairman Walton. July 1959 Chemical Engineering Progress Vol 55 No. 7
5) …. Parts 2 and 3. The operators report on safety in air and ammonia plants Chairman Walton. August 1958.
6) Safety aspects of reconstructed ICI tonnage Oxygen Plant. W. D. Mathews and GG Owen. Vol 5, 1962
7) Hazards Involved in handling Oxygen Rajesh Aggarwal, R K Bhatia. National Fertilizers India, Vol 52, 2011
8) Experience in optimizing ATR performance with high stability KATALCO catalyst. P W Farnell. IMTOF 2013
9) Delivery of world class OEE performance by JM ATR technology. M McKenna IMTOF (International Methanol Technology Operators Forum)2019
10) Max Appl. Ammonia principle and industrial practice.
11) Rupture of Process Air Vent Pipeline Incident and Lessons Learnt, Noelker, Uhde Ammonia Technology Users Group Meeting, September 2008;
12) Downgraded High Temperature Shift Converter and Explosion in Air Line to Secondary Reformer; Othman, Niwa & Bakkerud; AIChE Ammonia Technical Manual; 2002
13) Failure of a Process Air Line; Clark; AIChE Ammonia Technical Manual; 1981
14) Lessons learned in nickel catalyst handling in methanol plants. J R Brightling. IMTOF (International Methanol Technology Operators Forum) 2023
15) Linked in Post. https://www.linkedin.com/posts/alireza-safavi-477b151bb_its-not-a-mine-digging-its-catalyst-activity-6881928564510347264-axJs/
16) Comprehensive Guide to Mining Health and Safety. HSE network January 2024. https://www.hse-network.com/comprehensive-guide-to-mining-health-and-safety/

Primary Reformer Riser Transition Assembly Damage, Repair, and Future Risk Mitigation Plan

After accomplishment of a turnaround in 2023, during plant start up, a process operator reported knocking and bubble sounds in one of the primary reformer risers which led to thermal imaging of all the risers. Upon thermal imaging of risers, it was identified that one riser water jacket metal wall temperature was 115 °C [239°F], which was 25°C [77°F] higher than the other remaining 5 risers. This temperature rise and sound combined were the clear indication of an internal leak. Hence the metal jacket was cut locally for inspection and as expected a 30 mm [1.18in] hole was found on the pressure containment carbon steel shell. Later a thorough inspection was performed on the pressure shell and installed refractory. Upon inspection, localized loss of pressure shell thickness was observed which was caused by severe refractory channeling initiated at the junction point between the riser transition assembly and transfer line. Based on the above inspection results, a temporary repair was performed, which is sustainable until 2025-26. This incident resulted in 2 weeks of delay in plant startup and substantial production loss.

Mukesh Kumar
Yara Pilbara fertilizers

Romain Kabs
Yara Pilbara fertilizers

Marcel Markus
Yara International

Abhishek
Yara Pilbara fertilizers

Introduction

Yara Pilbara fertilizers (YPF) operates KBR purifier technology based 2450 TPD ammonia plant since 2006. After completion of a turnaround in 2023, a significant failure was observed in the riser transition assembly of the primary reformer (101MB).

This paper describes the Yara Pilbara experience and lessons learned as listed below:

- Discovery of failure
- In depth analysis and understanding of refractory and pressure shell damage mechanism
- Inspection results
- Repair methodology of pressure shell and refractory
- Future preventive measure and risk mitigating actions

Incident

After the completion of a turnaround in 2023, on 5th August 2023, during start-up of Primary Reformer, a process operator became aware of a newly observed bubbling noise in one of the riser's transition assembly in row D. Subsequently using thermal imaging, the skin temperature of that riser (on water jacket) was observed 115°C [239°F], which was higher than that of the other risers at 85-90°C [185-194°F].

Based on the above observation, it was decided to bring down the reformer from 650°C [1202°F] to minimum temperature at 350°C [662°F] and steam to be cut off. Subsequently on 7th August the reformer was taken off-line without any personnel injuries. Figure 1 below represents a visible hot spot at the water jacket metal surface of the riser transition assembly.

Figure 1: Hot spot visible with thermography camera on the water jacket (riser row D).

Design and Constructional features of Transfer line and Riser transition assembly

YPF primary reformer is a top fired reformer which consists of :

- 98 arch burners uniformly spaced into 7 rows (tagged Row A to G)
- 288 catalyst tubes uniformly spaced into 6 rows (tagged row A to F)
- 1 riser per catalyst tube row (6 risers in total)

Riser transition assembly and transfer line were supplied as separate units during original plant construction. Welding of these two assemblies and local junction casting of refractory was done on-site during plant construction in year (2005-2006). Below are the main design and construction features of transfer line (TL) and riser transition (RT) assembly.

TL pressure shell material: SA516 GR70
TL liner material: Incoloy 800H/HT
TL /RT assembly design pressure: 43.5bar(g) [630.9 Psig]
TL /RT assembly pressure shell design temperature: 205°C [401°F]
TL /RT assembly refractory: Bubble Alumina
RT assembly pressure shell material- SA106 GRB
RT assembly liner material: Incoloy 800HT
RT assembly pressure shell O.D : 406.4mm [16in]
Reformed gas temperature: 755°C [1391°F].

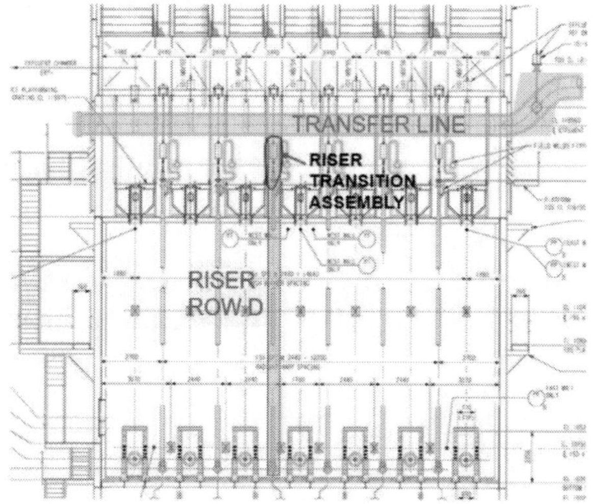

Figure 2: Section drawing showing the outlet manifold, risers, and transfer line. Failure occurred in Row r D riser.

The riser transition assembly is the intermediate connection element between Riser tube and Transfer line. The RT transition assembly is constructed from carbon steel and is refractory lined for protection against the high temperature of the primary reformer effluent. An internal Incoloy 800 HT liner protects the refractory from erosion by the gas. Additionally, the riser has an external water jacket to prevent overheating of the pressure bearing part of the riser in case of refractory failure.

Reformed gas from outlet manifold (at bottom) flows through the riser tube at temperature of 755°C [1391°F], through the riser transition assembly and finally enters transfer line as shown in Figure 3.

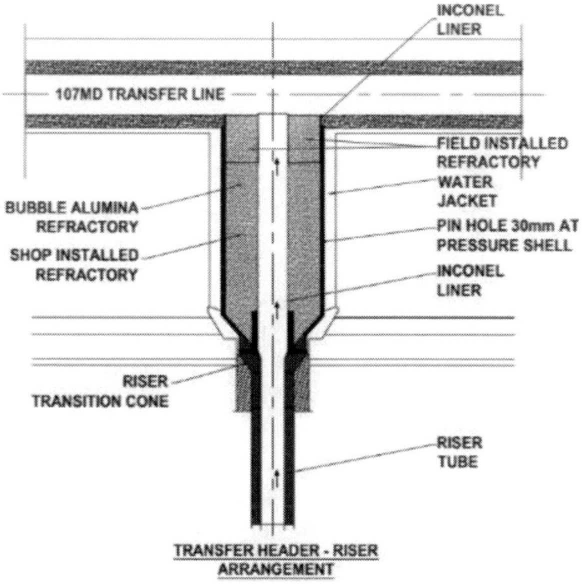

Figure 3: Simplified schematic of RT transition assembly and TL design.

Troubleshooting

After shutting down the reformer, a trouble shooting process started to find out the cause of bubbling sound reported by operator. A small section of water jacket was cut initially at the same location where higher temperatures were reported via the thermal imaging. Upon removal of this local section of the water jacket, a large hole of 30mm [1.18in] was found as shown in Figure 4 in the pressure shell.

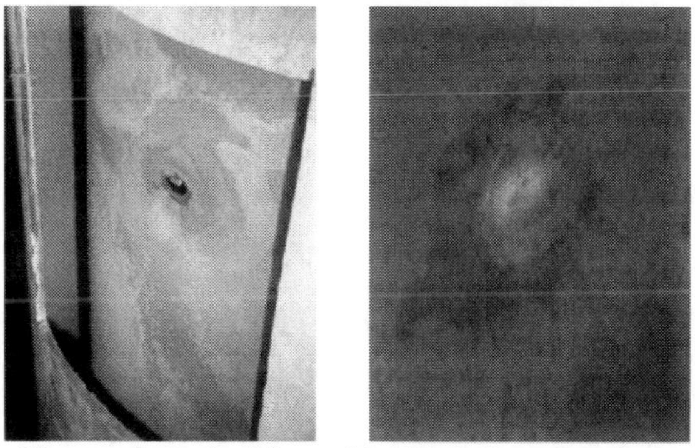

Figure 4: Left :30mm [1.18in] diameter hole in pressure part. Right: Erosion found on water jacket material at same location of hole in CS pipe.

Initially it was expected that the liner pipe at the same location should have been damaged, which would have created a path for the reformed gas to penetrate and consequently create conditions leading to pressure shell damage. But upon refractory removal of that local area, it was found that liner pipe was in good condition.

Then the pressure shell visual inspection was performed which highlighted heat impacted area and it was decided to cut the water jacket to do ultrasonic thickness measurement of pressure shell. Upon ultrasonic thickness measurement on the pressure pipe, heat impacted area was found to have a reduced thickness down to 5.5mm [0.22in] instead of 12.7mm [0.5in].

Hence a 180° semicircular window of approx. 1100mm [43.31in] height was cut on pressure shell. Upon removal of the shell section, refractory channeling was observed in upward direction towards the transfer line (107MD) as shown in below Figure 5

Figure 5: Left: Refractory channeling, Right: Heat impacted area found with associated loss of thickness during repair.

Once gas channeling in refractory was detected, the process of refractory demolition started, and precaution was taken that only the damaged refractory section was removed.

After careful demolition of approx. 190° semicircular refractory area as shown in Figure 6*Figure* ***6***, it was observed that the remaining refractory was in good condition, hence demolition work was stopped.

Figure 6: Incoloy 800H liner found intact upon inspection after refractory removal.

Visual inspection with borescope was performed to find out the origin of gas channeling and it was found that it was initiated from the transfer line side as shown in Figure 7 and Figure 8.

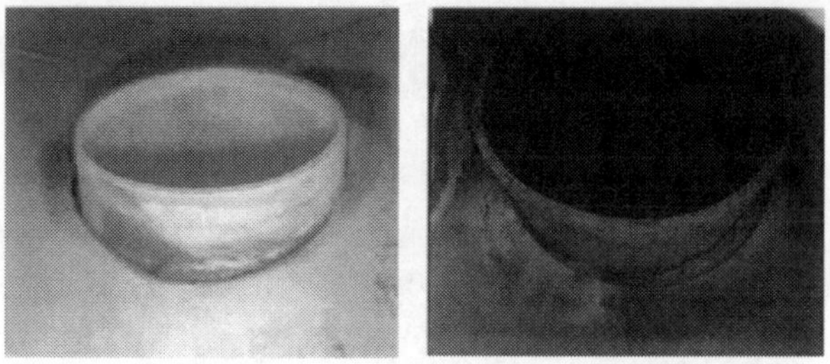

Figure 7: View of riser liner protruding into TL. There is a gap between riser and TL liners to allow for thermal expansion.

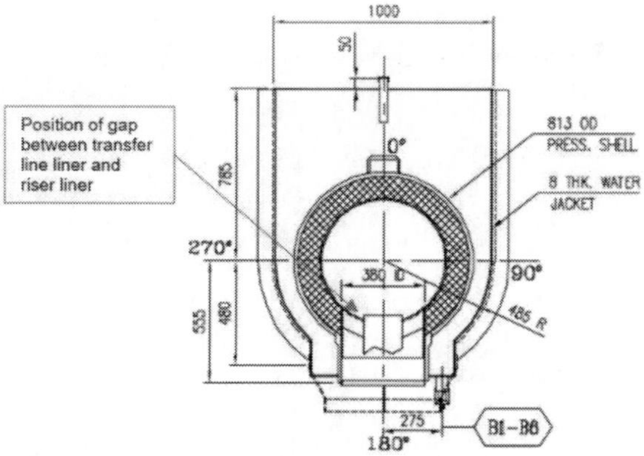

Figure 8: Section of transfer line drawing showing the transverse cross-section at the location of risers.

The joint between TL and riser Incoloy liners shown on Figure 7 is subject to thermal expansion in vertical (riser) and horizontal (transfer line) direction. This is the most susceptible area for refractory failure and channeling initiation.

Upon review of the manufacturer data report, no record was found of radiography performed on RT assembly by vendor. Radiography can be used to check that refractory is well poured and no cavity exists against the pressure boundary.

Damage Mechanism and Root Cause Analysis

Sample of the failed pipe was sent to Yara Porsgrunn where failure analysis was carried out of this serious failure to confirm the damage mechanism and link the root cause to the site observations.

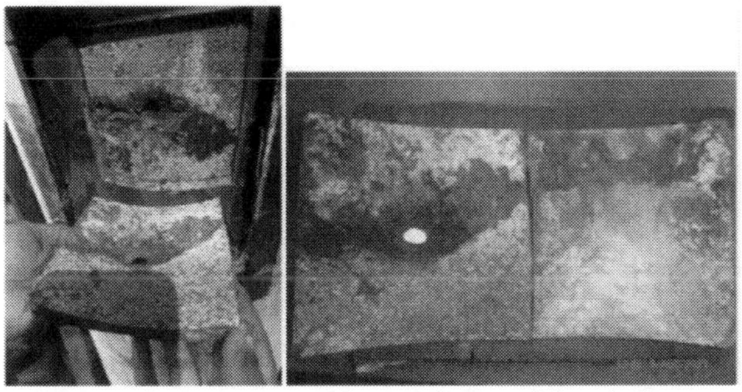

Figure 9: Location of sample used for failure analysis. Wall thickness loss on pipe internal surface and refractory damage are evident.

The samples were visually examined before performing XRF (X-ray fluorescence) and XRD analysis (X-ray diffraction).

The internal surface shows an area of localized corrosion with deep grooves around the area of the hole as shown in Figure 12. The grooves are rounded in shape and had a dark grey, dull, and smooth surface. The transition to the adjacent material is rather sharp.

The non-corroded parts of the internal surface had a dull and slightly rough appearance, with a thin layer of rust brown oxide as well as some white residue from the refractory.

The wall thickness measured at the cut surfaces of the sample is close to its nominal value of 12.3-12.6mm [0.5 in] measured with vernier calipers.

Figure 10: Sample external and internal surface

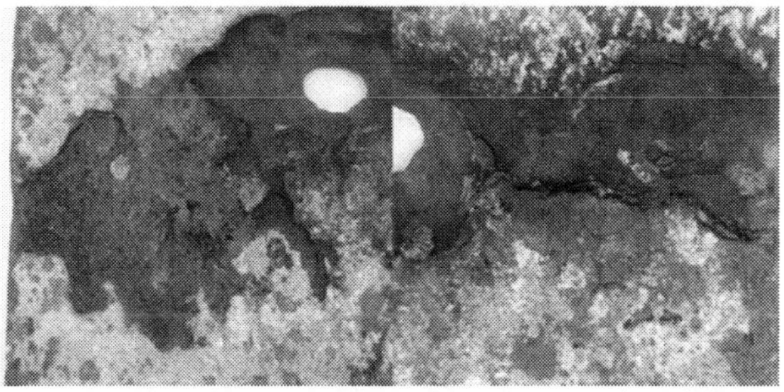

Figure 11: Close up view of groove and wall loss.

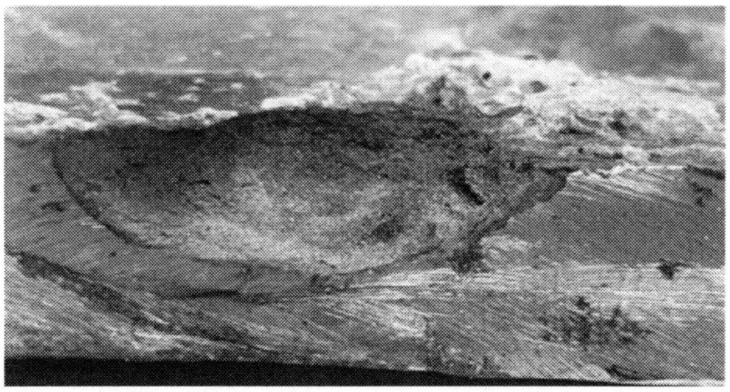

Figure 12: Corrosion groove as observed on the cross-section of sample.

Corrosion products were scraped of the sample and analyzed by XRF and XRD semi-quantitative analyses for detection of crystalline phases. The chemical composition as measured by XRF analysis is shown in Table 1 (reported as oxides). The crystalline phases as detected by XRD are shown in Table 2.

Element	*Wt(%)*
Al_2O_3	6.7
CaO	3.00
Co_3O_4	0.14
Cr_2O_3	0.14
Fe_2O_3	87
MnO	1.6
MoO_3	0.08
NiO	0.12
P_2O_5	<0.05
PbO	<0.05
SiO_2	0.65
SO_3	0.12

Table 1: Chemical composition of corrosion products as measured by XRF (reported as oxides).

Compound name	*Chemical formula*	*Semiquantitative amount [% wt.]*
Siderite	$FeCO_3$	65-85
Magnetite	Fe_3O_4	5-25
Calcite	$CaCO_3$	<10

Table 2: Crystalline phases as detected by XRD.

The observations during the examinations of the samples confirm that the riser suffered from localized corrosion and wall thickness loss due to condensation of gas behind the refractory. Damage and channelling of the refractory had allowed gas to encounter the pipe wall which is cooled by the water jacket, and condensation had occurred.

The primary reformer effluent contains about 12.51% (dry mole%) of CO_2, as well as steam. Due to the CO_2, acidic conditions will form upon condensation (carbonic acid) and corrosion of carbon steel is expected. Corrosion rates as high as 25 mm / year [1 in/year] have been observed upon condensation of effluent gas streams off the shift converters in hydrogen plants [1]. The degradation mechanism is confirmed by the presence of Siderite $FeCO_3$ in the corroded area. Siderite is one of the main corrosion products in this type of corrosion. The aluminium oxide (6.7%) and calcium oxide (3%) as detected by XRF are residues of the refractory on the surface.

Additional Information from Plant Operation

During the root cause analysis, the plant operations team highlighted that the main steam valve which is upstream of the reformer (feeding steam to the mixed feed supply to the catalyst tubes) was passing. Due to this passing valve during plant shutdown, when the valve is closed, a large amount of steam entered the system and condensed into the riser, before operations noticed it. Water levels up to the transfer line has been experienced several times during shutdowns between period of 2014-2018. Since 2018, extreme precautions being taken on draining during shutdown events.

The bottom part of the riser transition assembly is designed as a cone with draining holes 5mm [0.2 in] dia in the Incoloy liner to allow draining of water condensed into the refractory. Due to large amount of water present during restart together with refractory channelling, the drain was not sufficient and carbonic acid was able to be created.

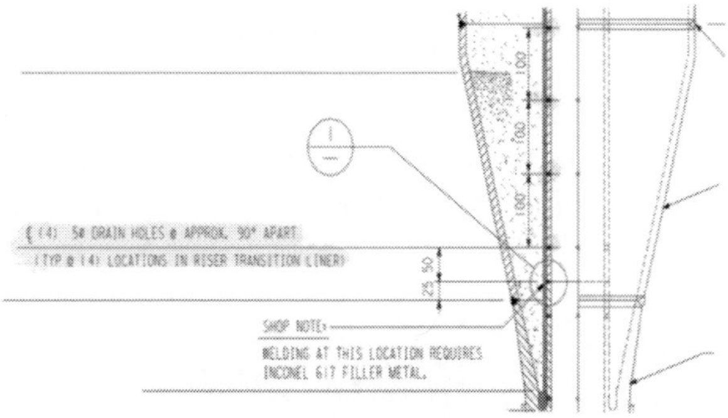

Figure 13: Draining holes in bottom cone of RT assembly.

After gathering of all this information, consensus on the root cause is as follow.

1. Fabrication defect could have been left from field refractory pouring on top of RT assembly to TL.
2. Water overflow events would have contributed to increase the defect into refractory channeling.

It is a thought that during restart, boiling water would create erosion of the refractory, increasing a small defect leading to a channeling.
3. Presence of channeling allows local pooling of steam during start-up. The steam condenses due to the cooling effect of the water jacket on pipe outer side. Presence of condensed steam with CO_2 of process gas leads to formation of concentrated carbonic acid which react with carbon steel at high corrosion rate.

Repair Methodology

Based on the following reasons it was decided to do temporary repair instead of a permanent repair and replace this RT assembly in the upcoming shutdown.

1. Spare RT assembly was not available in stock and lead time was approximately 6 months.
2. For permanent repair, in an ideal case, refractory pouring must be done via transfer line, which requires hot work and welding also, as there is a liner welded. It adds high personnel safety risk and team was not prepared for that.

Repair Procedure

Repair work started as defined in following sequence on 10th August 2023 and was accomplished on 16th August 2023.

- Prepared bevel ends of existing riser transition pipe and water jacket for welding and confirmed that they were acceptable by Penetrant test.
- Hand packed bubble alumina refractory in upper section (towards transfer line), before application 3-V anchors were welded, and shuttering was provided to support refractory. Shuttering was removed 8 hours after refractory works to provide sufficient time for settlement. Refer to Figure 14.

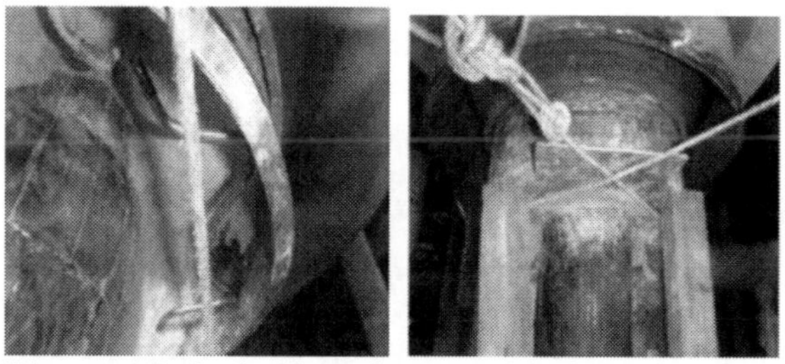

Figure 14: Left: V anchors welded, Right: Top refractory hand packed.

- Installed 3mm [0.12in] thick cardboard around exposed section of liner. Exterior of cardboard was wrapped with water-resistant tape.
- Left a 150mm [6in] mm wide refractory pouring window on the new riser transition pressure shell pipe 180° section as shown in Figure 15.
- Installed new 1100mm [44.04in] long riser transition pipe and weld in place using an approved weld procedure specification (WPS).
- Conducted 100% radiography testing and 100% Penetrant testing on all riser transition pipe welds.
- Poured bubbled alumina refractory through 150mm [6in] rectangular window and used small vibrator to avoid any cavity in refractory.

Figure 15: Left: 150mm[6in] rectangular groove for refractory pouring. Right: Refractory Pouring completion

- Welded SA 106 GR. B patch plate on the refractory pouring window using an approved WPS.

- Conducted ultrasonic testing and penetrant testing on 100% of welds.
- Installed section of water jacket by welding which had been prepared in two halves using the approved weld procedure specification.
- Conducted 100% Penetrant testing on all water jacket weld.
- Conducted general housekeeping and demobilized tools/equipment from primary reformer penthouse.

Recommendation and Risk Mitigation Plan

Due to the design of the RT assembly and temporary repair done, there is potential for recurring issues with refractory damage and subsequent corrosion. The risers must hence be followed up by inspection.

The pressure bearing part of the RT assembly is not inspectable from the outside due to the external water jacket. Inspection windows were cut on other risers during another opportunity shutdown (OPSD) to complete visual inspection with aid of borescope to identify any heat impacted area like the one observed as defined in above failure (as known to be associated to loss of thickness of the pressure shell).

In addition to above, following risk mitigation actions are taken and in progress.

- Perform thermal imaging of riser transition assembly(s) water jacket wall temperature on weekly basis.
- Keep 1 spare RT assembly in stock.
- Replace RT assembly in next Turnaround planned in 2026.
- Application of thermal sensitive paint on water jacket wall during next OPSD which changes color in case water jacket wall temperature increases to more than design due to pressure shell leakage.
- In new design of RT assembly, exploring the option to install a removable type of water jacket mechanism, so that during turnarounds, jacket can be removed easily, and inspection of RT assembly can be performed.

References:

[1] API RP 571: Damage Mechanisms Affecting Fixed Equipment in the Industry, p. 93-97; American Petroleum Institute; 3rd edition 2020.

E-208 Nitric Acid Cooler Condenser Failures

An unplanned outage occurred due to low pH in the cooling tower that resulted in the nitric acid condenser failures made of Zirconium alloy tubes and 304L stainless steel shell in March 2021. This paper examines the contributing factors to the failures, the experiences gained, including the effect of infrequent inspections and plant cycling. The practical changes made to continue operations, the importance of water chemistry control and the redesign of the condenser will be discussed.

Samuel Okulaja
Nutrien
Calgary, Alberta, Canada

Kafayat Adejare
Nutrien
Redwater, Alberta, Canada

Introduction

In March 2021, an unplanned outage occurred in the Nutrien Nitric Acid facility in Redwater, Alberta Canada due to low pH in the cooling water tower. Initial inspection revealed that the tubes upstream of the E-208 Nitric Acid Cooler Condenser failed and discharged hot nitric acid to the cooling water shell, leading to the contamination of the cooling water circuit. Further investigation shown through wall cracks on some of the tubes and thick calcium phosphate deposits on the affected tubes outer diameter (OD), which promoted the subsequent failure of the shell. This paper includes the destructive testing, the associated damage mechanisms, the contributing factors, the lessons learnt, and recommendations to prevent such failures in the future.

Background

The Nitric Acid plant was designed by Weatherly and began operation in 1969 with a capacity of 510 Metric Tonnes per day (MTPD). The plant produces 57% Nitric Acid, which is either converted into Ammonium Nitrate (NH_4NO_3) solution or combined with Urea to produce Urea Ammonium Nitrate (UAN).

The E-208 Nitric Acid cooler condenser cools the process gas (71%N_2, 14.2%H_20, 3.6%NO, 5.6%NO_2, 4.6%O_2, 0.84%Ar) from the preheater upstream of the heat train, flowing through the tube side, from 450°F (232°C) to 120°F (49°C) by cooling water on the shell side, condensing vapors to form weak nitric acid of 52% concentration. The weak acid then flows into the absorption tower to increase the concentration to 57% nitric acid. See Figure 1 for the layout of the E-208 Nitric Acid condenser.

The tubes, with an outer diameter (OD) of 0.75" (19.05 mm) and wall thickness of 0.035" (0.89 mm), are made of SB-523 Grade 702 alloy, which is a commercially pure zirconium (CP Zr). The shell material is constructed of SA-240 TP304L stainless steel, with an outer diameter of 63" (1600 mm) and a wall thickness of 0.5" (12.7 mm).

Subsequent visual inspection revealed a through-wall leak on the shell in the same vicinity as the failed tubes.

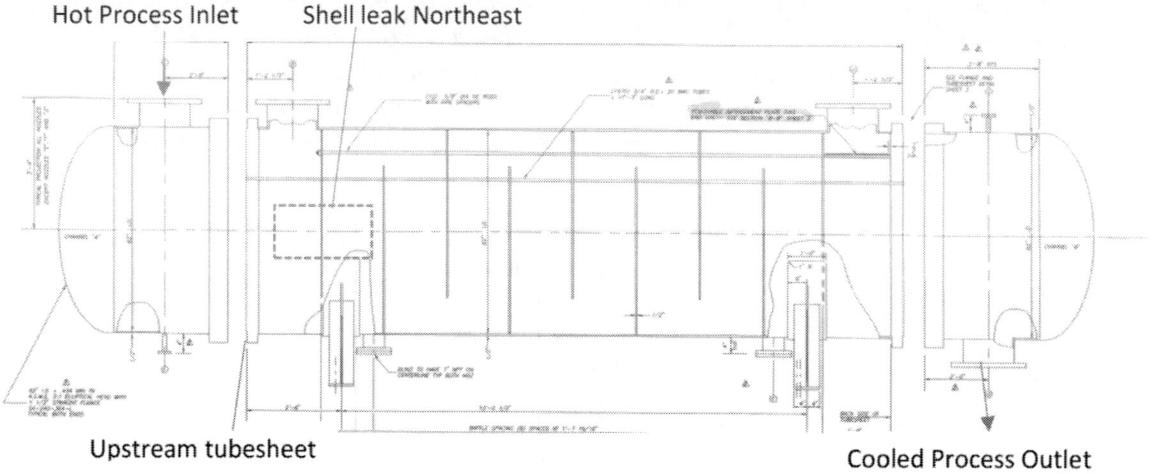

Figure 1: E208 Nitric Acid Cooler Condenser layout; Counter-current flow

Both the tubes and shell have been in service since 1997. Prior to the failures in 2021, the unit had experienced increased intermittent periods of operation due to product market demand since the previous outage in October 2020.

Failures Findings

Eddy current inspection was conducted in 2006 on 100% of the tubes (1975 tubes) and identified three (3) tubes being plugged due to leaks. The next full inspection completed prior to the 2021 failure was in 2012, and it showed no sign of wall loss on the tubes. These recent tubes failure (through wall leaks) occurred at the tube inlet (upstream), which is the hottest, with temperature of ~ 450°F (232°C). Other tubes away from this failure location did not show any damage. Visual examination of other internals, including the tubesheet and channel heads, did not exhibit signs of corrosion.

Tubes Failure Investigation

Upon removal of two of the damaged tubes, borescope imaging of the tube OD surfaces towards the upstream side showed thick external deposits identified as calcium phosphate, see Figure 2.

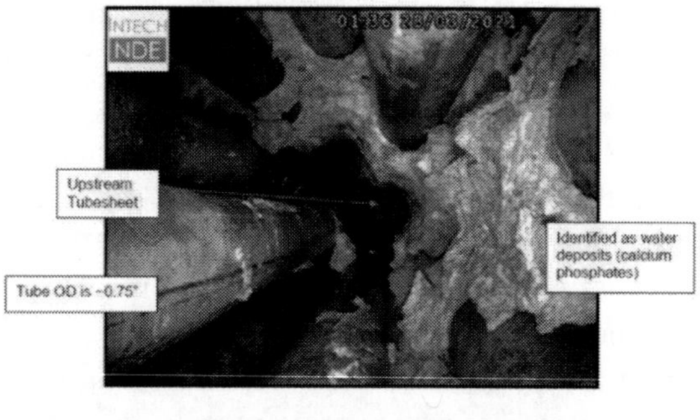

Figure 2: Borescope image of Tube OD surfaces near the upstream.

Failure analysis was conducted on three of the tubes, with two tubes from the areas near the tube inlet (upstream) and one tube from the tube outlet (downstream) for comparison. Wall holes with thin pale deposits were observed on the tube OD near the upstream as shown in Figure 3 while the downstream tube was in good condition.

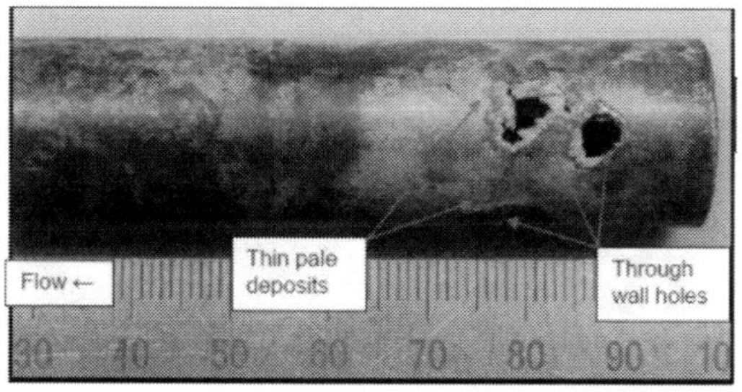

Figure 3: Tube OD, upstream failure.

Metallographic cross-sections of the damaged tubes showed corrosion losses, which were wider on the OD than the ID, with pits filled with scales. Zirconium oxide layers, about 2μm thick, were found on both OD and ID surfaces. However, the undamaged tube did not show any wall pit with a base microstructure comprising the expected polygonal alpha- Zr and various second-phase particles.

The tubes failed due to externally initiated pits from aggressive corrosion conditions created under deposits.

Shell Failure Investigation

Failure analysis was carried out on two cutouts from the Type 304L stainless steel shell, as shown in Figure 4. Initial Cutout A included the through-wall leak that was repaired by a flush patch. Cutout B was executed following additional pitting indication. Cutout B included part of the flush patch used to repair Cutout A. Visual inspection showed many deep pits on both cutout ID surfaces and visible cracks on the OD surfaces.

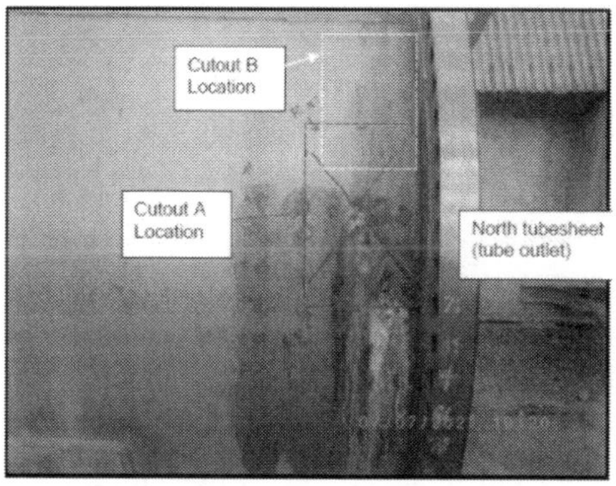

Figure 4: E208 Shell Cutout Locations.

Fracture surface examination using Optical microscopy was carried out on Cutout B which revealed that the cracks initiated on the ID surface and progressed outward to the OD. Metallographic examination of the cross sections also showed multiple cracks originating from the ID surface. Majority of the cracks initiated from the base of the pits on the ID surface. These pits were filled with deposits with the crack morphology as fine, branched, and transgranular consistent with chloride stress corrosion cracking as shown in Figure 5. The microstructure comprises of polygonal austenite and delta ferrite.

The damages on the shell ID traces the same path as the thick deposit formation.

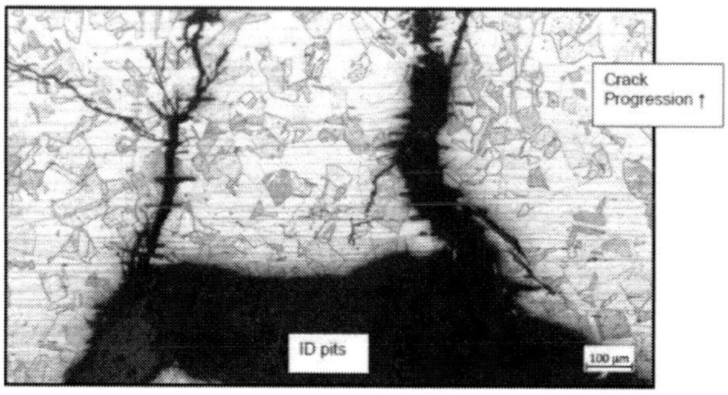

Figure 5: Cutout B, ID pits and cracks, 10% oxalic acid etch.

Process Changes & Contributing Factors

The increased intermittent operation of the Nitric Acid production led to aggregated conditions. Frequent startup and shutdown of the plant presented high occurrences of precipitate phosphate in the cooling water. Lower calcium concentration (as low as 100 ppm) in the cooling water during startups results in the addition of high amount (as high as 46 ppm) of phosphate, used as a corrosion inhibitor, to balance the water chemistry. This created high phosphate concentration in the water and assisted in the formation of deposits overtime when in contact with low pH and high operating nitric acid temperature.

Repair Strategy

Thirty-five (35) tubes, as seen in Figure 6, showed greater than 70% wall loss from the Eddy current test and were plugged. A 15-psi air test was conducted to ensure the plugs were adequately sealed.

Figure 6: Thirty-five (35) tube plugs on the northwest upper quadrant of E-208.

Following the repairs on the failed tubes, an attempt was made to wash the calcium phosphate precipitates formed inside the shell with formic acid, which was not completely effective. Large amount of deposit remained and was revealed when the damaged shell area was cutout for repairs. The deposit hardened, causing the formic acid solution to take the least resistant path while circulating in the shell. A mechanical cleaning method was eventually utilized to conduct the shell repairs, and Cutouts A and B on the shell were repaired with a flush patch.

Due to the depth of the deposit still present on the shell, the exchanger was rotated 180 degrees, since it was built symmetrically, to improve the performance and integrity. The existing north end, which received hot nitric acid, was flipped to the south end, where it discharges cold nitric acid to the absorber and vice versa for the cooling water piping.

Conclusions

Commercially pure zirconium alloys are widely used as tube materials in industry to resist nitric acid corrosion. However, external deposits can lead to aggressive corrosion due to depletion of the protective oxide layer. Adequate water chemistry program will prevent the development of precipitated calcium phosphate and localized chloride concentration formation.

Recommendations

The experiences gained from the failures have influenced the redesign of the exchanger to ensure an adequate flow path to the hottest side. It is very important to have an efficient cooling water flow in every corner of the exchanger. The new design is a combination of counter-current and co-current configurations including a dual strainer on the inlet of the cooling water using in-kind materials of construction.

Symmetrical exchanger design is very effective and to be considered in future designs.

Other recommendations to alleviate future failure include having a reliable cooling water treatment program, performing frequent acid washing of the exchanger to reduce the accumulation of solids before it solidifies and conducting a recommended scheduled inspection of the exchanger.

Synthesis gas compressor oil-seal failure with fire

In 2023, the Yara ammonia plant in Cubatão Brazil experienced an electrical power outage, resulting in limited steam availability. The syngas compressor tripped, and its lube and seal oil supply were impacted as well since both the steam turbine-driven and motor-driven oil pumps became unavailable. The oil sealing of the compressor stages could no longer be maintained. With the machine still partially pressurized, syngas passed through the seals towards the oil tank and vents, resulting in synthesis gas release and fire from the oil tank vent. The fire could be controlled and extinguished and, fortunately, did not result in any harm or damage. This paper analyzes the event and highlights the risks and preventive measures that need to be taken.

Carlos Dias, Marco van Graefschepe, Robson Grunow
Yara

Introduction

At the Yara Cubatão 2 site in Brazil an electrical blackout occurred, which resulted in limited steam availability. The syngas compressor train in the ammonia plant was impacted as the lube and seal oil pump drivers were no longer provided with steam and electricity. This resulted in a compromised oil seal system, with gas passing through the seals and eventually to atmosphere via the oil tank vent, where a fire occurred.

In the following paragraphs, the system will be described, the incident analyzed, and mitigating actions listed. The implemented and pending actions should ensure that this event will not repeat itself. There was no harm or damage caused by this incident, but this could have been very well the case if the fire further spread.

Description of the facility

The synthesis gas compressor is installed in the ammonia plant of the Cubatão 2 industrial complex, located in the city of Cubatão, within the state of São Paulo – Brazil. The plant has a Foster Wheeler design, with an original capacity of 450 MTPD, inaugurated in June 1970. The original feedstock was naphtha, which later was changed to refinery off-gas in 1983 and finally since 2018 to natural gas. Casale revamped the plant in 2001, and it currently has a production capacity of 600 MTPD.

Description of the event.

On March 6th 2023, at 7pm, an electrical blackout occurred at the Cubatão 2 production complex, causing the shutdown of all production plants simultaneously.

During the operational maneuvers following the blackout, a fire was observed in the ammonia plant at the atmospheric vent of the synthesis gas compressor oil reservoir. Firefighting actions were initiated and eventually succeeded after the compressor's gas intake valve could be closed.

Given the severity, it was decided to mobilize those accountable for turbomachinery engineering, process safety, process engineering and production and maintenance management after the incident to verify and understand the situation.

After evaluating the operational conditions, it was decided to initiate a cold production shut down for maintenance of the ammonia plant to correct the damage caused by the lack of electricity and assess the potential damage caused by the fire. Figure 1 shows the timeline of the event.

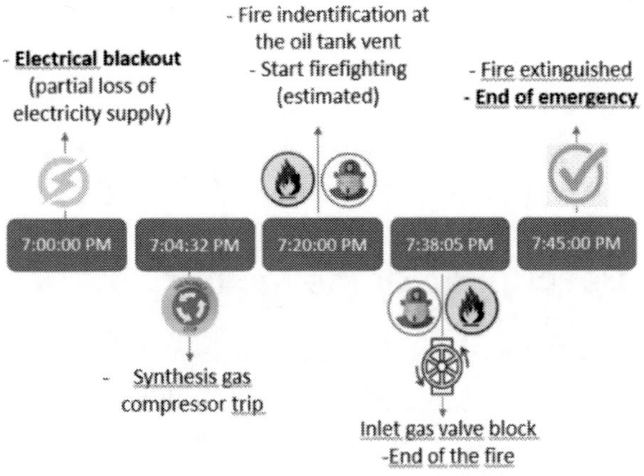

Figure 1. Timeline of the event.

After a detailed analysis of the reports collected from those involved in the occurrence and the careful evaluation of the equipment's process variables and construction conditions, a failure analysis study was carried out using the fault tree method. The results indicated there were no damages to the compressor and seals, and it became clear how the fire had occurred by comparing the operational data under normal conditions versus upset conditions and linking the main events from the blackout until the end of the emergency.

Design and layout of Synthesis gas compressor and related systems.

Synthesis compressor.

The synthesis gas compressor train consists of 3 compressor casings, of which the 3rd one also houses the recycle stage, manufactured by Clark/Dresser. Driven by an 11.2 MW (15,000 HP) Siemens steam turbine.

Figure 2. Sectional drawing of the synthesis gas compressor.

	1st Stage	2nd Stage	3rd Stage	Recycle
Model	2BC9	2BF8	272B7+2R	
Series	2-9-1812	2-8-1813	272-9-1814	
Inlet pressure	21 bar	55 bar	104 bar	125 bar
Discharge pressure	54 bar	105 bar	130 bar	147 bar
Flow	65000 Nm3/h			225000 Nm3/h
Seal type	Oil floating ring	Oil floating ring	Oil floating ring	
Construction	Vertically split	Vertically split	Vertically split	

Table 1. Design and operating data for the different stages of the synthesis gas compressor.

System configuration and normal operation.

Figure 3 shows the overview of the compressor and its lube and seal oil system. In normal operation, the lube oil pumps supply oil to the bearings and couplings and the suction of the seal oil pumps, which in turn supply oil for the compressor seals and guarantee the level of the overhead tanks.
The control valve (PDIC1338) ensures that the sealing oil pressure is 14 bar above the recycle discharge pressure and the control valves (LIC-1323/1324/1325) make sure that the oil sent to the seals is at a pressure of 400kPa (5.8 psi) above the reference gas pressure of the seals. This ensures hydrodynamic balance within the seal, preventing the synthesis gas to leak into the atmosphere.

One part of the oil that passes through the seal returns to the console along with the lubricating oil. The other oil encounters the gas and flows to the degassing system, after which the oil returns to the reservoir, and the released gas passes back to the compressor suction.

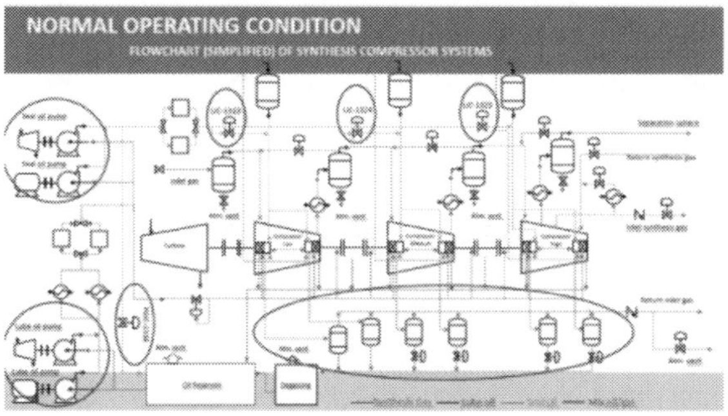

Figure 3. Layout of the synthesis gas compressor and its lube and seal oil system.

Details and functioning of the oil seal.

Figure 4 shows the lay-out and functioning of the oil seal system. Under normal operating conditions, the internal pressures of the seals are equalized in chamber "**D**", and compensated in chamber "**S**", ensuring that both seals (suction and discharge) have the same internal pressure.
This condition causes the gas arriving in chamber "**A**", where the pressure reference for the seal is taken, to have the same pressure in both seals.

In turn, the control valves use the reference pressure to ensure that the oil injected into chamber "**B**" is 400kPa (5.8 psi) above the gas pressure. The oil that flows from chamber "**B**" to chamber "**A**" ensures compressor sealing (oil pressure greater than gas pressure) and after contact with the gas it is directed to the degassing system.

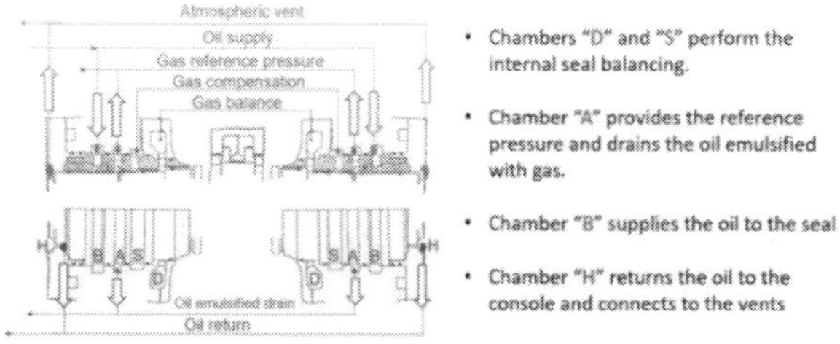

Figure 4. Lay-out of the oil seal system for one compressor stage.

Situation during the site event

Figure 5 shows the overview of the compressor and its lube and seal oil system during the event. After the site electrical shutdown, the energy source for the stand-by oil pumps' motor drives was compromised, making their operation unfeasible.

The lack of electricity also affected the steam generation of the utilities plant, which had its capacity reduced due to the trip of one of the two steam boilers and causing the cut-off of the steam supply to the ammonia plant. The steam generated internally by the ammonia plant became insufficient to keep the oil pump turbines running, causing the oil supply for the lubrication and the seals to cease.

Without a constant supply of oil caused by the lack of main and spare pumps, the elevated oil tanks began to supply the remaining oil to the seals, consequently lowering their levels.

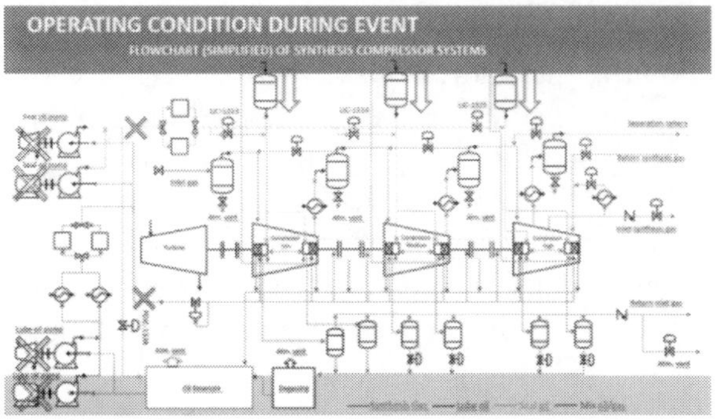

Figure 5. Lube and seal oil system during the event.

Figure 6 shows the oil seal flows during the event. When the overhead tank completely loses the oil level, the seals run out of barrier fluid. The gas contained inside the compressor (13.1 MPa (1900 psi) then invades the oil chambers ""A"" and ""B"" and the drain tubes ""H"" of the seal, which are shared between bearings and couplings.

Consequently, the gas pressurizes the oil return pipes that connect to the reservoir, to the coupling covers and the degassing system and escapes into the atmosphere through the system vents, entraining part of the oil along with it.

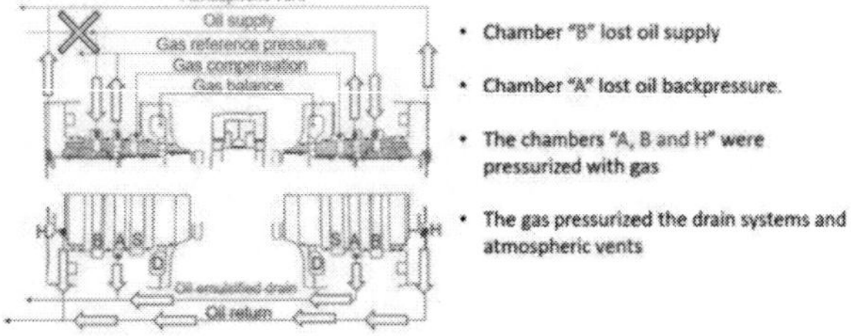

Figure 6. Oil seal flows during the event.

Timeline of the event.

After further analysis of the event, additional details in the timeline were discovered, which can be seen in Figure 7 below.

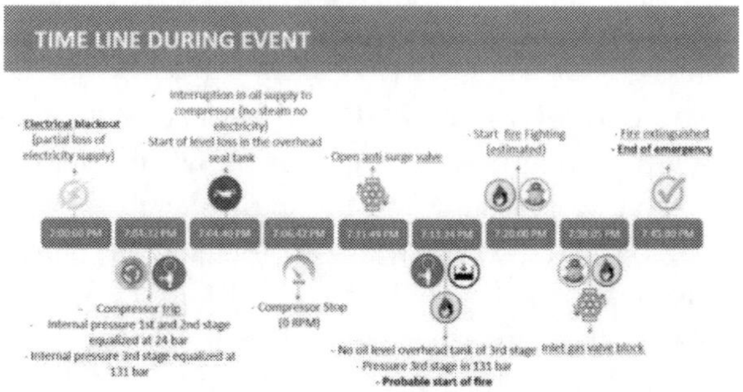

Figure 7. More detailed timeline during the event.

Evaluation and actions.

With the loss of the barrier fluid and the pressurization of the oil system with synthesis gas coming from the compressor's internal pressure, the gas escapes together with the oil through the atmospheric vents, creating conditions conducive to the start of a fire.

The high concentration of hydrogen in the synthesis gas coming into contact with any of the various heat sources and hot surfaces close to the vent of the oil reservoir (turbines and pipes with damaged insulation) likely caused the leaking gas to ignite, allowing the start of the fire.

As the facility does not have an automated firefighting system or nitrogen purge at the oil tank, the fire was only controlled and extinguished after closing the compressor suction valve, thereby removing the fuel.

Fortunately, the fire did not spread to other fuel sources, like oil residues on the pump skid or, in worst case, the oil from the oil tank itself. This would have led to a large fire and extensive damage to the plant.

Immediate/short-term actions after the event:

- Reroute atmospheric vents that may contain synthesis gas to suitable locations, i.e. from oil reservoir and coupling covers. See Figures 8, 9 and 10.
- Degas the entire oil inventory with thermo-vacuum equipment.

Figure 8. Oil tank venting under the pipe rack

Figure 9. The oil tank vent was rerouted to a safer location.

Figure 10. Coupling cover vents were also rerouted.

Mid/Longer-term actions:

- Installation of a water mist firefighting system connected to a fire detection system.
- A hydrogen detection sensor and a nitrogen smothering system should be provided in the oil tank.
- Automation of the manual compressor suction valves and interstage vessel drains for rapid depressurization of the gas system to a safe location in case of loss of oil seal supply (including logic modifications)
- Installation of an automatic surge control system.
- Re-evaluation of steam and power supply during different upset and emergency scenarios.

Conclusion

During the blackout of the unit, the oil sealing of the synthesis gas compressor was compromised due to the fact that both the electrical power and the steam as energy source for the lube and seal-oil pump drivers became unavailable.

The standard electrical power supply for the standby seal oil pump was no longer available, and there was no emergency backup from another electrical source.

During the blackout scenario, the steam to maintain the equipment was also undersupplied by the utilities area.

There was not enough time for the operator to perform manual depressurization during the abnormal blackout condition, leading to the flow of gas towards the oil tank and escape via the tank vent to the atmosphere, resulting in the fire.

A contributing factor was that the vent of the oil tank was in congested area and did not have a system to suffocate/dilute gases leaking into the atmosphere, using the supply of steam or nitrogen.

Fortunately, there was no harm or damage caused by the incident.

The implemented and pending actions should ensure that this event will not repeat itself.

Maritime Spills of Cold/Refrigerated Ammonia: Thermodynamic Model for Predicting Behavior

With an average global demand of nearly 100 million tonne per annum (MTPA), hydrogen remains a critical industrial gas. It has a very low boiling point and exhibits the highest energy density on a mass basis among all known substances, making it an efficient and lightweight energy carrier. Unfortunately, such characteristics also make the storage and transportation of hydrogen challenging and expensive. Ammonia provides a solution by readily reacting with hydrogen to act as a surrogate molecule with a much higher boiling point, capable of storing and transporting hydrogen within a liquid state. A pragmatic hydrogen economy would require large quantities of ammonia to be transported globally; while the safety of such transit is constantly improving, accidents remain an undeniable possibility. During marine transportation, semi-refrigerated barges are used for inland waterway transport of ammonia; and upon catastrophic perforation of a ship/barge tank, a small fraction of ammonia is expected to immediately flash, while the majority will spill into the ocean/river/lake waterbody. Liquid ammonia is fully miscible in water, but its low boiling point suggests that a small fraction will boil-off and/or evaporate. Herein, we present the fate of an ammonia spill within a riverbed through a thermodynamic model developed in Microsoft® Excel™. Our model yields the time evolution of the spill, capable of predicting the mass distribution of surviving ammonia within the liquid and vapor phases. Ultimately, the objective of our model is to aid in predicting the ecological impact of maritime ammonia spills from a health and safety standpoint.

Christopher Rzepa, Seshu Dharmavaram, Byung Hee Ko, and Vipul Parekh
Air Products, Allentown, PA, USA

Introduction

In 2022, global hydrogen demand had reached 95 million tonne per annum (MTPA) (105 short ton per annum STPA), exhibiting a nearly 3% increase year over year with an anticipated growth to 400 MTPA (441 STPA) by 2050. [1] This demand has been almost exclusively driven by global energy trends, with hydrogen being primarily consumed by the oil and gas industries. Hydrogen production is facilitated by fossil fuels, which accounted for as much as 900 MTPA (992 STPA) of CO_2 emissions to meet the hydrogen demand within 2022 alone. Mitigation efforts to meet climate goals has consisted of carbon sequestration projects [2], and net-zero (NZE) hydrogen generation through electrolysis. [1, 3] The feasibility of a low emission hydrogen economy heavily relies on international trade, with estimates of a NZE economy by 2030 requiring 20% of total merchant demand to be met by international trade. [1] A response to this prediction has been the announcement of approximately fifty terminals and related infrastructure for hydrogen and hydrogen-based fuels, striving to be operable by 2030. [1] Maritime transport therefore plays an integral role within a NZE hydrogen economy.

In addition to being a commercial feedstock, hydrogen is an established carrier of energy, exhibiting the highest energy density on a mass basis among all known substances. The other properties of hydrogen include being a colorless, odorless, and tasteless substance that is lighter than air and which has the second coldest normal boiling point of -252.78°C (-423°F), making the storage and transportation of hydrogen challenging, expensive, and highly demanding of expertise and procedures in process safety. Two-way liquid hydrogen carriers (LHC) are surrogate molecules with the capability to safely and efficiently store/release hydrogen on-demand. Ammonia is recognized as an exceptional good LHC, owing to its relatively high standard boiling point of -33°C (-27.4°F), ability to hydrogenate/dehydrogenate efficiently, and its versatility as a chemical commodity. Being a global fertilizer reinforces the presence of an already established ammonia transportation infrastructure. During inland waterway transport using existing barges within the USA, ammonia is typically stored as a refrigerated liquid at temperatures as warm as -22°C (-7.6°F) with relief valves set approximately at 0.68 atm (10 psig). While the safety of such transit is constantly improving, accidents remain an undeniable possibility. If a barge wreck and catastrophic failure occurred, under these conditions the entire contents of ammonia would be released. Approximately 4% of the ammonia will immediately flash upon discharge to atmosphere, leaving the remainder as a liquid spill. Liquid ammonia, with a density of about 0.684 g/cm^3 (42.7 lb/ft^3) at the normal boiling point, is buoyant in water, and fully miscible. Therefore, the liquid spill will rapidly spread across the surface of the river while it is boiling and mixing with the water.

In this paper, we present the evolution of an ammonia spill within a freshwater riverbed from a thermodynamic perspective. It has previously been shown that anywhere between 55% [4] - 70% [5] of the ammonia spilled into freshwater will be absorbed and dispersed into the water. A thermodynamic model computable with Microsoft® Excel™ was developed. It yields the time evolution of the spill and can predict the mass distribution of surviving ammonia within the liquid and vapor phases. Ultimately, the objective of our model is to aid in predicting the ecological impact of maritime ammonia spills from a health and safety standpoint.

Spill Model

A model for miscible buoyant spills, based on material balance equations for a control volume consisting of the ammonia remaining in the liquid phase plus the water which has been entrained into the spill by turbulent mixing, was developed for the United States Coast Guard. Called the HACS-R model [6]. It assumes that the spill control volume is well mixed and cylindrical in shape. The HACS-R model has been validated experimentally for buoyant spills using ethanol in water. Good agreement between the modeling and experimental results were obtained. [6] The HACS-R model includes the effect of vaporization of low volatility spills using a mass transfer coefficient. In the case of ammonia, the model must be enhanced to account for boiling and, after boiling stops, to account for rapid evaporation. Further, in case of an accidental release, the initial shape of the spill is unknown; and it is necessary for a given spill volume to run the model with different spill radii to examine the sensitivity. Once the volume and radius are specified, the initial depth of the spill is fixed. This paper presents an extensive enhancement of the HACS-R model to account for the boiling and rapid evaporation of ammonia from the mixture. Equations that are equivalent to those in the HACS-R documentation have an "H" affixed to the equation number. Equations used in the enhanced ammonia model do not. Table 1. defines the symbols used within the corresponding equations implemented within our model.

The control volume of the spill is represented as a cylinder whose volume increases with time as the river current perpetuates mixing. The depth of the spill "h_s", can therefore be computed at any given time from the total mass of the spill and the mixture density. The subscript 1 denotes water and 2 denotes ammonia.

Equation 1H

$$h_s = \frac{M_1 + M_2}{\pi \rho_m r^2}$$

The liquid-phase mixture density, ρ_m, is computed using the lever rule, neglecting any volume change upon mixing.

Equation 2H

$$\frac{1}{\rho_m} = \frac{x_1 V_{L1} + x_2 V_{L2}}{M_{wm}}$$

In this equation, the mole fractions are:

Equation 3

$$x_1 = \frac{\left(\frac{M_1}{M_{w1}}\right)}{\frac{M_1}{M_{w1}} + \frac{M_2}{M_{w2}}}$$

Equation 4

$$x_2 = \frac{\left(\frac{M_2}{M_{w2}}\right)}{\frac{M_1}{M_{w1}} + \frac{M_2}{M_{w2}}}$$

And the molecular weight of the mixture is:

Equation 5

$$M_{wm} = x_1 M_{w1} + x_2 M_{w2}$$

The spill radius increases with time as a result of the buoyancy driven flow in the gravity inertia regime.

Equation 6H

$$\frac{dr}{dt} = 1.3\sqrt{g h_s \left(\frac{\rho_w}{\rho_m} - 1\right)}$$

Equation (6H) shows how the spreading rate depends on the mixture density ρ_m, through its effect on the gravity wave speed. The mixture density changes with time as water is mixed into the spill. The rate of change of the mass of water in the spill is given by:

Equation 7

$$\frac{dM_1}{dt} = W_1 - W_{e1}$$

The HACS-R model gives the rate of entrainment of water into the spill through the bottom and side of the cylindrical spill as:

Equation 8H

$$W_1 = \pi \rho_w r^2 \left[u_{bot} + 2\left(\frac{h_s}{r}\right) u_{side}\right]$$

The entrainment velocities represent the rate of mixing at the spill boundaries caused by turbulent diffusion. The HACS-R model obtains these velocities from the following expressions:

Equation 9H

$$u_{side} = \frac{2\left(C_f \frac{dr}{dt}\right)^2}{\left(\frac{dr}{dt}\right)_{t=0}}$$

Where the friction coefficient, C_f is given by:

Equation 10H

$$C_f = \frac{3.807n}{R_h^{\frac{1}{6}}}$$

While the entrainment at the bottom of the spill is:

Equation 11H

$$u_{bot} = 1.3\frac{u^*}{6.0 + 0.25 N_{Ri}}$$

Where u^* is the friction velocity and N_{Ri} is the Richardson number, which quantifies the damping of vertical turbulent mixing by opposing buoyant forces. Because the light spill is more stable on top of denser water, turbulence at the boundary is damped. This damping is very strong in the case of an ammonia spill, with the friction velocity estimated from:

Equation 12H

$$u^* = \frac{3.807nu}{R_h^{\frac{1}{6}}} = C_f u$$

The Richardson number is:

Equation 13H

$$N_{Ri} = 0.59\left(\frac{\frac{dr}{dt}}{1.3u^*}\right)^2$$

A material balance on ammonia gives:

Equation 14H

$$\frac{dM_2}{dt} = -W_{e2}$$

In equations (7) and (12H) appear the rates of vaporization of water and ammonia, $W_{e1}$and W_{e2}, respectively. For ammonia spills, computation of these terms is more complicated than in the original HACS-R model. While the spill is boiling, the total boiling rate, W_e, is determined from an energy balance and the constraint that at all times the spill is perfectly mixed with its bubble point equal to one atmosphere.

Equation 15

$$W_e = W_{e1} + W_{e2}$$

This means that boiling occurs as fast as the water entrained into the spill can supply the enthalpy required for vaporization. Neglecting PV terms for liquids, the energy balance on the spill becomes:

Equation 16

$$\frac{dH_s}{dt} = W_1 h_w - W_e h_v$$

Where H_s is the total enthalpy of the spill, h_w is the enthalpy per unit mass of pure water, and h_v is the enthalpy per unit mass of the bubble point vapor which is boiling out of the spill. The vapor phase enthalpy was calculated by applying the departure function with the Peng Robinson equation of state. The enthalpies of pure components in the liquid phase were calculated by integration of the constant pressure heat capacity as a function of temperature, provided by the Design Institute for Physical Properties (DIPPR) v4.1.0. [7] Accordingly, the corresponding temperature of the spill was calculated by applying Newtons method on the following relation:

Equation 17

$$H_s = H_1 x_1 + H_2 x_2 + \Delta H_{mixing}$$

The enthalpy of mixing (or excess enthalpy) was calculated using the excess Gibbs energy.

Equation 18

$$\Delta H_{mixing} = -RT^2 \frac{\partial g^E}{\partial T}|_P$$

We elaborate on our calculation of the excess Gibbs energy within our discussion of the activity coefficients below.

During the integration of the time-dependent equations, the algorithm for finding the boiling rate W_e is a pair of nested Newton-Raphson iterations. A boiling rate is guessed and used in energy balance Eq. (16) to determine the total enthalpy of the spill at the end of the time step, and in the material balances to determine the composition and mass of the spill at the end of a time step. From the total enthalpy and composition, the temperature of the spill is computed by Newton-Raphson iteration using the known relations between pure component enthalpies and temperature, equation (17). Once the temperature is known, bubble point pressure and composition can be computed. Another Newton-Raphson procedure is used to adjust the boiling rate until the calculated bubble point pressure equals one atmosphere. Using thermodynamic models for ammonia/water mixtures, the bubble point pressure P is calculated to satisfy:

Equation 19

$$\sum_{i=1}^{2} P_i = P$$

The boiling vapor mole fractions are given by:

Equation 20

$$y_i = \frac{P_i}{P}$$

Where the partial pressure of ammonia and water are given by:

Equation 21

$$P_i = P_i^v \gamma_i x_i \Phi_i' EXP\left[\frac{V_{Li}}{RT}(P - P_i^v)\right] / \Phi_i$$

In equation (21), P_i^v is the vapor pressure of component "i", which was determined by the extended Antoine equation, with parameters given by DIPPR.

Equation 22

$$P_i^v(T) = EXP\left[A_i + \frac{B_i}{T} + C_i \log[T] + D_i T^{E_i}\right]$$

The activity coefficient, γ_i, within a binary mixture is related to the ratio of the liquid phase fugacity coefficient of the mixture, over the liquid phase fugacity coefficient of the pure component at the conditions of the mixture.

Equation 23

$$\gamma_i = \frac{\widehat{\emptyset_i^L}(T, P, x)}{\emptyset_i^L(T, P)}$$

The liquid phase fugacity coefficient of the pure component at the conditions of the mixture, $\emptyset_i^L(T, P)$, is provided by the Poynting corrected saturation pressure:

Equation 24

$$\emptyset_i^L(T, P) = \frac{P^{sat}}{P} \emptyset_i^L(T, P^{sat}) * EXP\left[\frac{v_i(T)(P - P^{sat})}{RT}\right]$$

The liquid phase fugacity coefficient of the pure component at its saturation conditions, $\emptyset_i^L(T, P^{sat})$, was fit to the DIPPR expression of the liquid phase saturation pressure using the data of Tillner-Roth and Friend [8]:

Equation 25

$$\log\left(\emptyset_i^L(T, P^{sat})\right) = A + \frac{B}{T} + Clog(T) + D * T^E$$

These expressions were used to calculate the activity coefficient γ_i of component "i" within the mixture; and were subsequently fit to the Redlich-Kister expression for the excess Gibbs energy:

Equation 26

$$g^E/RT = x_1 x_2 [A + B(x_1 - x_2) + C(x_1 - x_2)^2 + D(x_1 - x_2)^3]$$

Each Redlich Kister constant was fit against temperature based on the expression:

Equation 27

$$A = C_1 + C_2 * T + \frac{C_3}{T} + \frac{C_4}{T^2}$$

The pure component liquid saturation fugacity coefficient, Φ_i', as well as the mixture fugacity coefficients of water and ammonia in the vapor, Φ_i, were calculated using the Peng Robinson equation of state. Standard mixing rules were applied, and the interaction parameter was absent. The specific volume of pure component "i", at the conditions of the mixture were calculated using DIPPR expressions for the liquid phase density of each component:

Equation 28

$$\rho_1(T) = A + BT + CT^2 + DT^3 + ET^4$$

Equation 29

$$\rho_2(T) = \frac{A}{B^{1+\left(1-\frac{T}{C}\right)^D}}$$

We would like to emphasize that the density of pure liquid water was extrapolated below its standard freezing point. The trend within this hypothetical region should exhibit a decrease in the specific volume with temperature, which was ensured by fitting the functional form of equation (29) to the data of Wagner and Pruß [9]

Ultimately, equation (16) dictates the boiling rate within the Newton-Raphson scheme. Initially, a trial boiling rate for the mixture W_e is guessed; and a corresponding bubble point pressure is calculated based on equation (22). The boiling rate is then iteratively changed based on the Newton-Raphson formalism until a bubble point pressure criteria of $P = 1\ atm$ is reached. During the forward time integration, once a sufficient amount of ammonia has boiled off and/or enough water has been entrained in the control volume, the bubble point pressure of the mixture equation (19) falls below the ambient pressure $P < 1\ atm$ and boiling ceases. At this point, the vaporization of ammonia is now controlled by mass transfer across the gas-liquid interface at the top of the spill. Mass transfer controlled evaporation is much slower than boiling; however, it is important to estimate this rate to determine whether a significant airborne source term exists after boiling stops. Mass transfer across the interface occurs in two steps in series - from the bulk liquid to the interface and from the interface to the bulk air. Vapor liquid equilibrium exists only at the interface. Neglecting any evaporation or condensation of water, mass transfer coefficients are for transfer of ammonia in a non-transferring medium. The molar flux of ammonia (mol per unit time per unit area) from the bulk liquid to the interface is given by:

Equation 30

$$N_2 = k_{x0}/x_{1lm}\left(x_2 - x_2^i\right)$$

The symbol k_{x0} is the mass transfer coefficient at infinite dilution. It is estimated from the friction velocity in the river according to the Colburn analogy between mass transfer and momentum transfer:

Equation 31

$$St = \frac{f}{2} S_{C_L}^{-\frac{2}{3}}$$

Equation 32

$$\frac{k_{x0} M_{wm}}{u \rho_m} = \frac{f}{2} S_{C_L}^{-\frac{2}{3}} = \left(\frac{u^*}{u}\right)^2 S_{C_L}^{-\frac{2}{3}}$$

Equation 33

$$k_{x0} = \left(\frac{\rho_m}{M_{wm}}\right) \frac{u^{*2}}{u S_{C_L}^{\frac{2}{3}}}$$

The Schmidt number S_{C_L} for ammonia was calculated by the ratio of the viscous diffusion rate over the molecular diffusion rate:

Equation 34

$$S_{C_L} = \frac{\mu}{\rho D}$$

The kinematic viscosity μ_i for each component was taken from DIPPR:

Equation 35

$$\mu(T)_i = EXP[A_i + \frac{B_i}{T} + C_i log[T] + D_i T^{E_i}]$$

While the kinematic viscosity of the mixture was calculated by the method of Grunberg and Nissan [10]:

Equation 36

$$\mu(x_i, x_j) = EXP[x_i \log[\mu_i] + x_j \log\,[\mu_j]]$$

The kinematic viscosities were conservatively taken to be at room temperature, even though spill temperatures will be considerably colder. The diffusivity, D, of the liquid mixture was calculated by the method of Wilke-Chang [10]:

Equation 37

$$D = \frac{7.4 * 10^{-8} (\varphi M_{w1})^{\frac{1}{2}} T}{\mu_B v_2^{0.6}}$$

Where φ is the association factor of water (suggested value of 2.6). The flux from the bulk liquid to the interface must match that from the interface to the bulk air, given by:

Equation 38

$$N_2 = k_{GO}(P_2^i / P_{alm}/P)$$

The infinite dilution mass transfer coefficient was estimated from a turbulent boundary layer equation by Sherwood et al. [11] In the present notation, this equation becomes:

Equation 39

$$k_{G0} = 0.037 u_w Re^{0.2} S_{C_a}^{-\frac{2}{3}} / RT_a$$

Where Re is the Reynolds number based on the wind velocity u_w, the spill diameter $2r$, T_a is the ambient air temperature, and S_{C_a} is the Schmidt number for ammonia diffusing in air. Equation (39) is valid up to a Reynolds number of 300,000. Except in very calm air, the Reynolds number for air flowing over a large spill will be much greater than 300,000. In all the case study calculations, a wind speed of 10 ft/s (3.05 m/s) was used for evaporation estimates, which is conservative because less evaporates at lower speeds.

Analogous to the boiling scheme, a Newton-Raphson iterative procedure was used to find the unknown interface composition for which the fluxes calculated from equations (30) and (38) matched. Results showed that the interface composition was very close to the bulk spill composition, indicating that most of the mass transfer resistance is in the gas phase. Furthermore, the calculated evaporation rates were much lower than the boiling rates, typically two to three orders of magnitude in this study. Once boiling stops, the generation of significant airborne ammonia is essentially finished.

Emissions Estimate

Calculations of the fate of instantaneous spills from a barge wreck into a freshwater river were carried out using the model equations above. A Microsoft® Excel™ macro was written to solve the equations in a user-friendly format, where conditions of the barge (mass, pressure and temperature), riverbed (temperature, current, width, roughness), and atmosphere (wind velocity, ambient temperature and pressure) may be input by the user.

Herein we discuss the predictions of our model for a river with typical characteristics: 91.44 m (300 ft) width, 6.10 m (20 ft) depth, roughness of 0.008 $m^{1/6}$ (0.01 $ft^{1/6}$), current of 0.30 m/s (1 ft/s), and temperature of 10°C (50°F). The amount of ammonia was assumed to be 2500 metric ton (2756 short ton), which is typical for freshwater barges. Liquid ammonia was assumed to be stored at -22°C (-7.6°F) under 0.68 atm (10 psig) of pressure. Under these conditions, a catastrophic rupture of the ammonia tank would result in an instantaneous flash, releasing 4% of the ammonia to the atmosphere. The remaining ammonia will have spilled into the freshwater, now cooled to it's normal boiling point at -33° C (-28 F) but rapidly boiling as warm water mixes with the spill. The spill will of course never be perfectly cylindrical, and its effective initial radius depends on the manner in which the barge would break open. We assumed an initial spill radius of 30.48 m (100 ft).

Figure 1 shows the time evolution of the surviving ammonia spill as predicted by our simulation. The initial depth of the spill is 1.2 m (3.9 ft). The spill boils for about 8 min, during which time the spill radius has spread to a calculated value of 289 m (947 ft) and its depth has reduced to 0.018 m (0.06 ft). Note that the calculated radius is much larger than the width of the river, this is the result of the spill shape shifting from cylindrical to rectangular as the current stretches the spill downstream. To a first approximation, the spill would have a similar volume, concentration, and boiloff rate as the equivalent cylindrical spill. Constrained by the riverbank, the spill would actually entrain water at a lower rate than an unconfined cylinder. This means that the time period over which the ammonia boils off should be longer than calculated, but the total amount boiled off would be the same. When the boiling stops at about 8 min, the mass of ammonia remaining in the spill is 1837 metric ton (2025 short ton). The total mass of ammonia which either flashed off instantaneously or subsequently boiled off is 663 metric ton (731 short ton), about 27% of the initial 2500 metric ton (2756 short ton) spill. This prediction is consistent with the work of Dharmavaram et al. [5], which was the basis for our model, and who predicted 28% of ammonia had boiled off under the same conditions. After the boiling stops, ammonia continues to evaporate very slowly under mass transfer controlled conditions. Beyond that point, the evaporation rate is essentially zero. In addition to the mass of surviving ammonia within the riverbed, Figure 1 also shows the temperature of the spill, as well as the corresponding amount of ammonia to enter the atmosphere. One can see that the amount of ammonia released into the atmosphere is necessarily the amount of ammonia to escape the liquid phase, as dictated by the mass balance of equations (7), (14H) and (15). Moreover, as boiling ceases and freshwater entrains the control volume, the spill temperature is guaranteed to equilibrate at the temperature of the riverbed.

In a separate work, Dharmavaram et al. [4] had studied the fate of 136 kg (300 lb) of unpressurized ammonia at -34°C (-29.2°F) discharged into 1000 kg (2205 lb) of water contained within a static bund. The results of those experiments measured 45% of ammonia had boiled off, while our model predicted 22% under identical conditions. We would like to emphasize that within Dharmavaram et al. [4], the discharge of ammonia occurred above water, into a small 16 m^2 (172 ft^2) bund, with the release being characterized as intense: "ammonia puff initially deflects off the water surface and continues moving downwind as a puff" followed by vaporization from an aqueous ammonia pool. Such characteristics contrast the physics within our model, which simulates controlled boiling of a perfectly mixed, quasi-equilibrated solution in a flowing water stream. It is therefore conceivable that our model provides conservative estimates on the amount of absorbed ammonia.

The majority of maritime ammonia transport is anticipated to occur via long distances on oceanic routes. The consequence of which is the impact of ocean water on the solubility of ammonia; and thereby its vapor-liquid equilibrium within our model. Sodium and chloride ions predominate the composition of dissolved salts in ocean water, accounting for approximately 3.5% of the mole fraction. [12] The solubility of ammonia in aqueous solutions of single electrolytes have been studied by Singh et al. [13]; who showed a relatively small reduction in ammonia solubility in the presence of sodium chloride within the aqueous phase. Namely, the addition of 4 mol of sodium chloride into an 8 molar aqueous-ammonia solution increased the corresponding ammonia partial pressure by less than 0.02 MPa (2.9 psig) at 80°C (176°F). Although the application of our model was not intended for ocean water spills; we hypothesize the disparity to be limited.

Conclusions

A thermodynamic model was developed with Microsoft® Excel™ to yield the time evolution of the ammonia mass distribution within the liquid and vapor phases. The model was applied to predict the fate of an ammonia spill into a riverbed from a pressurized and semi-refrigerated tank, commonly used within freshwater barges. Model predictions were also compared against the experimental discharge of ammonia into a stagnant freshwater bund. [4] The model underpredicted the quantity of vaporized ammonia by 23%, likely due to the vigorous discharge of ammonia within the experiment, which contrasted the perfectly-mixed, quasi-equilibrated boiling assumed within the model. Because most maritime ammonia transport is anticipated to occur via long distances on oceanic routes, the impact of ocean water salinity on the model was examined, hypothesizing that the disparity should be minimal.

r Spill radius.
t Time.
h_s Spill height.
ρ_w/ρ_m Density of water/mixture.
M_1/M_2 Mass of water/ammonia.
x_1/x_2 Liquid mole fraction of water/ammonia.
V_{L1}/V_{L2} Liquid phase specific volume of water/ammonia.
$M_{w1}/M_{w2}/M_{wm}$ Molecular weight of water/ammonia/mixture.
g Gravitational constant.
R Universal gas constant.
T Temperature.
W_1 Rate of water entrainment.
W_{e1}/W_{e2} Rate of evaporation of water/ammonia.
C_f Chèzy coefficient in the standard Manning formula for open-channel flow.
N_{Ri} Richardson number.
R_h Hydraulic radius of the river.
u River current.
$H_s/H_1/H_2$ Total enthalpy of spill/water/ammonia.
h_w Enthalpy per unit mass of water.
h_v Enthalpy per unit mass of the bubble point vapor boiling out of the spill.
g^E Excess Gibbs energy.
P Pressure.
P^{sat} Saturation pressure.
y_i Vapor mole fraction of species "i".
P_i^v The vapor pressure of species "i".
P_{alm} Log-mean partial pressure of air between the interface and the bulk air.
P_2^i Partial pressure of ammonia in air at the interface.
γ_i Activity coefficient of component of species "i" within the binary mixture.
$\widehat{\phi}_i^L$ Liquid phase fugacity coefficient of the mixture.
ϕ_i^L Liquid phase fugacity coefficient of the pure component "i" at the conditions of the mixture.
v_i Liquid molar volume of species "i".
Φ_i Pure component liquid saturation fugacity coefficient of species "i".
N_2 Molar flux of ammnia.
x_{1lm} Log mean mole fraction of water between the bulk liquid and the interface.
x_2^i Mole fraction of ammonia at the interface.
k_{x0} Mass transfer coefficient at infinite dilution.
St Chilton and Colburn analogy.
f Friction velocity.
Sc_L Schmidt number.
μ_i Kinematic viscosity of species "i".
D Diffusivity of the liquid mixture.
φ Association factor of water.
k_{G0} Mass transfer coefficient at infinite dilution.
Re Reynolds number.

Table 1: Symbol definitions

Note: The Microsoft® Excel™tool used to implement the thermodynamic model within this work will be available at a later time.

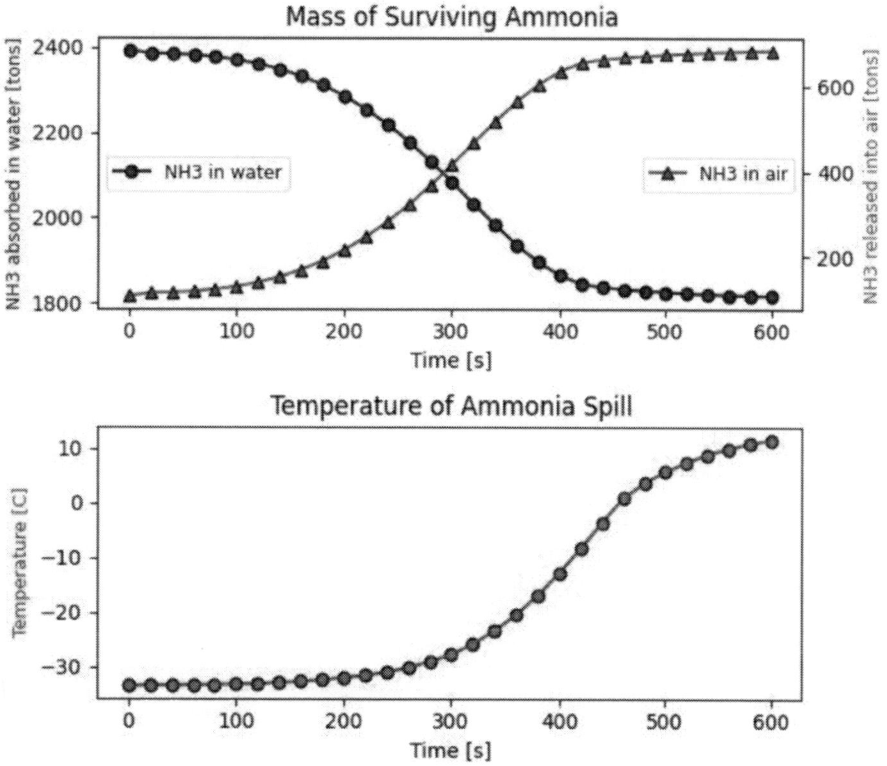

Figure 1: Mass of ammonia absorbed by the riverbed (water) and released into the air; and the temperature of the ammonia spill within the riverbed as a function of time as predicted by the model

References

[1] IEA, "Global Hydrogen Review 2023," IEA, Paris, 2023.

[2] Air Products & Chemicals, "https://www.airproducts.com/energy-transition/louisiana-clean-energy-complex," 11 March 2024. [Online]. Available: https://www.airproducts.com/energy-transition/louisiana-clean-energy-complex.

[3] Air Products & Chemicals, "https://www.airproducts.com/energy-transition/neom-green-hydrogen-complex," 11 March 2024. [Online]. Available: https://www.airproducts.com/energy-transition/neom-green-hydrogen-complex.

[4] Dharmavaram, S., Carroll, M. J., Lutostansky, E. M., McCormack, D., Chester, A., & Allason, D. (2023). Red Squirrel Tests: Air Products' ammonia field experiments. *Process Safety Progress*, *42*(3), 481-498.

[5] Dharmavaram, S., Tilton, J. N., & Gardner, R. J. (1994). Fate and transport of ammonia spilled from a barge. *Journal of hazardous materials*, *37*(3), 475-487.

[6] Dodge, F. T., Buckingham, J. C., & Morrow, T. B. (1982). *Analytical and Experimental Study to Improve Computer Models for Mixing and Dilution of Soluble Hazardous Chemicals* (p. 0274). US Coast Guard Report CG-D-1-83.

[7] American Institute of Chemical Engineers (AIChE), "Design Institute for Physical Properties v4.1.0".

[8] Tillner-Roth, R., & Friend, D. G. (1998). A Helmholtz free energy formulation of the thermodynamic properties of the mixture {water+ ammonia}. *Journal of Physical and Chemical Reference Data*, *27*(1), 63-96.

[9] Wagner, W., & Pruß, A. (2002). The IAPWS formulation 1995 for the thermodynamic properties of ordinary water substance for general and scientific use. *Journal of physical and chemical reference data*, *31*(2), 387-535.

[10] Prausnitz, J. M., Lichtenthaler, R. N., & De Azevedo, E. G. (1998). *Molecular thermodynamics of fluid-phase equilibria*. Pearson Education.

[11] Sherwood, T. K., Pigford, R. L., & Wilke, C. R. (1975). *Mass transfer* (Vol. 23). New York: McGraw-Hill.

[12] Millero, F. J., Feistel, R., Wright, D. G., & McDougall, T. J. (2008). The composition of Standard Seawater and the definition of the Reference-Composition Salinity Scale. *Deep Sea Research Part I: Oceanographic Research Papers*, *55*(1), 50-72.

[13] Sing, R., Rumpf, B., & Maurer, G. (1999). Solubility of ammonia in aqueous solutions of single electrolytes sodium chloride, sodium nitrate, sodium acetate, and sodium hydroxide. *Industrial & engineering chemistry research*, *38*(5), 2098-2109.

Evaluation of Safety Instrumented System in Ammonia Plant using HAZOP, LOPA, and RBD Analysis

The Interlock system is the heart of any petrochemical/gas processing plant, and catastrophic failure could occur due to system failure. In the last two years, there have been several interlock system failures in the Ammonia Plant of Pupuk Indonesia (PIHC), resulting in serious damage and long shutdown periods. This paper aim to verify that the intended safety integrity level (SIL) of a safety instrumented system (SIS) is achieved. The methodology followed to achieve the objective of this study starts with risk analysis, then allocates a SIL using a layer of protection analysis (LOPA) and finally, validates the SIL using a reliability block diagram (RBD). Of the five ammonia plants, there are a total of 501 SIL classified SISs. 443 SISs (88%) meet the SIL target, while the other 58 SISs (12%) do not meet the SIL target. Some of the recommendations made to maintain plant reliability include replacing and adding several sensors, the final elements and making improvements to architectural constraints to meet SIL requirements, with an estimated cost of around USD 720,000.

Helmi Qosim, Muhammad Azmi, Sigit Abdurrakhman, Rifqah Azzahra
PT Pupuk Indonesia (Persero), Indonesia

M. Hilal Amsori
PT Pupuk Sriwidjaja Palembang, Indonesia

Introduction

PT Pupuk Indonesia (Persero) is one of the largest fertilizer producers in Asia. Our group consist of fertilizer, pesticides, chemical products, industrial support, logistics, and trades companies. We currently own a production capacity of 22 million tons per year which consists of 14 million tons of fertilizers per year, and 8.6 million tons of ammonia and other products.

Ammonia, one of the largest-volume synthetic chemicals produced in the world, has a complex and dangerous production process [1]. To avoid hazardous events, ammonia plants must be equipped with safety devices to reduce the risk to an acceptable level. One of the most important protection layers is the safety instrumented system.

A Safety Instrumented System (SIS) is an independent instrument system that is installed to reduce the potential risk of the process. The SIS will include sensors, transmitters, logic solver and final control elements [2]. It is important to ensure the reliability of SIS.

To describe the reliability degree of the safety instrumented system, the SIS-related standard, IEC 61508, puts forward an important concept, safety integrity level (SIL). In IEC 61508, the SIL means the possibility of a safety instrumented function (SIF) implemented by safety-related systems at a certain time and under certain conditions. The safety function will decline due to the influence of the environment and usage of SIS, so it is necessary to formulate corresponding improvement measures after regularly carrying out SIL evaluation [3].

The main purpose of this paper is to evaluate a safety instrumented system starting with risk analysis, then allocate a safety integrity level (SIL) to the SIS and finally, validate of the SIS.

Methodology

The methodology followed to achieve the objective of this study is summarized as follows:

- Risk Analysis
- Classification
- Verification

A. Risk Analysis

Risk Analysis techniques are used to identify hazards and evaluate risk, also determine the appropriate ways to eliminate hazards. Several methods can be applied, one of them is hazard and operability analysis (HAZOP).

HAZOP is a qualitative hazard analysis technique that can be used at any time during the process life cycle, starting from the process development to plant shutdown and for any process modification that needs hazard evaluation.

The advantages of this method are systematic and inclusive. It is used to differentiate between any available deviations from normal operations and check if the appropriate safeguards are placed correctly to mitigate or prevent the occurrence of accidents. To be provided by an inclusive evaluation of the process, many guide words typically no/not/none, more, less, part of, reverse, other than, as well as, are used with different parameters (flow, pressure, temperature, reaction, level, composition), and applied to each pipe, equipment, instrumentation & control of the process [4].

B. Classification

The SIL classification process, which is conducted as part of the Safety Life Cycle, involves assessing the reliability of Safety Instrumented Systems (SIS) according to risk assessment. This classification step is essential for defining the instrumentation function. SIL classification is performed on all interlock systems identified in the Cause and Effect Diagram, excluding functions related to push buttons, fire & gas alarm systems, control, and electrical functions.

The Layer of Protection Analysis (LOPA) is a semi-quantitative technique used for assessing the frequency of undesirable consequences that are preventable [5]. The LOPA method is utilized in this paper. There are observable advantages and disadvantages. On the positive side, LOPA presents a simple approach, enhancing scenario identification through the causal linkages derived from HAZOP studies. It aids in determining the necessary Safety SIS and SIL rating, serving as a tool for Cost and Benefit Analysis during the selection of process safety instrumentation. Moreover, it facilitates risk-informed decision-making across various stages such as design and management of change. LOPA also ensures proper acknowledgment of all protective layers, providing an accurate assessment of unit/equipment risk levels, while simultaneously supporting compliance with process safety regulations. However, drawbacks exist; LOPA necessitates predefined risk tolerance criteria before initiation, and it does not prescribe specific Independent Protection Layers (IPL) [6].

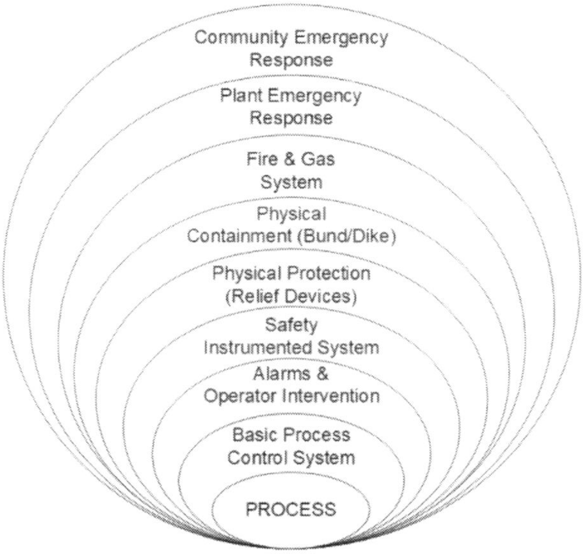

Figure 1. Layers of Protection Analysis

The LOPA method analyzes each protection layer within the instrument process based on the IEC 61511 standard [7]. It identifies the risk probability by calculating the severity consequence of an initiating event with the likelihood of hazardous events. The consequences of an initiating event are analyzed based on three parameters: Safety, Environment, and Asset Loss. The LOPA method in the classification stage will yield Initial SIL results assessed using the risk matrix in Figure 2, alongside the likelihood and consequences criteria shown in Table 1 and 2.

		Consequence / Impact				
		5	4	3	2	1
Likelihood	5	SIL 4	SIL 3	SIL 2	SIL 2	SIL 1
	4	SIL 3	SIL 3	SIL 2	SIL 1	NR
	3	SIL 3	SIL 2	SIL 1	SIL 1	NR
	2	SIL 2	SIL 2	SIL 1	NR	NR
	1	SIL 2	SIL 1	NR	NR	NR

Figure 2. SIL Risk Matrix

Table 1. Criteria of Hazardous Likelihood

Likelihood of Hazard (1/year)	
5	> 0.1
4	0.01 - 0.1
3	0.001 - 0.01
2	0.0001 - 0.001
1	<0.0001

Table 2. Criteria of Consequences Severity

	Personal (S)	Environment (E)	Asset Loss (A)
5	Multiple Fatalities	Leakage exceeds the TLV at high intensity outside the Plant area	>$45.81 M > 3 months
4	Single Fatality	Leakage exceeds the TLV at low intensity outside the Plant area	$29.01 M - $45.81 M 1 - 3 months
3	Permanent Disability	Leakage exceeds the TLV in the Plant area and site	$9.68 M - $29.01 M 12 - 36 days
2	Injuries requiring medical treatment and hospitali-zation	Leakage exceeds the TLV in the Plant area	$32,258 - $9.68 M 3 - 12 days
1	Injuries requiring first aid medical treatment	Leakage within the TLV in the Plant area	< $32,258 < 3 days

The results of the LOPA method are affected by Independent Protection Layers (IPL). An IPL is a device, system, or action, designed to prevent consequences resulting from an initiating event. The initial SIL obtained previously is affected by every protection layer owned by the Safety Instrumented Function (SIF). Each protection layer will provide credits to reduce the probability of hazardous events. Table 3 shows the IPL credit. The more IPLs are identified, the lower the targeted SIL. Sufficient identification of IPLs may result in a SIF not being required.

Table 3. Independent Protection Layer Credit

No	IPL	Credit
1	Pressure Relief Device Non-Potential of Solidification, Polymerization (RDN)	2
2	Pressure Relief Device Potential of Solidification, Polymerization with Measure (RDPM)	2
3	Pressure Relief Device Potential of Solidification, Polymerization without Measure (RDP)	1
4	Safety Instrumented System 1 (SIL 1)	1
5	Safety Instrumented System 2 (SIL 2)	2
6	Safety Instrumented System 3 (SIL 3)	3
7	Basic Process Control System 4 (BPCS)	1
8	Double Check Valve (DCV)	1
9	Dikes	1

C. Verification

The SIL verification process is a calculation involving the Probability of Failure on Demand (PFD) and the architecture of a SIS to ensure the achievement of the target SIL. Verification is conducted to analyze whether each SIF's PFD, classified in the previous stage, meets the target SIL and architectural constraints. Additionally, the determination of the optimum proof test interval for the Classified SIFs is carried out. Any SIF that does not meet the SIL requirements will be recommended for modification. There are two stages in the SIL verification process: verification according to the PFD Requirement and Architectural Constraint Requirement.

To demonstrate that the Probability of Failure on Demand (PFD) meets the target SIL, a reliability analysis is conducted. The SIL verification process for each Safety Instrumented Function (SIF) is carried out by modelling each SIF into a reliability block diagram (RBD). RBD is a reliability system analysis technique used to assess the likelihood of system failures. The primary focus of RBD is a diagrammatic representation that illustrates system components interconnected through a network, where the failure probability of a system can be identified and calculated based on specific conditions.

Figure 3. Reliability Block Diagram

To calculate the RBD, it is important to determine the system's configuration by identifying its components and connections, which may exist in series or parallel arrangements. Subsequently, the RBD diagram should be drawn, utilizing blocks to represent components and lines to signify connections. Parallel blocks denote redundant subsystems or components that contribute to a lower failure rate, while series blocks indicate that a failure in one component will result in the entire system failing.

The RBD method serves as a crucial tool in the verification of SIL due to its numerous advantages. RBD models offer inherent simplicity and ease of establishment, making them highly intuitive. Additionally, these models possess the capability to accurately depict the reliability structure of the system under scrutiny, providing a comprehensive understanding of its operational dynamics. Moreover, the numeric accuracy of the average probability of failure on demand within RBD models is deemed satisfactory, further enhancing their utility in SIL verification processes. Furthermore, the method aligns closely with the principles outlined in IEC 61508-6, offering detailed explanations that contribute to its robustness and reliability in safety assessment endeavors [8].

Case Study

The risk analysis case study in this paper examines the air compressor node within PT Pupuk Sriwidjaja Palembang Plant 2B. The air compressor serves multiple functions including supplying normal plant air, instrument air, providing process air for the secondary reformer to maintain the H2:N2 ratio in the converter to supply heat for chemical reactions, and exporting air to the off-site battery limit (OSBL). This process air compressor operates through four stages of centrifugal compression (101-J), with intercoolers between stages and a condensate separator within the interstage cooler

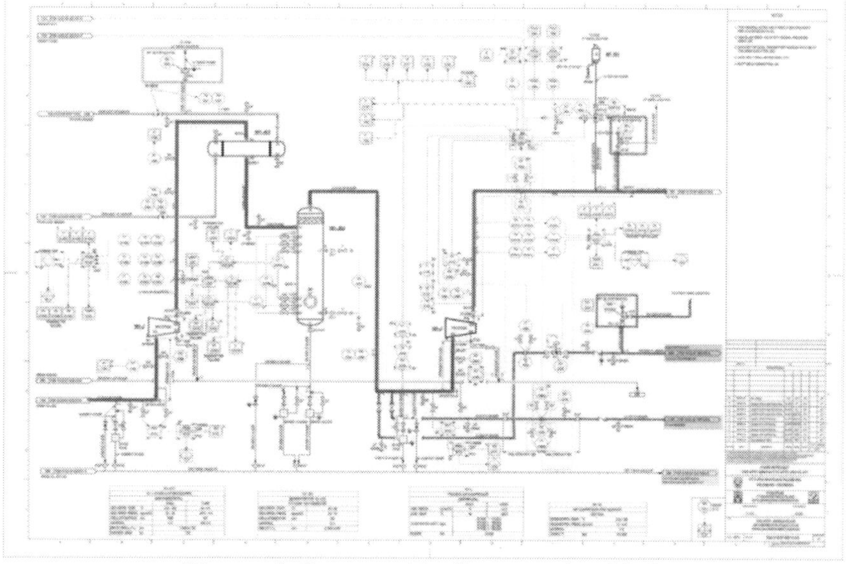

Figure 4. Process Air Compression Node

A. Risk Analysis

Table 4. HAZOP Risk Analysis

Deviation	Causes	Consequences	Safeguards
No/Low Flow	No flow of condensate from 101-JD1	Potential high level in 101-JD1. Potential liquid carryover to 2nd stage 101-J, Compressor damage.	LSHH-1235 (2oo3)
	No flow of condensate from 101-JD2	Potential high level in 101-JD2. Potential liquid carryover to 3rd stage 101-J.	LSHH-1236 (2oo3)

Deviation	Causes	Consequences	Safeguards
	No flow of condensate from the Separator for 101-JD3	Potential high level in 3rd stage separator 101-JD Potential liquid carryover to 4th stage 101-Jand damage to Compressor. Also, air coil damage due to less flow.	LSHH-1237 (2oo3)
	XV-1212 malfunctions and closes	Potential overheating of the air preheat coil 101-BCA	FSLL-1203 (2oo3)
Reverse/ Misdirected Flow	FIC-1004 malfunctions	Potential loss of airflow to the Secondary Reformer	FSLL-1203 (2oo3)
		Potential overheating of the air preheat coil 101-BCA	FSLL-1203 (2oo3)
	HIC-1020 malfunctions	Potential loss of airflow to the Secondary Reformer	FSLL-1203 (2oo3)
		Potential overheating of the air preheat coil 101-BCA	FSLL-1203 (2oo3)
	101-J trip	Potential reverse flow of process gas into Process Air Compressor and damage.	FSLL-1203 (2oo3)
Others	Tube leak 101-JC1/2/3	Potential leakage of CW	LSHH-1235 (2oo3)

According to the HAZOP analysis performed of the Process Air Compressor node in Table 4, consideration of potential causes and consequences of deviations has resulted in significant probabilities of risk. Consequently, further measures including a comprehensive risk ranking process are needed to mitigate these risks.

B. Classification

The risk assessments acquired from HAZOP are subsequently followed by the determination of SIL targets in the Classification process that implemented the LOPA method. A case study examining the SIF within the Process Air Compressor node is illustrated in Table 5.

Table 5. SIL Classification - LOPA Risk Analysis

No	SIF	Consequences	F	C	L	Initial SIL	IPL	Required SIL
1	LSHH - 1235 (2oo3)	2	1	0.3	0.30	SIL 2	-	SIL2
		2	0.1	0.3	0.03	SIL 1	-	SIL1
2	LSHH - 1236 (2oo3)	2	0.1	0.3	0.03	SIL 1	-	SIL1
3	LSHH - 1237 (2oo3)	2	0.1	0.3	0.03	SIL 1	-	SIL1

No	SIF	Consequences	F	C	L	Initial SIL	IPL	Required SIL
4	FSLL - 1203 (2oo3)	3	1	0.3	0.30	SIL 2	-	SIL2
		3	0.1	0.3	0.03	SIL 2	-	SIL2

Based on the SIL classification steps using the LOPA worksheet, the analysis identified that, within the Process Air Compressor Node at Ammonia Plant P-IIB, out of the four SIFs loops, two were classified as SIL 1, while the remaining two were classified as SIL 2.

C. Verification

The SIL verification process begins with the calculation of the Probability of Failure on Demand (PFD), which is determined based on factors including hardware and software failures of the SIS components, demonstrates that the PFD meets the requirements, as shown in Table 6.

Table 6. SIL Verification - PFD Requirement

No	SIF	SIL Target	PFD				Proof Test Interval			Note
			Sensor	Logic Solver	Final Element	Tota l	Sensor	Logic Solver	Final Element	
1	LSHH - 1235 (2oo3)	SIL 2	0.0000 36759	0.0000 0970	0.00030 939	0.00 036	3	3	3	Meet the PFD Req.
2	LSHH - 1236 (2oo3)	SIL 1	0.0000 36759	0.0000 0970	0.00030 939	0.00 036	3	3	3	Meet the PFD Req.
3	LSHH - 1237 (2oo3)	SIL 1	0.0000 36759	0.0000 0970	0.00030 939	0.00 036	3	3	3	Meet the PFD Req.
4	FSLL- 1203 (2oo3)	SIL 2	0.0000 36759	0.0015 3302	0.00200 031	0.00 357	3	3	3	Meet the PFD Req.

The SIL verification process is continued by verifying with Architectural Constraints, which involves assessing the number of installed SIF components and comparing them with the Minimum Hardware Fault Tolerance (HFT) required by IEC 61511 standards.

Table 7. SIL Verification - Architectural Constraint

No	SIF	SIL Target	SFF Actual (%)		Req. Minimum Hardware Tolerance		Actual Hardware Tolerance		Note
			Sensor	Final Element	Sensor	Final Element	Sensor	Final Element	
1	LSHH-1235 (2oo3)	SIL 2	73.8	60	0	0	1	0	Meet the Architectural Constraint.
2	LSHH-1236 (2oo3)	SIL 1	73.8	99	0	0	1	0	Meet the Architectural Constraint.
3	LSHH-1237 (2oo3)	SIL 1	73.8	99	0	0	1	0	Meet the Architectural Constraint.
4	FSLL-1203 (2oo3)	SIL 2	73.8	60	1	0	1	0	Meet the Architectural Constraint.

As shown in Table 7, the verification of Architectural Constraints shows that all four SIFs of the process air compressor node meet the requirement that is aligned with the respective SIL classifications.

Furthermore, expanding upon the detailed analysis, Table 8 below provides consolidated results of SIL Evaluation for five Ammonia Plants within Pupuk Indonesia (PIHC).

Table 8. SIL Evaluation Results

No	Company	SIL Classified	SIL Not Required	SIL Compliance	SIL Non-Compliance	Estimated Cost
1	PKC -1B	116	15	87	14	$161,472.68
2	PKG-1B	167	83	80	4	$329,032.26
3	PSP-2B	192	68	124	0	-
4	PIM-2	113	12	70	31	$177,361.29
5	PKT-1A	118	27	82	9	$48,959.32
	Total	**706**	**205**	**443**	**58**	**$716,441.86**

The SIL evaluation within five ammonia plants resulted that among the 706 classified SIFs, 205 are categorized as SIL Not Required. The remaining 443 are deemed SIL Compliance, while 58 are classified as SIL Not Compliance. Following the evaluation, recommendations were issued for the 58 SIL Not Compliance. As a result, modifications are required following the recommendations, necessitating the replacement of sensors and final elements, with a total cost of $716,441.86.

Conclusion

The main objective of this work was to evaluate a safety instrumented system using Risk Analysis, LOPA, and RBD. We have first introduced the main steps of the proposed methodology. Then, we have briefly described the system on which we have illustrated this approach. Of the five ammonia plants, there are a total of 501 SIL classified SISs. 443 SISs (88%) meet the SIL target, while the other 58 SISs (12%) do not meet the SIL target.

SIL analysis is critical to ammonia plants. Therefore, we have to continue our work at the other plant. For the next project we will roll out this study to the Urea Plant.

Nomenclature

Nomenclature	
HAZOP	Hazard and Operability Study
IPL	Independent Protection Layer
LOPA	Layer of Protection Analysis
PFD	Probability Failure on Demand
RBD	Reliability Block Diagram
SIF	Safety Instrumented Function
SIL	Safety Instrumented Level
SIS	Safety Instrumented System
TLV	Threshold Limit Value
PKG	PT Petrokimia Gresik
PKC	PT Pupuk Kujang
PKT	PT Pupuk Kalimantan Timur
PIM	PT Pupuk Iskandar Muda
PSP	PT Pupuk Sriwidjaja Palembang

References

[1] Pattabathula, V., & Richardson, J. (2016). Introduction to ammonia production. AIChE. https://www.aiche.org/resources/publications/cep/2016/september/introduction-ammonia-production

[2] Mitchell, J. K. (2010). Kenexis Safety Engineering Handbook. Columbus: Kenexis Consulting Corporation.

[3] Yang, Y.J., Huang, C., Zhong, Q.Y., Chen, H.J., & Huang, H.-Z. (2021). A Case Study on Safety Integrity Level Analysis for Shale Gas Station. Journal of Mechanical Science and Technology, 35(12), 5445–5452. https://doi.org/10.1007/s12206-021-1115-x

[4] Dunjó, J., et al. (2010). Hazard and operability (HAZOP) analysis: A literature review. Journal of Hazardous Materials, 173(1–3), 19–32. https://doi.org/10.1016/j.jhazmat.2009.08.076

[5] Ouazraoui, N., Nait-Said, R., Bourareche, M., & Sellami, I. (2013). Layers of protection analysis in the framework of possibility theory. Journal of Hazardous Materials, 262, 168–178. https://doi.org/10.1016/j.jhazmat.2013.08.042

[6] Darwish, A. Salaheldine, Mansour, M. Salem, Farag, H., & Ezzat, K. H. (2020). Applying LOPA and Fuzzy Logic to Identify SIL Requirement for Safety Critical Functions in a Direct Reduction Iron Industry. Alexandria Engineering Journal, 59(5), 3575–3585. https://doi.org/10.1016/j.aej.2020.06.003

[7] ISA (International Society of Automation). (2004). IEC 61511-1 Functional Safety: Safety Instrumented System for the Process Industry Sector - Part 1. North Carolina: Author.

[8] Guo, H., & Yang, X. (2007). A simple reliability block diagram method for safety integrity verification. Reliability Engineering & System Safety, 92(9), 1267–1273. https://doi.org/10.1016/j.ress.2006.08.002

Prioritizing Safety Culture for Ammonia as a Fuel in Energy Transition

Proposed and in-construction expansions in ammonia production and usage are gaining ground quickly. "Green" and "blue" projects are being undertaken by people at varying levels of experience—from long-term ammonia producers and energy companies to Private Equity companies new to chemical processing. Using ammonia as a hydrogen transport medium or as a fuel is driving the effort. With this rapid advance comes the challenge of producing, transporting and using ammonia safely at all points in the distribution chain, down to the end users.

On the flip side, the industry still faces the public perception of "toxic ammonia" as being an H2 transport mechanism and faces serious risks if even one major incident occurs, such as Bhopal, Toulouse, or the Waco AN explosions that impacted public areas.

With an industry that is transitioning from primarily fertilizer production into fuel production, clear guidance for safe design and operations is critical to ensure that the drive for ammonia-as-fuel succeeds. This paper addresses the various facets of ammonia production, transport, and deployment, along with any preventive and mitigation system requirements, existing guidance, and various standards pertaining to these facets of the ammonia lifecycle, and highlights the gaps so that the industry can respond. Lessons from the Ammonia and Fertilizers Joint Industries Project (JIP) can be used in developing approaches for the various stakeholders in this booming industry.

Murtaza I. Gandhi, P.E., Dorothy Shaffer, Daniel J. Benac, P.E., CFEI
Baker Engineering and Risk Consultants, Inc. (BakerRisk®)
3330 Oakwell Court, Suite 100
San Antonio, Texas 78218

Introduction

Energy transition is occurring at a rapid rate with hydrogen being seen as the cleanest carbon free molecule. Anhydrous ammonia is in high demand, as it provides a low carbon energy source with high hydrogen density, with viability in a huge variety of industries including shipping, storage of thermal energy, power plants, and others. This transition has resulted in many legacy companies expanding from traditional ammonia manufacturing for the fertilizer and chemical industries to these energy related uses. In addition, ammonia distribution terminals are being added to handle the transport of materials. The prospective business growth opportunities are also attracting many new companies to this space. However, many of those newer companies do not have experience dealing with ammonia process safety, and even some of the established companies are struggling to find sufficient resources. These issues are exacerbated by the lack of ammonia specific standards and lesser known or unpublished safety practices.

This paper summarizes some of the standards that are applicable to the ammonia manufacturing industry and how they can be applied to different use cases. The paper also discusses the Ammonia and Fertilizers Joint Industries Project (JIP) started by BakerRisk in 2019 as a model that can be used to bring knowledgeable stakeholders together in developing a set of resources that supports safe ammonia manufacturing and use processes during this period of high growth.

Recognition of ammonia as energy source

A few headlines listed below show how ammonia is being developed as an energy source:

1. "Japan, Singapore to establish green shipping corridor. As part of the new agreement, a "Green and Digital Shopping Corridor" will be established between the two countries."
2. "Mazda includes ammonia fuel in future energy roadmap."
3. "Mitsubishi Heavy Industries has successfully combustion-tested its ammonia single-fuel burners."
4. "Project Cormorant: 8 Rivers will develop the 880,000 tonnes-per-year ammonia production facility in Port Arthur, TX."
5. "Neom's 2.2GW green hydrogen and ammonia complex will meet 'high bar' EU definition of renewable fuel"
6. "Yara and ACME signed a binding agreement for supply of green ammonia"

Figure 1 shows the projected growth of clean ammonia export by various regions. As shown in the figure, clean ammonia energy production/export is projected to increase from 1 Million tonnes in 2024 to 75 Million tonnes by 2035.

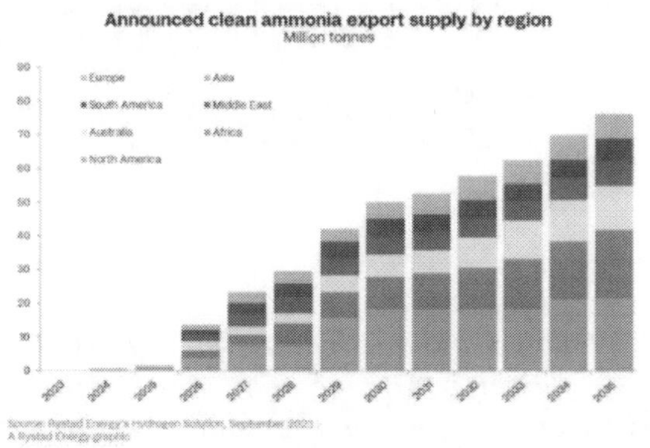

Figure 1. Clean Ammonia Growth Chart

Figure 2 shows the projected growth of green ammonia in US. According to Grandview Research, the global green ammonia market size was valued at $152 million in 2022 and is expected to grow at a Compound Annual Growth Rate (CAGR) of 116.5% from 2023 to 2030. The US CAGR in the same period is expected to be 113.9%.

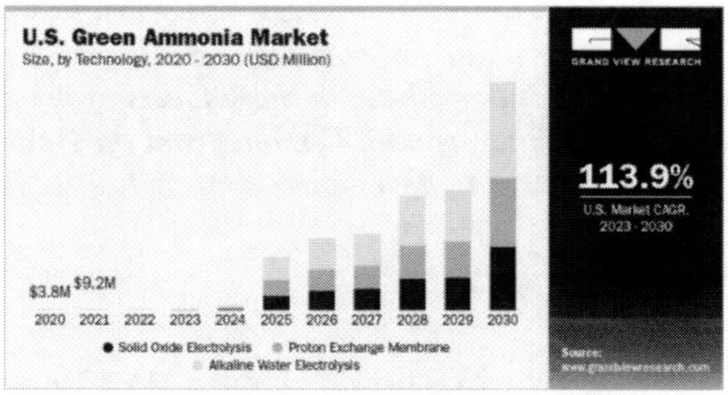

Figure 2. US Green Ammonia Market

This exponential growth potential in the industry will be a huge driver of technological innovations in the production and use of ammonia as a fuel. This level of growth cannot be achieved by the companies currently producing ammonia, and will require newer producers and technologies to meet the demand requirements of the market. This significant and exponential growth brings with it a lot of growing pain and challenges that organizations might find hard to manage. In such a challenging environment, there is a need for a set of Recognized and Generally Accepted Good Engineering Practices (RAGAGEP) documents that are widely adopted by global authorities and can help this industry grow safely at the pace that is needed.

Current standards in the industry

As is the case for many other chemicals, the ammonia industry also relies heavily on general standards such as the API RP (Recommended Practices) and standards for fabrication of its piping and vessels. The well tested broad spectrum of standards such as these is adequate in most circumstances. However, specific guidance is critical for knowing when exceptions should be recognized.

For example, within the US, anhydrous ammonia is shipped as a Division 2.2 non-flammable compressed gas, but for international transport, it is designated instead as Division 2.3 D poison gas. Copper-bearing materials are often attacked if they come into contact with ammonia, and the differences between soft materials have caused issues when propane transport equipment is put into ammonia use without modification. Ammonia-specific damage mechanisms must be understood as evidenced by the lessons learned in the AIChE Ammonia Safety Symposium and the BakerRisk Ammonia JIP efforts.

A literature search on the existing industry standards pertaining to ammonia generally point towards anhydrous ammonia refrigeration systems which operate in a much different environment than what is present in anhydrous ammonia manufacture. A paper published by IChemE in 2016 [1] summarizes the various international standards pertaining to this industry. Ignoring those specifically focused on refrigeration systems, it was found that the global standards for ammonia primarily revolve around the hazardous material regulations. Other organizations such as Fertilizer Europe and the North American The Fertilizer Institute have valuable guidance documents that have been generated for use that avoid some of the time and authorization issues inherent in developing a defined "standard". Additionally, new standards are being generated by organizations and regulatory agencies in response to the specific needs of the ammonia as fuel transition. For example, the 2023 very recent issue by the American Bureau of Shipping[4] and ongoing work of the Bureau Veritas of France[5] provide guidance on ammonia fueled, ocean-going vessels.

Examples of key regulations and standards (other than those specific to refrigeration systems) that apply to the manufacture, storage, handling and use of ammonia are stated below, separated by various regions.

USA

Regulation / Standards	Title
ANSI K61 / CGA 2.1 - 2023	Requirements For The Storage And Handling Of Anhydrous Ammonia - 7th Edition
CGA G-2	Anhydrous Ammonia
ISO 5771: 2008	Rubber hoses and hose assemblies for transferring anhydrous ammonia
OSHA 29 CFR 1910.111	Storage and handling of anhydrous ammonia
US DOT 49 CFR Parts 171-180	Transportation of hazardous materials
US EPA- RMP	Risk Management Plan
US OSHA 29 CFR 1910.119	Process Safety Management of Highly Hazardous Chemicals

EU

Regulation / Standards	Title
ATEX 94/9/EC	Equipment Directive - Equipment and protective systems intended for use in potentially explosive atmospheres
ATEX 99/92/EC	Workplace Directive - Minimum requirements for improving the safety and health protection of workers potentially at risk from explosive atmospheres.
Chemical Industry Association (UK)	Guidance for the large scale storage of fully refrigerated anhydrous ammonia in the UK, June 1997
SEVESO Directive 2003/105/EC	Control of Major-Accident Hazards involving dangerous substances
EN 14620	Design and manufacture of site built, vertical, cylindrical, flat-bottomed steel tanks for the storage of refrigerated , liquefied gases with operating temperature between 0°C and -165°C
Fertilizers Europe Meeting	Guidance For Inspection Of Atmospheric, Refrigerated Ammonia Storage Tanks, Guidance For Inspection Of And Leak Detection In Liquid Ammonia Pipelines

Australia/New Zealand

Regulation / Standards	Title
AS/NZS 2022:2003	Anhydrous ammonia—Storage and handling
AS/NZS IEC 60079.10.1:2022	Explosive atmospheres, Part 10.1: Classification of areas - Explosive gas atmospheres

China

Regulation/Guidance	Title
Decree - 591	Regulations on Safe Management of Hazardous Chemicals in China

India

Regulation/Guidance	Title
IS 4544 (2000), ICS 71.060;13.300	Ammonia - Code of Safety [CHD 8: Occupational Safety, Health and Chemical Hazards]

As is evident from the list of resource documents above, most of the documents follow the storage and handling of anhydrous ammonia or follow general process safety requirements. A general lack of a coherent set of RAGAGEP may be of concern, especially if designs for new ammonia equipment and production operations are performed by companies that may not have the extensive experience that traditional ammonia production companies typically have. This includes an understanding of ammonia tank design, mechanical integrity issues, and the use of effective spill mitigation strategies, to name a few.

Capital Projects Lifecycle and some key regulations/guidance

The paper aims at identifying some key regulations / standards from a US perspective and how those requirements can be met by aligning them with the lifecycle of capital projects.

While this paper focuses on US-based regulations / standards, each project should have its own lifecycle aligned safety studies that follow the pertinent regulations / standards for that particular location. Typically, a well-established ammonia process and designers with significant ammonia tank experience will be assigned for the project.

In the early 1900s, the need for ammonia production both to feed growing populations and create weapons of war initiated the development of high pressure synthesis designs. First developed using the Haber-Bosch process by BASF, the basics were quickly incorporated into process designs offered by several ammonia industry process design companies.

While process design background is important, the operator must also satisfy safety requirements to ensure design, operations, and maintenance practices are grounded in good safety practices. The intent of the paper is not to outline all the design and safety studies but to provide guidance on how the alignment can be done according to the project lifecycle, to ensure the appropriate regulations and guidance are followed, and to highlight the gaps in the standards regime to inform the readers about various efforts needed to close those gaps.

Figure 3 shows a typical capital project lifecycle for a facility producing ammonia and other materials. As shown in the figure, the project starts with a conceptual design, site selection, preliminary and detail design, construction, commissioning, operations/maintenance and eventually either expansion or demolition.

Figure 3. Typical project lifecycle

All these stages have their unique needs in terms of safety analyses. Not all studies need to be performed at each stage, and many of these are built over time as more data becomes available.

A Stage Gate Review (SGR) is conducted at the end of each stage, which helps to verify that the intent of the particular stage has been met and that inputs are developed to move to the next stage of the project. The overall goal of a SGR process is to minimize personnel exposure, asset damage, and business interruption while also controlling costs and ensuring ease of project implementation [2].

Figure 4 and Figure 5 show an example matrix of various safety studies that are typically performed for a typical blue or green ammonia facility. As shown, some of the studies are started at preliminary stages and then are developed incrementally over time.

General Technical Safety Study for H2 NH3 CCUS						
	Front End Loading (FEL)			Project Execution		
General Technical Safety Studies	FEL-1 (Appraise)	FEL-2 (Select)	FEL-3 (Define)	Detailed Design	Construction	Operation
HSE plan	Updated continuously throughout all project phases					
Risk Register						
	TASK(S)	TASK(S)	TASK(S)	TASK(S)	TASK(S)	TASK(S)
Hazard Identification	Prelim HAZID	HAZID	Prelim HAZOP	HAZOP		HAZOP
		Impact identification (ENVID)				
					Pre-startup safety review (PSSR)	
				Noise Study		
Consequence Assessments		Screen FSS	Prelim FSS	FSS		FSS
		HCA				
			Prelim HAC	HAC		HAC
			SIP	SIP Design		Shelter Testing
		Prelim WSA	WSA			
Safety Assessments	Prelim ISD	Inherently Safer Design (ISD)				
		Screen HFA	Prelim HFA	HFA		
			Prelim BD/PR/Flare	BD/PR/Flare		Inspection of PR
			Prelim RAM	RAM		
			Prelim SCE	SCE		
			Prelim SCEVA	SCEVA		
			Prelim EERA	EERA		
				DOA		
				Safety Report		
				FSA		
	Concep EIA	Prelim EIA		EIA		
			SVA			
			Prelim Simultaneous Operations (SIMOPS)		SIMOPS	
						Audits

Figure 4. Matrix of safety studies – 1

General Technical Safety Study for H2 NH3 CCUS						
	Front End Loading (FEL)			Project Execution		
General Technical Safety Studies	FEL-1 (Appraise)	FEL-2 (Select)	FEL-3 (Define)	Detailed Design	Construction	Operation
HSE plan	Updated continuously throughout all project phases					
Risk Register						
	TASK(S)	TASK(S)	TASK(S)	TASK(S)	TASK(S)	TASK(S)
Risk Assessments	Concept RA	Prelim RA				IRE
			Preliminary Quantitative Risk	QRA		QRA
		HRA				
				LOS		
				RMP (Detailed Design)		RMP
Risk Mitigation				LOPA		
						Emergency Response Program (ERP)
				SIS/SIL		Testing of Safety Instrument
			Prelim F&G	F&G		
			Prelim FHA	FHA		
			Prelim FW	FW		

Figure 5. Matrix of safety studies - 2

For each of the studies listed above, it is a good practice to define the appropriate regulations / standards, or other RAGAGEP resources and a basic scope and deliverables. This can help ensure that the entire project team is aware of the various requirements, and can plan accordingly to prevent impacts to schedule and costs.

Similarly, mechanical integrity programs will include a set of standards used for design, maintenance and inspection of equipment for ammonia production and storage.

An example for a Facility Siting Study (FSS) task at various stages is discussed below.

FSS study progression

- Screening Level
  - Intent: High level risk study to compare locations where site could be developed
  - Typically performed: FEL 2 (Site Selection)
  - Regulation: OSHA 1910.119
  - Standards: API752/ 753Scope of work:
    - Model selected release scenarios based on preliminary process data.
    - Calculate discharge, flammable and toxic dispersion, thermal radiation.
    - Generate overpressure contours and maximum blast loads on building targets.
    - Generate flammable, toxic, and thermal radiation contours.
  - Deliverables:
    - Report showing flammable and toxic dispersion contours, thermal radiation contours for different thresholds, overpressure contours.
- Preliminary FSS
  - Intent: Perform risk assessment for a specific location, with intent to be able to change the plot plan layout based on risk and operability.
  - Typically performed: FEL 3 (Define)
  - Regulation: OSHA 1910.119
  - Standards: API752/ 753
  - Scope of work:
    - Update release scenarios based on the process data available.
    - Update discharge, flammable and toxic dispersion, and thermal radiation.

- Update overpressure contours and maximum blast loads on building targets.
- Update flammable, toxic, and thermal radiation contours.

o Deliverables:

- Updated report showing flammable and toxic dispersion contours, thermal radiation contours for different thresholds, overpressure contours.

- Detailed Design
  - o Intent: Perform risk assessment for a specific location with a locked in plot plan, with the ability to make small changes to the layout and help move people out of occupied buildings near the main release sources.
  - o Typically performed : Detailed Design
  - o Regulation: OSHA 1910.119
  - o Standards: API752/ 753
  - o Scope of work:
    - Review the detailed design process flow diagrams, heat and material balance, and P&IDs to update release scenarios.
    - Update release scenario locations based on updated plot plan.
    - Update zones of congestion and confinement based on updated plot plan and 3D model (if available).
    - Model site buildings' blast capacity based on structural drawings or construction types with conservative assumptions.
    - Update discharge, dispersion, thermal radiation, overpressure calculations.
    - Generate building damage levels, flammable and toxic concentrations at buildings, thermal radiation levels and occupant vulnerability of buildings.
    - Generate flammable and toxic dispersion contours, thermal radiation contours, and overpressure contours.
    - Generate API RP 753 contours for portable building siting.
  - o Deliverables:
    - Report that documents input and output (discharge rates, dispersion distances, thermal radiation levels on buildings, toxic concentrations predicted at buildings and occupant vulnerabilities).
- Operations (typically updated every 5 years or when a Management of Change (MOC) occurs).

Working toward RAGAGEP

Ammonia companies in the US must meet the performance-based Process Safety Manage (PSM) regulatory requirements for RAGAGEP. However, as identified above, these regulations provide general guidance for ammonia facility, but do not specifically provide inputs or standards for ammonia hazards. This may create inconsistent designs for similar facilities, based on the EPC or consulting companies that are designing and analyzing such plants. These designs are based on an individual organization's experience, which means industry experience as a whole is not incorporated in the overall design. Also, with the various global regulations, similar facilities in different countries can have very different design and mitigation requirements. This can create huge risk gaps and design/operational issues for organizations that have no background in ammonia production. Hence, it is important to set up a consistent resource that can be used for guidance and is available to ammonia producers and distributors globally.

A set of RAGAGEP similar to IEC61511 for functional safety that have been widely adopted by global regulations and the set of documents developed by The Chlorine Institute are both good examples of an industry coalescing around a common set of practices to support industry safety.

An example structure for developing such guidance was initiated by BakerRisk ammonia and mechanical subject matter experts at the 2019

AICHE Safety in Ammonia Plants & Related Facilities Symposium [3]. Nine well-known ammonia producers joined the Ammonia-Fertilizer Damage Mechanism and Mitigation Joint Industry Program (A&F JIP). The first studies were done on the ammonia process, ammonia storage and Urea process. [3,4] While the initial thought was to create an ammonia document similar to the structure of API RP 571 Damage Mechanisms Affecting Fixed Equipment in the Refining Industry, the A&F JIP members quickly made it clear that additional information regarding industry experience along with prevention and mitigation of damage were desired to enhance industry safety and reliability.

The collaboration inputs are based upon the past 20+ years of industry experience in ammonia related fields, including AIChE NH3 Safety Symposium papers, Urea KnowHow resources, National Transportation Safety Board reports of ammonia incidents, NASA reports, ANNA presentations, designer and fabricator inputs, along with other public industry resources, and finally, BakerRisk's extensive experience and knowledge of the industry. In addition, a framework was created in order to share the JIP members' unpublished incidents and learnings confidentially. JIP members choose roundtable topics for each year of the effort, participate in review meetings, ask questions on current issues that drive the content, and are invited to provide input on the drafts.

The JIP structure (first started by BakerRisk as a collaboration of refineries and petrochemical companies to address explosion risks) is readily available to tackle industry concerns and issues arising from the fast-paced growth of the industry, new industry participants, and new facilities to identify and compare existing standards and guidance, identify gaps and develop RAGAGEP resources for the industry that may indeed be adopted as accepted practice if not standards. This method helps ensure sensible industry input is incorporated into practices where gaps could potentially invite development of overly reactive standards by agencies or bodies with less practical experience.

A key challenge the group has faced has been some reluctance by several companies to participate, which could be seen by their management as a loss of competitive advantage. This has been addressed by BakerRisk protecting information shared by companies. The clear benefits are seen as the JIP membership group has grown from the original nine participating companies to a total of sixteen (16) companies [5]. The A&F JIP includes almost all North American producers, some major producers overseas, and several major chemical and petroleum producers that are new to the industry.

While JIP participants have been heavily represented by mechanical integrity personnel to this point, feedback is very positive from those reviewing the work who are active in reliability and process safety. Ensuring the learning opportunities and JIP structure can be fully exploited by the Asset Integrity (AI) and Mechanical Integrity (MI) field and beyond is a challenge for member companies. Since the information provided by the JIP includes 200+ page documents, ensuring the knowledge is easily accessible to different sectors of company personnel, and can be readily absorbed is a challenge. Finding new learning tools to broaden the scope of learning includes holding progress review meetings and utilizing online roundtables on discussion topics as well as webinars. Filters and searchable listings of references provide easier access to topics of interest.

The learnings from the A&F JIP can be used toward developing comprehensive resources available for creating RAGAGEP documents for ammonia production and distribution. A proposed structural framework for such a body can be developed using the following approach:

1. Identify the most significant industry risks and review for those with gaps in industry practices that pose a threat.
2. Bring various stakeholders together to coalesce around building standards and any associated technical reports.

3. Create appropriate working groups to build and review the RAGAGEP documents, incorporating industry inputs.
4. Pursue promotion of the RAGAGEP documents and any associated technical reports to gain widespread acceptance.
5. Once an organizational structure is proven, the industry can choose to tackle various problems with further efforts as needed.

Examples could include safe design and operating mitigation practices for ship loading and unloading facilities; improved guidance for specifications on water and for operating practices to avoid ammonia stress corrosion cracking in ammonia tanks (e.g., Fertilizers Europe [12] and distribution systems; or considerations for co-locating ammonia, oxygen, and hydrogen facilities in close proximities. Developing a comprehensive listing of standards and resources in place or in development would be a first step.

Conclusion

The ammonia industry has a bright future with the developments in energy transition. However, the current limited set of standards around safe ammonia practices, particularly for those who are new to the field, can pose a critical risk that can affect the entire industry. Regulatory agencies with limited knowledge and lacking practical experience could attempt to fill the gap. These conditions could hinder the industry's ability to reach its full potential and hinder its development.

By intentionally developing a consistent and accessible set of RAGAGEP resources with widespread international input and buy-in, the ammonia as energy transition can progress while increasing the safety of these production, distribution, and co-located facilities, allowing innovation to flourish.

References

[1] Rystad Energy's Hydrogen Solution, September 2023

[2] Grandview Research, grandviewresearch.com

[3] Review of Global Regulations for Anhydrous Ammonia Production, Use, and Storage, 2016 IChemE Symposium Series No 161, Feck, et al

[4] ABS Annual Review 2023

[5] BUREAU VERITAS PUBLISHES RULES AND GUIDELINES ON AMMONIA, May 4, 2021

[6]The Weitz Company 360° Facility Life Cycle

[7] Leveraging Process Safety Techniques in Early Phases of Capital Projects to Optimize Risk Reduction and Project Costs, 16th Global Congress on Process Safety, 2020, Karen Vilas, Mike Broadribb.

[8] D. J. Benac, D. Shaffer, "Damage Mechanism Assessment and Methodology for Ammonia Equipment, Vessels, and Piping," 64th AIChE, Safety in Ammonia Plants & Related Facilities Symposium, 2019, Vol. 60, pp 165-176.

[9] [2020-2021 Ammonia & Fertilizer JIP BakerRisk Project No. 01-07250-001-20 Identification of Damage Mechanisms and Mitigation Options for the Ammonia Process, April, 2021.

[10] 2021-2022 Ammonia & Fertilizer JIP BakerRisk Project No. 01-07250-002-20 Identification of Damage Mechanisms and Mitigation Options for the Ammonia Storage Tanks and the Urea Process, April 2024, October 2024.

[11] 2023-2024 Ammonia & Fertilizer JIP BakerRisk Project No. 01-07250-004-20 Identification of Damage Mechanisms and Mitigation Options for the Nitric Acid and Ammonium Nitrate plants, April 2024.

[12] Fertilizers Europe, Guidance for Inspection of Atmospheric, Refrigerated Ammonia Storage Tanks, 2008 Edition, Issue 2014

Glossary

BD	Blowdown
DOA	Dropped Object Assessment
EERA	Evacuation, Escape and Rescue Analysis
EIA	Environmental Impact Assessment
ENVID	Environmental Impact Identification
FSS	Facility Siting Study
HAC	Hazardous Area Classification
HAZID	Hazard Identification
HAZOP	Hazard and Operability Study
HCA	High Consequence Area
HFA	Human Factors Analysis
HRA	Health Risk Assessment
IRE	Insurance Risk Evaluation
ISD	Inherently Safer Design
LOPA	Layer of Protection Analysis
LOS	Loss of Stability Assessment
MI	Mechanical Integrity
PR	Pressure Relief Study
PSSR	Pre-startup Safety Review
QRA	Quantitative Risk Assessment
RAM	Reliability, Availability and Maintainability
RMP	Risk Management Plan
SCE	Safety Critical Elements
SCEVA	Safety Critical Elements Vulnerability Analysis
SIMOPS	Simultaneous Operations
SIP	Shelter-in-place
SIS	Safety Instrumented Systems
SIL	Safety Integrity Level
SVA	Security and Vulnerability Assessment
WAS	Water Suitability Assessment

Steam Methane Reformer Outlet System – Mechanical Integrity

Steam Methane Reformer outlet systems come in various designs, each susceptible to damage mechanisms specific to the design. The outlet system designs range from "cold wall" designs that utilize insulating refractories to reduce pressure boundary skin metal temperature to "hot wall" high alloy designs whose pressure boundary experiences process conditions. The primary damage mechanisms include creep, metal dusting, thermal aging and fatigue, excessive stress, stress relaxation cracking, and high-temperature hydrogen attack. Inspection for these damage mechanisms can be invasive and requires detailed planning and a thorough inspection and test plan. The paper will discuss these details, share examples of actual plant experience, and review relevant industry documents such as CGA H-12 Mechanical Integrity of Syngas Outlet Systems and the forthcoming API 561 Reforming Furnaces for Hydrogen and Syngas Production.

Chad Texter, Seshu Dharmavaram, Jason Cassell
Air Products and Chemicals

Introduction

Steam methane reformer (SMR) outlet systems utilize several fundamentally different designs based on the plant's vintage and technology provider. Designs range from a cold wall design that utilizes internal refractory to minimize skin metal temperature to hot wall designs that are externally insulated and whose skin metal temperature operates at process conditions. The applicable damage mechanisms are a function of the design and can occur slowly over a prolonged period or relatively fast. Loss of containment can be limited to small leaks or result in ruptures based on the damage mechanism and how long the damage is allowed to progress. Past studies discussed below review several past failures in reformer outlet systems. These failures have resulted from small releases to catastrophic failure of components, resulting in jet fires and/or significant energy releases. Given the range of damage mechanisms and possible outcomes, it is important for the SMR operators to understand their design and applicable damage mechanisms and to have inspection and monitoring programs in place to manage them.

Currently, four industry documents provide holistic guidance on SMR outlet systems [1, 2, 3,4]. The documents provide detailed information on outlet system design, applicable damage mechanisms, inspection options, and industry experience.

This paper draws on Air Products' vast experience as the world's largest, independent hydrogen supplier to provide a brief overview of SMR outlet system design, discuss past incidents, provide insights into particular vulnerabilities, and discuss applicable industry literature. The intent is to share both industry and internal experience so others are aware and may learn from past history.

Incidents

An API study [1] published in 2014 summarises reformer outlet system issues affecting the industry. At the time of publication, survey respondents reported that 13 of 21 reformer operators reported problems with their outlet system pigtails, with eight of the thirteen reporting significant, major, or severe issues.

A catastrophic refractory lined piping failure [6] occurred as recently as 2009 at a PraxAir Port Arthur, TX, facility designed in 2003. The line rupture resulted in extensive damage to the site. Other failures of alloy components have occurred repeatedly in industry, resulting in fires under the reformer structure, jet fires extending 30 feet (9 meters) from the reformer structure, equipment damage and downtime.

Cold Outlet Systems

Cold outlet systems utilize lower alloy refractory lined components to contain the hot syngas stream. The refractory lining is critical to protecting the pressure boundary from the high-temperature process fluid. Figure 1 depicts a cold outlet system reproduced from CGA H-12, *Mechanical Integrity of Syngas Outlet Systems* (1st ed.), with permission from the Compressed Gas Association, Inc. For more information, please visit www.cganet.com.

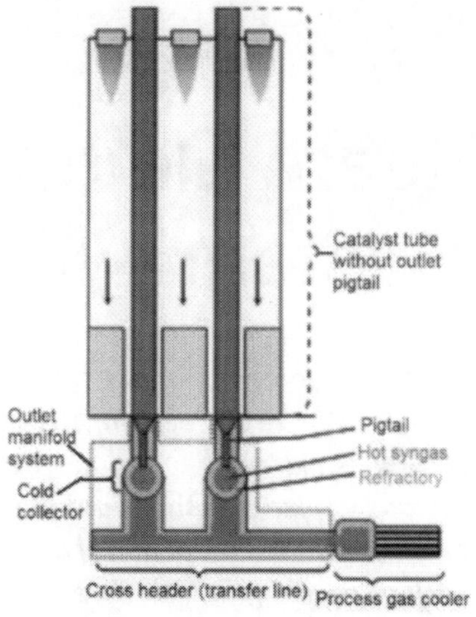

Figure 1. Cold outlet system design [2]

Depending on design details and operating conditions, these systems are susceptible to multiple damage mechanisms, including Metal Dusting, High High-Temperature Hydrogen Attack (HTHA), Hydrogen Embrittlement (HE), Material Deformation, and Carburization. The system can suffer stress rupture if the refractory lining significantly degrades or is otherwise compromised, exposing the pressure boundary to the elevated process temperature. Refractory degradation can occur from improper design, improper application, and plant thermal cycles. Less obvious but equally important is that maintenance activities can significantly impact cold outlet systems. An outlet system catastrophically failed [5] in the early 2000's at an SMR that Air Products acquired later when an insulation blanket was left on the refractory lined Cross Header with the plant in operation, as shown in Figure 2. The Cross Header had hot spots from existing refractory degradation, and adjustments were being made to an external steam lance system to provide localized cooling. The insulation blanket was placed for personnel protection while the steam lance was adjusted and not subsequently removed. The insulation blanket on the Cross Header's outside prevented the steel pressure boundary from effectively cooling to the ambient environment. The insulation blanket coupled with degraded refractory allowed the pressure boundary to heat up, weaken, and fail catastrophically.

Figure 2. Refractory line cross header failure [5]

Plants designed before 2016, especially plants designed before 1990, using cold outlet systems, are particularly vulnerable to HTHA attack because their designs used P1 material such as carbon steel or carbon ½ moly steel for pressure containment. At the time, industry did not fully understand the vulnerability of P1 materials to HTHA. As the industry gained more experience, API 941 "Steels for Hydrogen Service at Elevated Temperatures and Pressures in Petroleum Refineries and Petrochemical Plants" [6] was updated in 1990 to remove carbon ½ moly steel from its Nelson curves, effectively treating carbon ½ moly the same as carbon steel. API 941 was further updated in 2016 to create a derated "Carbon steel with no PWHT" curve in response to the 2010 Tesoro Anacortes Refinery incident, as shown in Figure 3.

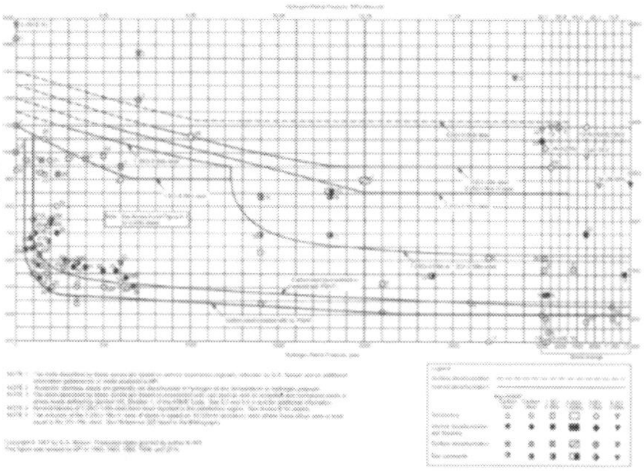

Figure 3. API 941 "Nelson" curves [6]

Many existing operating assets have had portions of their plants upgraded over time. It is critical to understand the details of cold outlet systems, including past upgrades that may retain older material that may effectively govern the system's overall design, and have effective inspections and controls in place to manage HTHA.
Figure 4 illustrates how portions of legacy P1 material can remain in cold outlet systems as other components are upgraded to P11 chrome-moly steels.

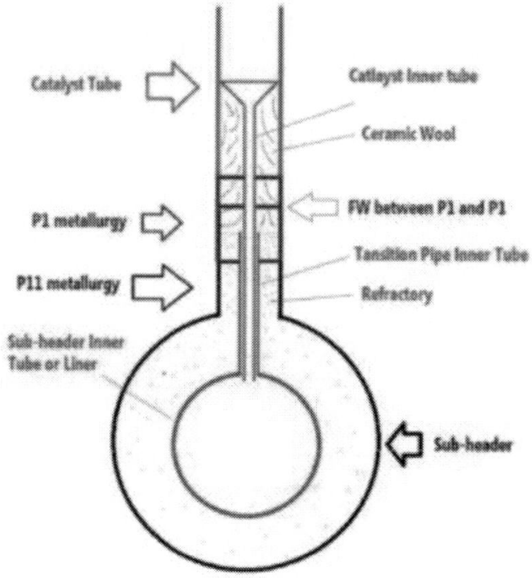

Figure 4. Cold outlet system design with P1 material [5]

These remnants of P1 material represent vulnerabilities that must be accounted for in the mechanical integrity of the entire system.

Cold Wall systems require periodic monitoring to detect and manage refractory degradation. Infrared thermography is a powerful tool to assess refractory performance while the plant is operating, and internal visual inspection can be executed when the plant is in a scheduled Turn Around (TAR). Figure 5 shows an example of IR thermography performed on a cold outlet system with areas of interest highlighted.

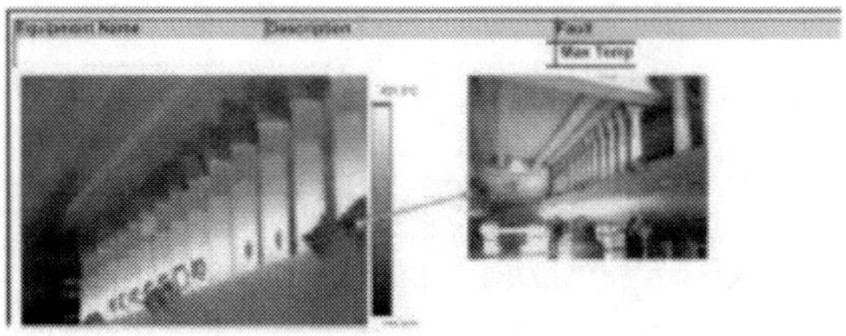

Figure 5. Example of IR survey [5]

Hot spots can form where refractory breaks down, subjecting the metal pressure boundary to elevated temperatures. Hot spots can form anywhere on a refractory-lined system; therefore, infrared thermography or other refractory monitoring systems must fully cover all applicable equipment. Specific areas of risk include joints where thermal expansion occurs in two different directions, domes where thermal cycles and gravity can cause the refractory to degrade and areas that operate below the process dew point. Figures 6 and 7 demonstrate refractory damage that was caused by differential thermal expansion. Figure 7 shows how the Cross Over refractory's thermal expansion damaged the Waste Heat Boiler's refractory.

Figure 6. Hot spot on a refractory line Cross Header at a joint [5]

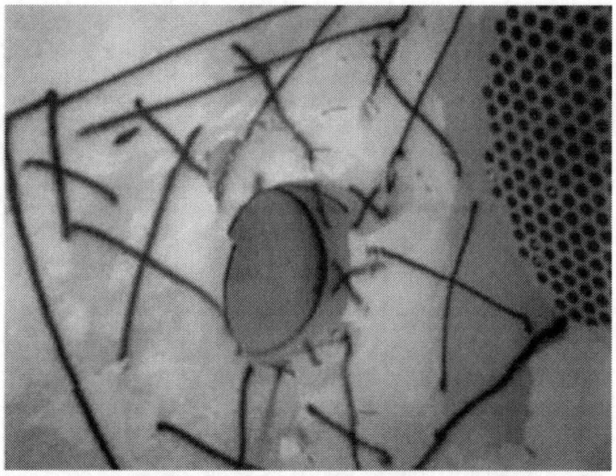

Figure 7. Refractory damage at a joint [5]

If the refractory design allows process condensation, the fluid can attack either the refractory or the pressure boundary. Process condensate often attacks the refractory, leading to a hot spot, as shown in Figure 8.

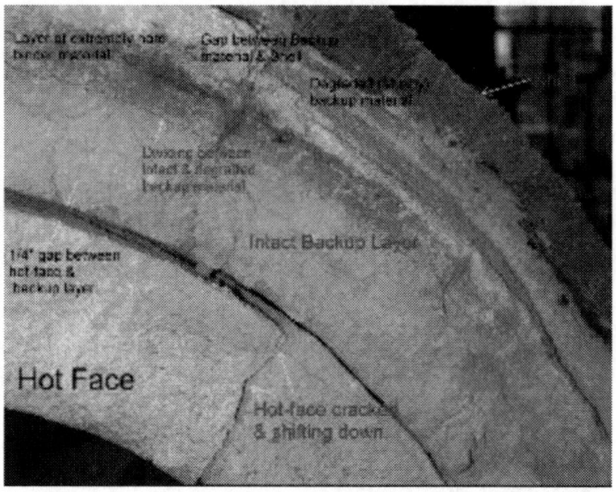

Figure 8. Degraded refractory on a Cross Header that initially operated below the dew point and then transformed into a hot spot [5]

If hot spots on the pressure boundary are detected, they must be acted upon. Potential actions include applying localized cooling and executing a Fitness For Service (FFS) to assess hot spots. Localized cooling is often used by blowing air and/or steam over a hot spot. Applying a magnetic thermocouple allows for continuous monitoring of skin metal temperature. Figure 9 provides an example of a cooling system.

Figure 9. Degraded refractory on a Cross Header with steam cooling applied and magnetic thermocouples attached to monitor skin metal temperature [5]

The FFS should evaluate the potential for accelerated HTHA damage and the pressure boundary's load carrying capability.

Hot Outlet Systems

Hot Outlet Systems use high alloy material that operates at process temperatures. The outlet components are externally insulated and must be properly designed and supported to accommodate thermal expansion and minimize stress at these challenging operating conditions. Figure 10 shows an example of a Hot Outlet System using a long pigtail design.

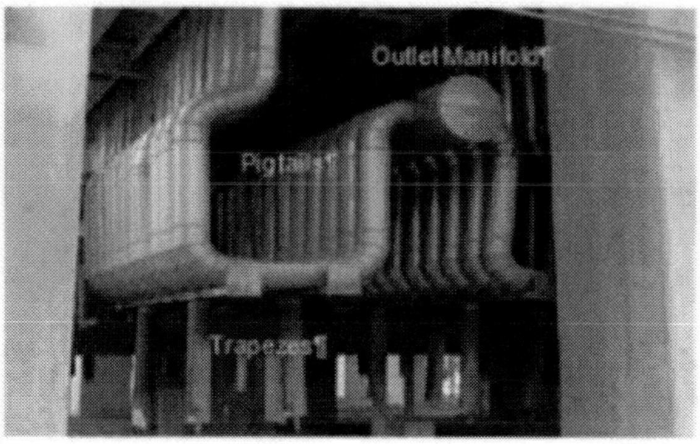

Figure 10. Hot outlet system [5]

Hot Outlet Systems are susceptible to multiple damage mechanisms, including Metal Dusting, Thermal Fatigue, Creep, Stress Relaxation Cracking, and external Nitridation, depending on design details and operating conditions. Metal dusting is a particularly aggressive form of corrosion that occurs at specific temperature ranges based on process fluid composition and metallurgy. When metal dusting is active, carbon penetration rates can reach up to 25 mm (1 inch) per year. Figure 11 shows an example of a metal dusting attack.

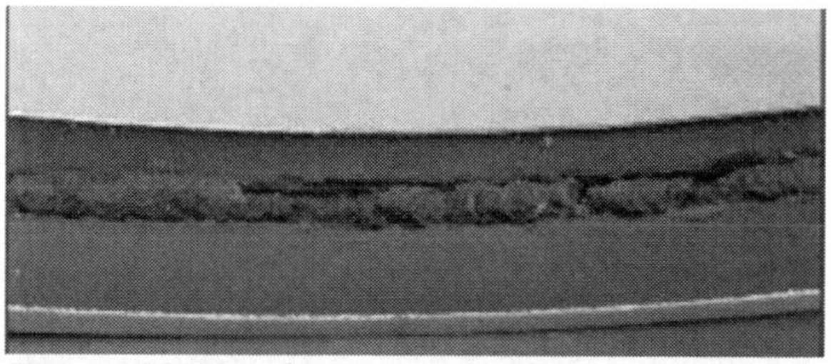

Figure 11. Metal Dusting attack at weld root [5]

Thermal fatigue can occur when insulation gaps allow localized cooling of the hot outlet system. Figure 12 shows an insulation gap forming at the joint between structural calcium silicate at the pigtail trapeze support and the softer insulation material used on the remainder of the length. Insulation designs often use layers with a staggered joint to prevent direct line-of-sight gaps from forming at joints.

Figure 12. Insulation gaps on a Hot Outlet System [5]

Thermal fatigue is particularly an issue in Short Hot Outlet System designs where the high-temperature outlet material penetrates a header box. In these cases, the header box runs at a slight vacuum, drawing cold air through any insulation gaps over the hot outlet system. Figure 13 shows thermal fatigue-driven failure on a Short Outlet System at a weld in the high alloy piping. Insulation gaps should be addressed, and particular attention should be paid to Short Outlet System Header box penetrations.

Figure 13. Thermal fatigue failure on Short Hot Outlet System penetration to header box [5]

Hot Outlet Systems operate in the creep range. Creep degradation rates are strongly influenced by operating temperature and stress. Creep damage can be managed through a combination of operating controls and inspection. Reformer balancing to reduce or eliminate hot zones in the furnace is important to managing the creep damage of individual pigtails. Online thermography, as shown in Figure 14, can balance the reformer to eliminate hot zones.

Figure 14. Examples of reformer thermography to balance the furnace [5]

During planned TARs, the outlet system should be periodically inspected for creep damage. Pigtails can be checked for diametrical growth to infer the degree of creep damage. Figure 15 illustrates outlet pigtail creep damage.

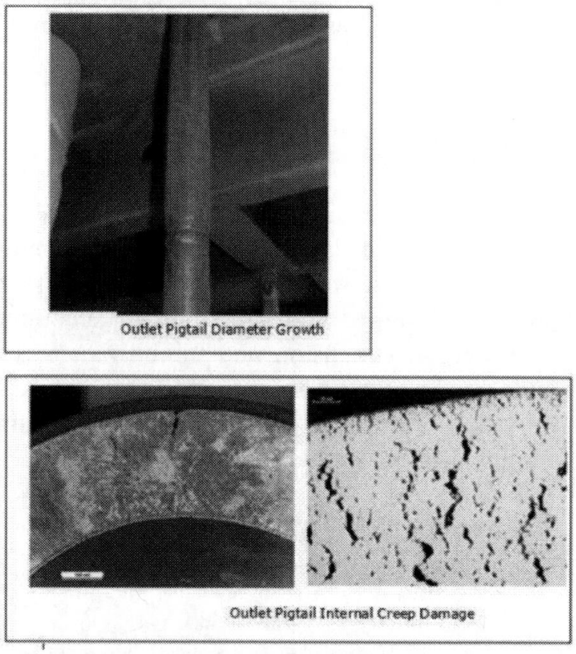

Figure 15. Outlet pigtail diametrical growth caused by creep damage [5]

Outlet system welds represent a vulnerable area for creep damage and potentially stress relaxation cracking. The process contents are hot enough that leaks may result in autoignition. Figures 16 illustrates an outlet system weld failure.

Figure 16. Long Hot Outlet System loss of containment [5]

Inspections for cracking at the weld and weld toes should be conducted via surface examination and radiography. Figure 17 illustrates an outlet system weld failure.

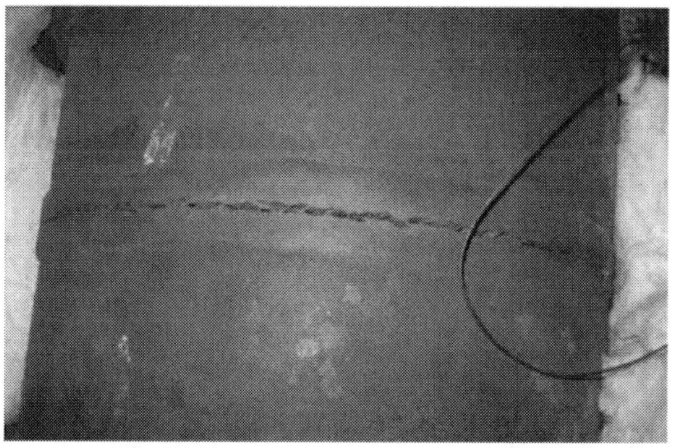

Figure 17. Failed Long Hot Outlet System weld [5]

The tenacious oxide layer must be removed to a bright metal finish to achieve good results before conducting the surface examination. Figure 18 illustrates the as-found and cleaned conditions for an outlet header weld.

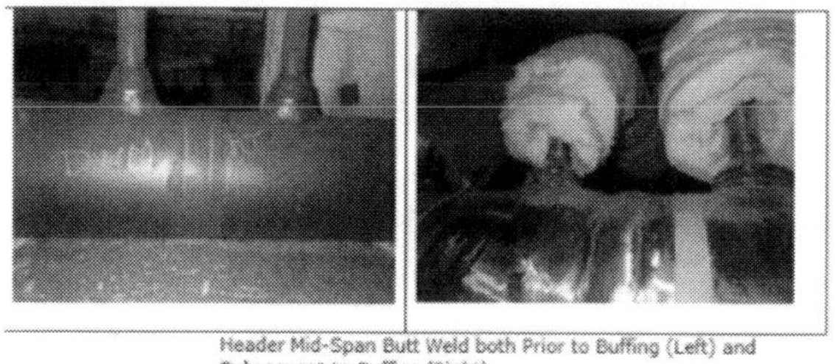

Figure 18. Outlet system weld before and after surface oxide removal [5]

As outlet systems age, their support systems can degrade as well. When the support system fails to function correctly, the outlet system components can potentially be overstressed, leading to physical deformation and/or accelerated creep damage. Structural calcium silicate insulation is often used at support points for the outlet system. The structural calcium silicate can either crush over time, or the support point can shift off the structural insulation, as shown in Figure 19. Either case will impact the support of the outlet system.

Figure 19. Support trapeze shifting off structural calcium silicate insulation [5]

Support components exposed to high temperatures can deform over time, impacting the functionality of the support system. Support components can also lift off or otherwise disengage. Examples are shown in Figure 20.

Figure 20. Support lifting off support trapeze and support deforming due to high-temperature exposure [5]

Short Hot Outlet Systems are particularly prone to support issues. Catalyst Tubes support the entire outlet system. Because of the high operating temperatures, the outlet system must effectively float, leaving a small margin for error for Catalyst Tubes support issues. Both excessive support providing a net upward lift or insufficient support are detrimental. Figure 21 and Figure 22 show different Short Hot Outlet System failure modes caused by improper support. Special attention must be provided when changing catalyst as changes in catalyst weight can impact the support of the outlet system. Visual inspection for deformed components and weld inspection via surface examination and radiography recommended, along with visual inspection of the supports.

Figure 21. Severed Short Hot Outlet System weld caused by improperly supported outlet system [5]

Figure 22. Deformed pigtails caused by improperly supported outlet system [5]

When necessary, adjustments to the supports to account for catalyst weight changes or load testing may be required.

The importance of managing penetrations through Short Hot Outlet System Header Boxes is not as obvious. Items such as thermocouples need to accommodate the outlet system's thermal expansion. If the thermowell penetration is not managed during outages, the thermowell can bind during the subsequent plant restart and cause excessive stresses at the thermowell penetration, leading to loss of containment as illustrated in Figure 23.

Figure 23. Bound up thermowell penetration through header box leading to failure [5]

Industry Documents

CGA H-12 [2] was developed in response to the recurring issues in the industry SMR outlet systems. The document provides detailed discussion on steam methane reformer outlet systems damage mechanisms and inspection strategies. CGA H-12 provides in-depth descriptions of outlet system designs, applicable damage mechanisms and inspection techniques. Table 1, shown below, was reproduced from CGA H-12, *Mechanical Integrity of Syngas Outlet Systems* (1st ed.) with permission from the Compressed Gas Association, Inc. For more information, please visit www.cganet.com.

	Hot Outlet System					Cold Outlet System				Common Components	
	Pigtail	Hot outlet collector	T-piece	Reducer cone	Cross header	Cold pigtail	Hot pigtail	Cold outlet collector	Cross header	Support members	Cross header thermowells
Metal dusting		X [1]				X	X				X
Thermal aging [2]	X	X	X	X		X	X				
Thermal fatigue	X	X	X	X			X				
Creep	X	X	X	X			X			X	
Excessive stress	X	X	X	X							
Stress relaxation cracking	X	X	X	X			X			X	
High temperature hydrogen attack					X	X		X	X		
Hydrogen embrittlement					X			X	X		
Material deformation					X	X		X	X		
Dissimilar weld cracking				X [3]	X [3]	X [3]		X [3]			
Nitridation	X [1]	X [1]					X [1]				
Carburization					X			X			

[1] Externally insulated components only.
[2] Applies to cast components only.
[3] Dissimilar weld cracking occurs in the reducer cone to the cross header, and the cold pigtail to the cold collector welds.

Table 1: CGA H-12 outlet system damage mechanism summary [2]

API TR 942-A [1] was issued in 2014 to document an API project on materials, fabrication, and repair issues associated with steam methane reformer outlet systems. The industry had encountered multiple occurrences of in-service degradation of outlet systems, and the goal of the technical report was to document why some reformer operators experienced issues while others did not. The paper provides an overview of design and material selection, welding practices, and results of industry surveys for several specific issues, and provides discussion on seven case studies.

API Standard 561 [3] is currently under development. The document will focus on Steam Methane Reformer design and construction considerations and will provide Informative Annexes on in-service monitoring and inspection strategies. Publication is anticipated in 2024 or 2025.

API 941 [4] is the definitive industry document on High Temperature Hydrogen Attack (HTHA). API 941 current design recommendations and discusses the assessment of legacy systems containing carbon steel and carbon-1/2 moly steels (P1 material) operating in potential HTHA at-risk zones.

All four documents are excellent references for any Steam Methane Reformer Operator.

Summary

Steam methane reformer outlet systems have a variety of designs, each with specific applicable damage mechanisms. These damage mechanisms can progress slowly over time or rather quickly, depending on the specifics of the mechanism. Failures can start as small leaks that can progress to more significant releases, while other damage mechanisms may proceed directly to catastrophic ruptures. The industry has experienced multiple consequential loss of containment events that resulted in significant equipment damage. Operators must understand the specific details of their SMR outlet systems. This is particularly important for older reformers whose design may no longer be appropriate for original design conditions. Several specific failure cases were discussed in this paper, and references to four industry documents were provided that every SMR Operator should be familiar with.

References

[1] API Technical Report 942-A "Materials, Fabrication, and Repair Considerations for Hydrogen Reformer Furnace Outlet Pigtails and Manifolds", 2014

[2] CGA H-12 "Mechanical Integrity of Syngas Outlet Systems, 1st edition", 2015
[3] API Standard 561 "Reforming Furnaces for Hydrogen and Synthesis Gas Production in General Refinery and Petrochemical Services" 2024 (Draft).
[4] API Recommend Practice 941 "Steels for Hydrogen Service at Elevated Temperatures and Pressures in Petroleum Refineries and Petrochemical Plants Eighth Edition, Errata 2, Addendum 1", 2020
[5] Air Products & Chemicals Reformer Failure Internal Incidents, personal communication.
[6] Kelly Holleran, "Praxair blames Linde, others for energy release at Motiva facility", Southeast Texas Record, December 14, 2011

High Temperature Equipment Service Damage Identification and Inspection

High temperature alloys are essential for fertilizer facilities. They resist the high temperatures and corrosive fluids demanded by the service. They are carefully fabricated with specifications and recommended practices. Thus, they can provide the high temperature strength and corrosion resistance combinations required for long term reliable service. Despite these precautions, the industry still has challenges.

This article discusses some mechanical integrity challenges for equipment in high temperature service: the consequences of fabrication flaws and various forms of damage. Where applicable, inspection methods to identify and characterize the damage before failure occurs are detailed.

Keywords: Fabrication defects, Stress Relaxation Cracking (SRC), liquation cracking, laser scanning dimensional analysis, temper embrittlement, field Post Weld Heat Treatment (PWHT), metal dusting, cyclic carburization and oxidation, nitriding, High Temperature Hydrogen Attack (HTHA).

Matthew Bell, Ana Benz
IRISNDT

Introduction

The fertilizer industry has many equipment components operating at high temperature. Alloys of all kinds are used to resist high temperature damage. However, these services cause specific damage mechanisms, and they can alter the materials in long-term service. The case studies discussed further below in this paper illustrate various high temperature damage mechanisms and the challenges. These case studies are from investigations done by IRISNDT for various customers in various industries.

- Fabrication defects such as tube piercing or rolling defects leading to leaks, the various pitfalls of fabricating heavy-wall Alloy 347H materials often used in hydrogen reforming equipment, characterizing the extent of deformation due to restrained thermal expansion, and insufficient Post-Weld Heat Treatment (PWHT) in the field causing premature decommissioning of a vessel.
- Consequences of composition changes from service, such as losses due to metal dusting, losses due to cyclic oxidation-carburization, and welding challenges due to nitriding.
- Consequences of long-term service including Creep damage, monitoring for High Temperature Hydrogen Attack (HTHA) and Stress Relaxation Cracking (SRC) due to long-term high temperature service.

In most cases, these issues can be detected with inspection or prevented with proper operating procedures. Vigilant inspections are essential since fertilizer processes can develop dangerous leaks and potential explosions; some examples are highlighted in the reference that is updated annually.[1]

Fabrication Defects found by Pneumatic Testing

A system of small diameter, thin walled, seamless UNS S31603 austenitic stainless-steel tubing used for instrument air failed during a pneumatic test after construction and before service. The pneumatic test did not reach the intended test pressure. A visual examination found axial splits on the OD surfaces of some tubes. The tubes were examined with Penetrant Testing (PT) and sectioned in half to examine the ID surfaces.

The tube ID surfaces had multiple instances of rippled flakes of metal, like a warped skin, near the axial splits found on the OD surfaces, see Figure 1. A metallographic cross section through the feature is shown in Figure 2. The seam is not through-wall in this plane. The austenite grains on either side of the seam do not match; this area did not rupture. The tube never had an intact wall after fabrication.

Figure 1. Rippled, folded metal found on ID surface coincident with axial splits

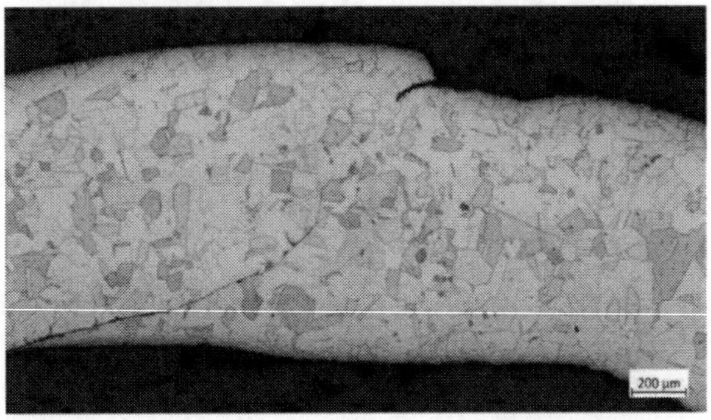

Figure 2. Portion of a fabrication Seam in cross section. Note the mismatch in grain structure on either side of the seam, 10% Oxalic Acid Etch

The warped skin – also sometimes termed inward folding – on the ID surfaces can be associated with imperfections in the casted tube blank or improper tube piercing procedures.[2,3] The folds reduced the effective wall thickness so that the pneumatic test pressure caused a ductile rupture.

Eddy Current (ET) and Internal Rotary Inspection Systems (IRIS) after fabrication can identify these seams but may require special calibration and test probes. Pneumatic or hydrostatic testing can also identify if these seams would develop rupture at service pressure. Finally, these seams could act as stress concentration or crevice corrosion points for in-service failures.

Heavy Sections of UNS S34709 (347H) Stainless Alloys

Alloy 347H with cross-sections heavier than 20 mm have had many reports of Heat Affected Zone (HAZ) damage after a short operational time (one to two years).[4] Within the 300 series stainless steel grades (including 316, 304, 321, and 347), Alloy 347 is the most susceptible to SRC. [5] 347H SRC failures have been reported in heavy wall hydrogen service piping.[6] The alloy has been assessed as having a shorter high-temperature fatigue service life than the 304H parts it has replaced.[7] In laboratory examinations, it is prone to grain-boundary liquation in the HAZ.[8] Liquation cracks have been identified in 347H fitting welds with large grains as they were welded before commissioning.[9]

As listed above References 5 and 6, have described failures of heavy 347H sections. They detail measures followed to reduce the failure incidence. The measures include optimizing welding processes, removing external reinforcement, designs minimizing secondary stresses, optimizing the PWHT, and others.[10] Users should consider the industry's findings when using heavy sections of 347H.

Monitoring the Movement of Piping Subjected to Significant Thermal Stresses

Piping systems' thermal stresses are complex. They depend on support movement and design, transient loads, constraints, and other variables. Monitoring the movement is of utmost importance, especially in reformer lines. Today, metrology-based laser scans give quick and reliable maintenance information as illustrated below. [11]

When a newly commissioned plant was started for the first time, an NPS 12 [324 mm] pipe expanded, pushing an NPS 2 [60.3 mm] drain into an I-Beam (see Figure 3). An operator noted that the NPS 2 [60.3 mm] drain had deflected, so operations immediately shut down the plant to investigate further. One question was whether the NPS 12 [324 mm] and NPS 2 [60.3 mm] pipes were plastically or elastically deformed. In other words, how prone were they to fail? Based on this information, the mitigation and redesign had to be assessed.

The NPS 12 [324 mm] piping did not appear to have local deformation, such as dents or bulges. The ovality was within the manufacturer's ±1%OD tolerance based on the measurements extracted from the laser scans – it had not deformed. Laser scanning showed the NPS 2 [60.3 mm] drain was at an angle of ~88° (instead of 90°) to the normal axis of the NPS 12 [324 mm] piping.

From the laser data acquired and supporting information, a repair plan was made, and the NPS 2 [60.3 mm] drain was repaired. Further measurements were not required after the plant continued to operate. The NPS 12 [324 mm] line remained intact without requiring further changes or additional NDT. Ideally laser scans before (though not in this instance) and after operating newly commissioned high-temperature piping can ease piping management.

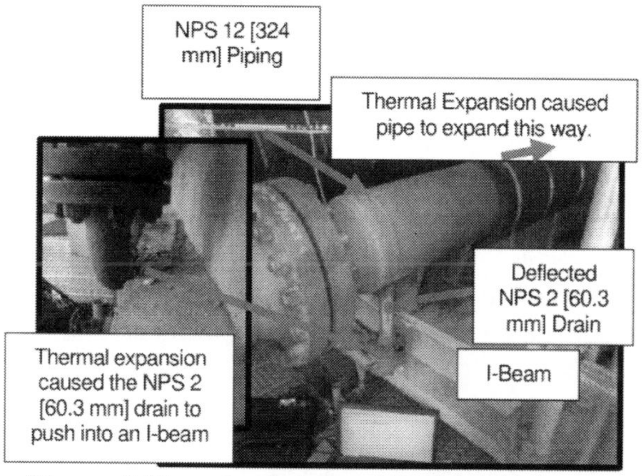

Figure 3 Note the deformed NPS 2 [60.3 mm] drain deflected away from normal

Monitoring Field PWHT of Cr-Mo Vessels

Local PWHT is now commonly used instead of the furnace PWHT used in the past: it is economical and efficient. However, it requires more diligence than furnace PWHT. The diligence importance is illustrated by a small leak from the top tubesheet to the channel weld of a 1¼Cr – ½ Mo [UNS K11789] ammonia converter boiler feed water exchanger detailed below.

The 2.125-inch [54 mm] thick-walled heat exchanger operated at 700°F [371°C]. The vessel welds were PWHT locally and in the field. The vessel developed cracks after seven years of service. [12] Only the top channel to the tubesheet butt weld cracked, not the bottom one. The previous exchanger had operated for fifteen years without developing these cracks.

Based on the fracture, metallography, and the service cracking was due to temper embrittlement. Based on chemical analyses and API RP 934-C (Materials and Fabrication of 1 ¼ Cr- ½ Mo Steel Heavy Wall Pressure Vessels for High-pressure Hydrogen Service Operating at or Below 825°F [440°C], Second Edition), the top and bottom tubesheet were susceptible to temper embrittlement. Neither the top nor bottom tubesheet met the required impact properties at -20°F [-29°C].[12]

Field portable ultrasonic contact impedance (UCI) hardness tests on meticulously prepared surfaces solved the puzzle. The HAZ of the top channel's inside diameter (ID) had hardness values exceeding the 235 HV API 934-C recommended maximum. The ID of the bottom channel and the outside diameter (OD) of the top and bottom channels had hardness values below 235 HV.

The heat treatment of the top channel's ID had not been rigorous. The Welding Research Council (WRC) Bulletin 452 (Recommended Practices for Local Heating of Welds in Pressure Vessels) should be followed to prevent damage.

Losses due to Metal Dusting

A dissimilar metal weld in a High-Pressure steam superheater developed localized losses at a dissimilar metal weld; the base metals are undamaged. The shell side has hot syngas with a high mol% of steam. Notably, the location of the tube bundle baffles makes this location difficult to access and inspect.

The base materials include a UNS S34700 tube, a UNS K21590 [2 ¼ Cr – 1 Mo] tube, and a UNS N06600 [ERNiCr-3] Deposited Weld Metal (DWM). A cross section through a less affected area is shown in Figure 4. The damage is contained entirely within the DWM; the sharp damage follows the fusion line. The base materials are essentially undamaged.

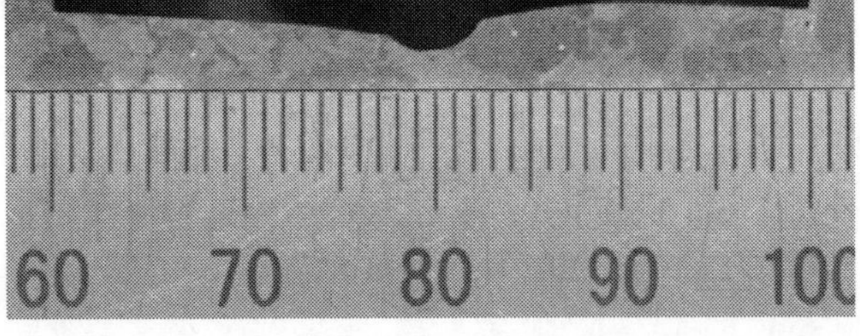

Figure 4. Cross section through a mostly intact portion of the losses. Note the damage is entirely within the DWM, no etchant.

The DWM is shown in Figure 5. The recessed areas of the DWM contain a carbon rich scale with entrained metallic compounds, consistent with metal dusting. The DWM has a pale layer and cross hatching visible after etching; targeted microhardness measured these areas at over 750 HV and 550 HV, respectively, compared to 210 HV in the unaffected area. This is consistent with the metallurgical changes caused by carburization.

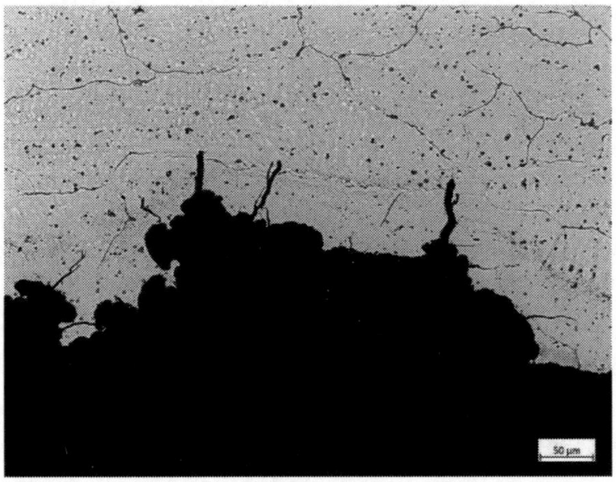

Figure 5. Metallographic image of the affected DWM. Note the cross hatching in the damaged DWM, oxalic acid etchant.

The nickel-rich DWM was particularly susceptible to metal dusting damage, especially in a system with high steam content. [13,14] Details surrounding this failure, repairs, and potential solutions have been previously reported.[15]

Losses Due to Cyclic oxidation and Carburization

A carbon steel manway cover with a refractory filled plug was observed glowing red in service. The system was taken offline to investigate; the refractory inside the plug was missing. The UNS N08810 [Alloy 800HT] shroud containing the refractory had failed. The shroud had pits all over the surface and the weld affixing the refractory support in place had also failed. The piping carries high temperature syngas. The operator reported that the process has alternated between carburizing and oxidizing conditions in the 20-year service life.

The pitted shroud is shown below in Figure 6. The pits are round and hemispherical and join to form general wastage. The pits are clean with no visible scale. Delta-ferrite readings on the shroud ranged from 1 away from the pits to over 130 near the pits.

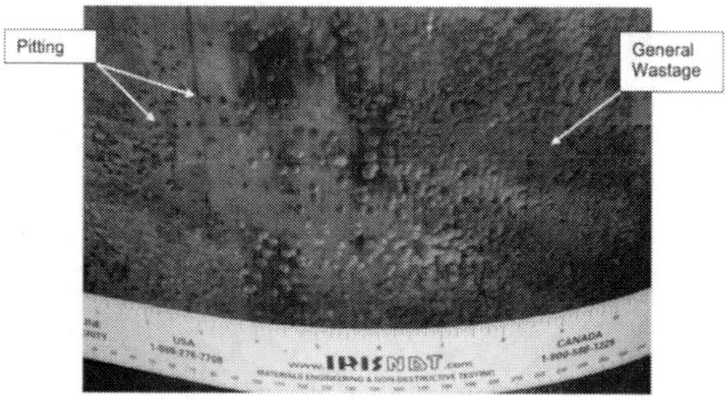

Figure 6. Shroud outer surface exhibiting clean, scale free pits.

A metallographic examination of the pits is shown in Figure 7. The pits have layers of bright white material and dark material. In some areas, grains have fallen out of the structure. Detailed Energy Dispersive X-ray (EDX) analysis shows areas of decarburized material and carburized material. In the examined DWM, the material also has thick oxide scales.

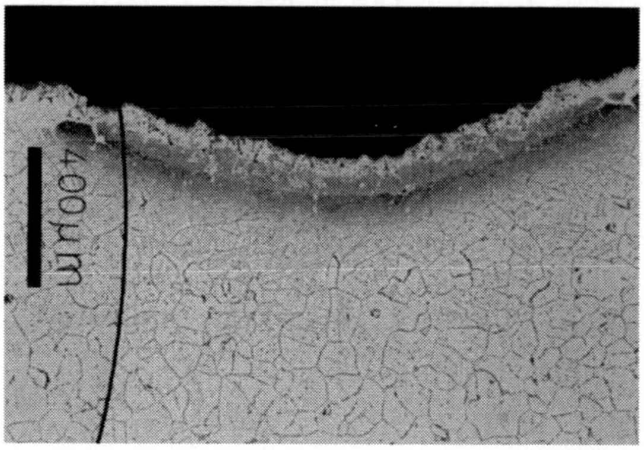

Figure 7. A typical shroud pit, showing both decarburized and carburized layers around pitted areas, 10% Oxalic Acid Etch

The shroud is suffering from cyclic carburization-oxidation damage. The carburization cycles can cause metal dusting damage; the oxidation cycles are developing oxide scales. The oxide scales spall and expose fresh material for carburization.[16] The carburization can also develop high internal stresses that can promote cracking, especially during temperature cycles.[17] Higher nickel alloys such as UNS N06601 [Alloy 601] may have better resistance; however, no alloy is immune to metal dusting conditions. The process must be carefully controlled.

Weldability Challenges due to Nitriding

Successful weld-tie ins on NPS 10 [273 mm] SCH 120 UNS S34700 (Type 347) stainless steel piping were frustrated by the development of severe porosity in the weld root. The piping was from an ammonia converter operating at high temperature for many decades. Field hardness testing indicated a substantially higher hardness value on the ID surface than at the OD surface.

A test coupon was extracted to characterize the stainless steel for nitriding. A metallographic cross section without etching is shown in Figure 8. A detailed hardness traverse showed values of 680 HV – 824 HV in the first 600 μm from the ID surface. The hardness values decrease to 175 - 190 HV at 700 μm to 2000 μm from the ID surface.

Figure 8. Metallographic image of the nitrided Type 347 piping. The unetched material has a visible band of affected material at the ID surface, approximately 650 µm deep to the unaffected material, no etchant.

A sample of the pipe was submitted for detailed wt.% nitrogen analysis by LECO. The ID material was milled out in 400 µm thick bands; each band was tested, and the results are plotted below in Figure 9. The nitrogen wt.% is as high as 5 wt.% at the ID surface and decreases with increasing depth into the material. EDX Line Scans showed a similar trend with an abrupt decrease in wt.% nitrogen at the boundary.

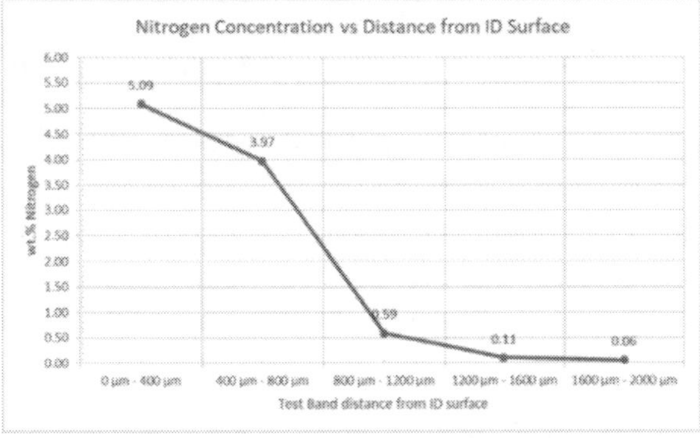

Figure 9. Plot of nitrogen content determined by LECO versus sample depth from ID surface.

The depth of the affected material was consistent in all examined samples. The weld repair was completed successfully by milling the ID to a specified depth, thereby removing all the nitrided material.

Creep Damage

Creep damage must be mentioned since it is a prevalent damage mode in the ammonia industry's high temperature equipment. An image of advanced creep load from a 50-year-old UNS J94204 (HK40) reformer tube is shown below in Figure 10. The tube did not fail; the damage was flagged by inspection with a combination of Ultrasonic Testing (UT) and Eddy Current (ET). In the worst area, more than two-thirds of the tube wall thickness had macrocracks.

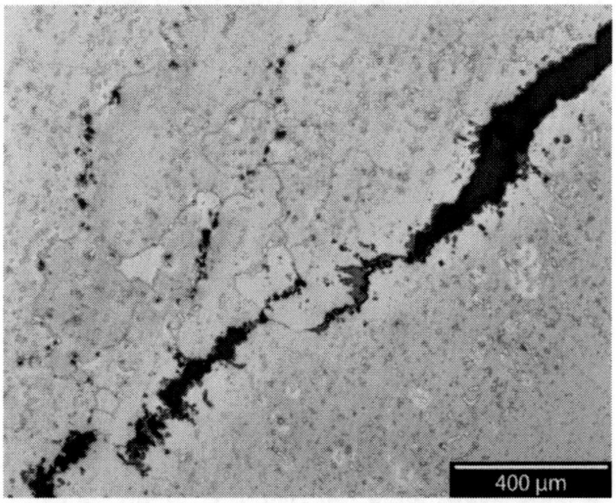

Figure 10. Advanced creep damage in a 50-year old UNS J94204 reformer tube, Vilella's Reagent etch

In a previous Ammonia Symposium Proceeding, creep was defined as "A process whereby the material slowly and progressively deforms with time under the action of stress, for steels usually at elevated temperatures [752°F, 400°C]."[18] The proceedings have detailed information from design, Non-Destructive Testing (NDT), operation, life prediction, to repair practices. The subjects cover from overall creep failures[19] to those of methanators,[20] tubes,[21] and many other reformer components.

New Developments in Monitoring and Predicting HTHA

Monitoring HTHA continues to be challenging. The Non-destructive Testing (NDT) guidelines in API RP 941 (Steels for Hydrogen Service at Elevated Temperatures and Pressures in Petroleum Refineries and Petrochemical Plants, Eighth Edition) are routinely used. Previously in 2016, new curves for non-PWHT Carbon Steels were added. This development necessitated re-evaluating existing piping for the possibility of damage.[22]

An image of incipient HTHA damage, including fissuring and decarburization, is shown below in Figure 11. This HTHA damage was found in the weld root of 55-year old non-PWHT carbon steel piping in Reform Gas service. Inspecting previously ignored piping was triggered by the published advances in API RP 941. The damage was not widespread but piping replacement with UNS K11789 (1.25 Cr-0.5Mo) was scheduled. [23]

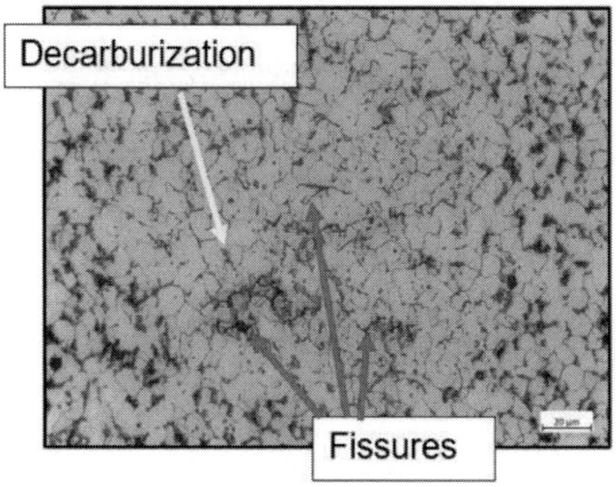

Figure 11. Incipient HTHA damage in a 55-year old non-PWHT carbon steel in Reform Gas. 2% Nital Etch

New advances in predictive assessments are developing time-dependent Nelson Curves for carbon steels and C-½Mo steels.[24,25] These new time-dependent curves may appear in future versions of API RP 941. The Materials Technology Institute (MTI) loans ex-service samples with HTHA damage to facilitate inspection regime testing for member companies. The samples are used during technician demonstrations/qualifications, to ensure that the inspection regime and operator can detect the defects present.[26]

Stress Relaxation Cracking in Valve Stem and Pins

A UNS N07750 (Alloy X-750) ammonia syngas control valve stem failed in service; the valve could not be closed. The valve plug was secured to the stem with taper pins; the taper pins also failed. The valve stem separated at the taper pin holes. The failed stem and pins were in service for over 45 years at a typical operating temperature of 1100 °F [590 °C]. An overview of the failed components is shown below in Figure 12.

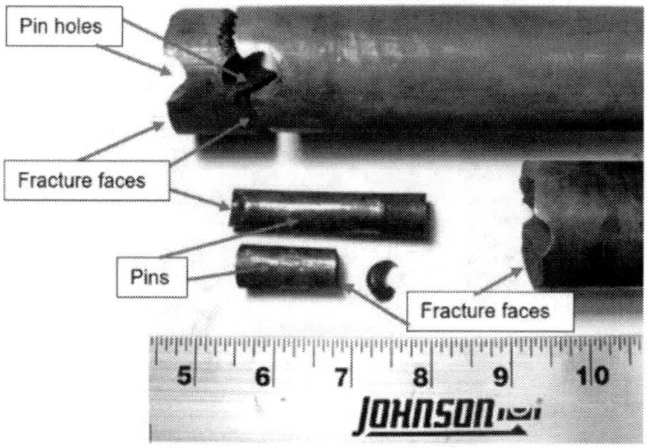

Figure 12. Overview of fractured stem (larger diameter) and pins (smaller diameter).

The stem and the pin fracture surfaces had a rock-candy macroscopic appearance. The stem fractures initiated in multiple locations within the pin bore and progressed through diameter. The taper pin fractures initiated in multiple locations around the entire circumference and progressed through the diameter. Fractography by Scanning Electron Microscopy (SEM) revealed intergranular fracture in all locations.

Metallography on the failed parts confirmed the intergranular crack modes. The cross section of the valve stem is shown below in Figure 13; an image of the intergranular cracks is shown below in Figure 14. The intergranular cracks have a filament of metal down the centre of the crack. EDX analysis of the filament shows it is almost pure nickel and the scale is iron, chromium, and oxygen rich and nickel poor. The hardness values of the stem and pins range from 300 – 330 HV.

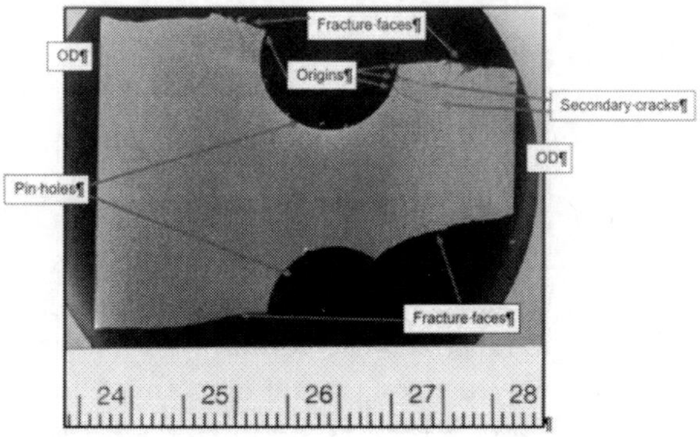

Figure 13. Metallographic cross section of the valve stem through the pin holes showing extensive secondary cracking.

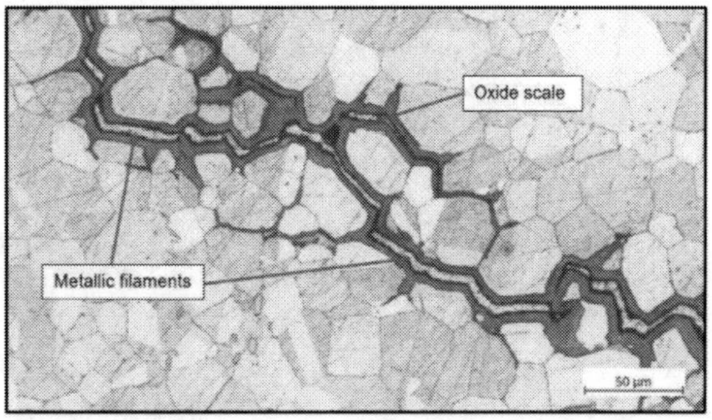

Figure 14. Stress Relaxation Cracks typical of all the examined valve stem and taper pin main and secondary fractures. The filament is nickel rich; the scale is chromium, oxygen, and iron rich and nickel poor, 10% oxalic acid etch.

The valve components failed due to SRC. The intergranular cracks, a nickel-rich filament, and a chromium rich oxide scale are all consistent with SRC. [27,28] Precipitation hardened alloys, such as UNS N07750 are susceptible to SRC especially after long-term high-temperature service causing aging of precipitates. The stress causing the failure may be from operating stresses and internal stresses. The pins likely failed first followed by the stem, since the pins would not be loaded if the stem had broken first. The long service life (over 45 years) may be sufficient and pre-emptive replacement may prevent reoccurrence. SRC can also occur in austenitic piping components. Proper design, heat treatments, and material selection can address the mechanism.[29]

Summary

The fertilizer industry requires materials with long-term reliability to prevent incidents. These materials require careful fabrication, consistent operation, and detailed inspections to prevent failures. New industry learnings should be incorporated into asset integrity programs. After failure occurs, failure investigation is useful to characterize the damage and contributing factors.

Acknowledgements

The authors would like to acknowledge our understating customers for sharing their learnings through this article. We would also like to thank our Metallography Laboratory Technologists for their excellent work producing the images shown herein. Finally, we express our sincere thanks to the IRISNDT management for their support.

References

[1] Marsh JLT Specialty. "100 Largest Losses in the Hydrocarbon Industry." March 2020. www.marsh.com. Accessed September 3, 2021.

[2] Nan-Steel, "Reasons for inward folding of seamless steel pipes," https://www.nan-steel.com/news/reasons-for-inward-folding-of-seamless-steel-pipes.html. Accessed October 10, 2023.

[3] Landee Pipe. "The Causes of Warped Skin of Seamless Steel Pipes (Part One)" https://www.landeepipe.com/the-causes-of-warped-skin-of-seamless-steel-pipes-part-one.html . Accessed October 10, 2023.

[4] Y LEE, H.. "Comparison of Stress Relaxation Cracking Susceptibility of Austenitic Stainless Steels." Supplement to the Welding Journal, September 2022. Electronic.
Sponsored by the American Welding Society

[5] Van der Westhuizen, C.E. "Stress Relaxation Cracking of Welded Joints in Thick Sections of Type-347 Stabilized Grade of Stainless Steel," Corrosion 2008, NACE-2008-08454. Electronic.

[6] Fahrion; M. E. *et al.* "Technical Basis for Improved Reliability of 347H Stainless Steel Heavy Wall Piping in Hydrogen Service, *CORROSION 2003*. NACE 2003—03647. Electronic.

[7] Ngonyoza, N., *et al.*, "Failure of a Residue Fluid Catalytic Cracker (RFCC) Catalyst Cooler's Aeration Piping System (Cat Cooler Internals)," *CORROSION 2018*, NACE-2018-11086, Electronic.

[8] Tamura, H, *et al.*, "Study on the Behaviour of Grain Boundary at Heat-Affected Zone of SUS 347 and SUS 321 Austenitic Stainless Steels." Transactions of the Japan Welding Society, Vol. 5, No. 1, April 1974.

[9] Benz, A.; Bell, M.; "Combatting Corrosion in the Fertilizer Industry," Proceedings of Ammonia Plants and Related Facilities Symposium, Vol. 62.

[10] Messer, B. *et al.* "Optimized Heat Treatment of 347 Type Stainless Steel Alloys for Elevated Temperature Service to Minimize Cracking." Corrosion 2004. NACE 2004 – 04640. Electronic.

[11] Sjerve, C. *et al*, "Using Laser Scanning Creatively in a Plant Environment." Inspectioneering, Volume 30, Issue 1. January – February 2024.

[12] Benz, A., *et al*, "Abnormal Cracks Led to Premature Decommissioning of Boiler Feed Water Exchanger – What Happened and Why?" Inspectioneering, Volume 20, Issue 5. September – October 2014.

[13] Hoffman, J.J., *et al.* "09161 - Proposed Metal Dusting Mechanism in Lower Temperature, High Steam Syn Gas." Presented at *Corrosion 2009*. NACE International. 2009. Electronic.

[14] Britton, S.L., Cooper, G.A., "15614 – New Trends in Alloy Resistance to Metal Dusting in High Steam, Low Temperature Syngas." Presented at *Corrosion 2015*. NACE International. 2015. Electronic.

[15] Butlin, C., Okulaja, S. "Inconel Weld Damage and Metal Dusting in NH_3 Front End Equipment." Ammonia Technical Manual. pp 87 – 94. 2019.

[16] Harper, M.A., *et al.* "Cyclic Oxidation Plus Carburization of Heat-Resistant Alloys," CORROSION–Vol. 51, No. 3, p191

[17] Lai, G.Y., High Temperature Corrosion and Materials Applications. Materials Park, OH: ASM International. 2007. Print.

[18] Decker, C.J., "Lifetime Assessment of NH_3,-Plants: Approach for Lifetime Extension." Proceedings of Ammonia Plants and Related Facilities Symposium, Vol. 62.

[19] Hester, M. W., P.E, "Mechanisms in Ammonia Plant Equipment - Reducing the Risk of Failure

[20] Babwah, R., "Methanator vessel overheating Why still a "Hot Topic?" Ammonia Technical Manual. pp 251 – 263. 2019.

[21] Okulaja, S., "Primary Reformer Air-Steam Coil Alloy 800HT Tube Failure." Ammonia Technical Manual, 2022.

[22] AFPM Q&A Database. "Question 15: What practices and modifications have you implemented in response to the new High Temperature Hydrogen Attack (HTHA) guidelines and updated Nelson curves?" Available at https://www.afpm.org/data-reports/technical-papers/qa-search/. Accessed April 1, 2024.

[23] Bell, M. "Planning to Fail: Some Unexpected Developments." MTI Global Symposium. Baton Rouge, LA. February 26 – 29, 2024.

[24] Dewees, D., Buchheim. "Breakthroughs in Predictive HTHA Life Assessment: Time Dependent Nelson Curves and Inspection Planning Enhancements." Inspectioneering, Volume 28, Issue 1. January – February 2022.

[25] Sutton, N., Swindeman, M., "HTHA Time Dependent Nelson Curves & Assessment of Equipment." API Inspection & Mechanical Integrity Summit. San Antonio, TX. January 23 – 25, 2024.

[26] Young, N., "Unexpected HTHA Susceptibility and Path Forward." January23 – 25, 2024 Inspection and Mechanical Integrity Summit. Electronic.

[27] Van Wortel, H., "Control of Relaxation Cracking in Austenitic High Temperature Components." TNO Science and Industry. NACE International, Paper No. 07423. Electronic

[28] Shoemaker, L.E. *et al.*, "Fabricating Nickel Alloys to avoid Stress Relaxation Cracking." 2007 Corrosion Conference and Expo. NACE International, Paper No. 07421. Electronic.

[29] API TR 942-B. "Material, Fabrication, and Repair Considerations for Austenitic Alloys Subject to Embrittlement and Cracking in High Temperature 565 °C to 760 °C (1050 °F to 1400 °F) Refinery Services. May 2017. Electronic.

Analyzing Ammonia Dispersion Under Varying Atmospheric Conditions Using DRIFT

Atmospheric conditions, such as ambient temperature and relative humidity, can influence dispersion of toxic chemicals. Ammonia is hygroscopic and therefore has complex interactions with water vapor present in the atmosphere. The integral model DRIFT has been utilized to predict ammonia dispersion and downwind concentrations for a range of temperatures and humidities. We have simulated ammonia dispersion for two types of release: (i) long-duration, typical of a leak from a hole in a vessel; (ii) instantaneous release, typical of a catastrophic vessel failure.

The two cases studied in this paper are somewhat idealized representations of what can happen during loss of containment. However, both release scenarios contribute knowledge to how a release of ammonia interacts with the environment, and how this affects downwind dispersion.

Rory Hetherington, Alison McGillivray, Simon Gant
Health and Safety Executive (HSE), UK

Gemma Tickle
GT Science & Software, UK

Introduction

A release of a pressure-liquefied industrial chemical, such as ammonia, can result in the formation of a large cloud of vapor. Depending on the release and atmospheric conditions, a cloud of ammonia can travel great distances and pose significant risks to public health and the environment.

The thermodynamic interactions of ammonia with water potentially make ammonia dispersion in a moist atmosphere sensitive to humidity and temperature. Anhydrous ammonia is hygroscopic, it therefore absorbs moisture from the surrounding air and forms a non-ideal solution, releasing heat in the process. This effect is more pronounced than with other substances that either interact weakly or not at all with water. Understanding and predicting the behavior of such releases requires accounting for both thermodynamic interactions and the non-ideal nature of the solution.

Experimentalists investigating the release and subsequent dispersion of a chemical such as ammonia, will often aim to control or at least characterize the storage and source conditions. However, the source conditions of an accidental ammonia release can be difficult to characterize. There are many factors involved, such as location, storage conditions, vessel type, and circumstances relating to the accident. Due to the variable nature of incidents, it is a challenge in modeling to specify conditions which can represent a wide variety of release scenarios. For instance, containment loss might occur suddenly due to a catastrophic failure, or gradually through a prolonged leak from a hole or crack.

In 1992 at a peanut oil mill in Dakar Senegal, a road tanker ruptured, releasing approximately 22 tonnes of liquid ammonia [1]. Overfilling led to overpressures which caused a previously repaired weld to fail, and the tanker burst open. A two-phase flow of ammonia liquid and vapor was produced. The cloud engulfed neighboring offices and restaurants, which were relatively quiet at the time due to the accident occurring during the Ramadan holidays. Even so, there were 129 fatalities and over 1000 injuries. This incident was reviewed and modeled in [2] using the DNV Phast integral model. Two scenarios were modeled: instantaneous release of 22 tonnes; and a continuous release from a hose failure. Dispersion modeling predicted concentrations of 1500 ppm up to 1.5 km away from the initial tanker position.

In 2019, a farm worker driving a tractor was towing a fertilizer applicator and a trailer carrying two 1000 gallon nurse tanks of ammonia on a public road in Beach Park Illinois. The applicator hose disconnected from the bulkhead adapter, releasing approximately 650 gallons of ammonia. A dense cloud of ammonia vapor was produced, leading to a one mile shelter in place order. A total of 83 people were evaluated at hospital, of which 14 were subsequently hospitalized [3]. Luckily there were no deaths. The time of release (early morning at 04:24), was probably a factor in this, in addition to the size of release (approximately 1.5 tonnes). In an investigation report, the National Transportation Safety Board in the United States concluded that a worn applicator coupling disconnected from the adapter, releasing ammonia through a 1 inch diameter orifice [4]. ASOS weather data for the nearest airport (Waukegan National Airport) indicates a low wind speed of ~0.5 m/s and relative humidity of 85% during the time of release. The orifice was pointing towards the tractor cabin, suggesting a momentum jet could have been impinging onto the tractor.

This brief review of two previous incidents (Dakar 1992 and Illinois 2019) highlights the difficulty in prescribing initial conditions for models. In this study we simulate two different types of release:

- Long-duration release, typical of a hole in a vessel;
- Instantaneous release from a catastrophic failure.

These simulations are not tailored to replicate specific accidents or incidents. Instead, they aim to explore a range of conditions and variables that could potentially occur during an ammonia incident. By doing so, the simulations provide insights into fundamental behaviors of ammonia dispersion in various conditions.

We model the problem using the integral model DRIFT. As one of the few operational models equipped with a sub-model specifically designed to account for ammonia-water interactions, DRIFT is appropriate for this investigation. In addition, it is relatively fast to run, allowing us to comprehensively explore the input space.

The main objective of this study is to examine the impact of atmospheric conditions, such as humidity and temperature, on the dispersion of ammonia.

Methodology

Long-duration release

Due to the difficulty in characterizing the initial conditions for a release such as the Beach Park incident, we chose to model a large-scale pressure-liquified experimental release of ammonia. The Desert Tortoise field trials conducted in 1983 are chosen as a suitable test case. Conditions for the selected Desert Tortoise trials are mostly taken from the SMEDIS database [5]. Temperature and pressure were recorded at the orifice. The release rate of 80 kg/s was estimated from the initial mass of ammonia in the vessel and the time taken to empty (126 s). For this release we increase the release duration to 10000 s, to produce a long-duration type release.

For the purpose of this study, 100% liquid has been assumed at the exit nozzle which is consistent with the normal assumption for a padded release, but does not account for flashing that may have been induced by pressure losses, e.g. due to the presence of a knee-bend immediately upstream of the exit orifice in Desert Tortoise.

Instantaneous release conditions

A source-term calculation of the expansion from storage to atmospheric conditions is not carried out in this study. This allows us to focus on the downwind dispersion behavior. We model the release from the point at which the ammonia has reached atmospheric pressure (101325 Pa or 14.7 psi) and temperature is at its boiling point (-33.34 °C).

The total mass of ammonia is 20 tonnes, similar to the Dakar incident which was 22 tonnes. To simulate different fills of a vessel, we take the liquid fraction to vary from 0 to 0.3. This might be typical of a road or storage tanker, where there is a vapor space above the liquid fill line. Ammonia vapor at 20 °C and atmospheric pressure has a relative density compared to air of approximately 0.6 [6]. Therefore when the liquid fraction is zero, the cloud is expected to be initially buoyant, and remain buoyant as it disperses into the atmosphere. The cloud is initially stationary and released from ground level. The aspect ratio of the initial cloud is assumed to be 1, i.e. height equal to width.

Problem overview

Table 1 summarizes the source conditions for the two release configurations.

Parameter	Unit	Long-duration	Instantaneous
Source type	-	Momentum jet	Catastrophic
Orifice diameter	m	0.081	-
	inch	3.19	-
Temperature	°C	21.5	-33.34
Release pressure	Pa	1.01E+06	101325
	psi	146.5	14.7
Liquid fraction	-	1	0.0, 0.1,..., 0.3
Release rate	kg/s	80	-
Release duration	s	10000	-
Inventory	tonne	-	20
Release location	m	(0, 0, 1)	(0, 0, 0)

Table 1 Geometry and source conditions for the two types of release.

We wish to study the dispersion of ammonia from two different release types. Therefore, initial conditions which are inputs to the dispersion model will vary between the two different release scenarios. However, where possible, conditions have been matched between the two scenarios. This allows for a comparison between the two cases. A summary of the atmospheric conditions is provided in *Table 2*.

Parameter	Unit	Value
Temperature*	°C	0, 3, ..., 30
Pressure	Pa	101325
	psi	14.7
Relative humidity*	%	0, 10, ..., 80
Reference height	m	2
Roughness length	m	0.003
Pasquill stability class	-	D
Friction velocity	m/s	0.442
Wind speed	m/s	2

Table 2 Atmospheric conditions. Parameters marked with an asterisk are sensitivity inputs to the model.

The atmospheric stability was assumed to be neutral (Pasquill D), with a wind speed of 2 m/s. We assume zero rainout because this leads to the formation of a pool and subsequent vaporization. This is considered to be a separate problem which will not be considered in this study, but could be included in future investigations. In addition, there is no heat transfer to or from the ground.

Numerical model

Both release scenarios are modeled using DRIFT version 3.7.19. DRIFT (Dispersion of Releases

Involving Flammables or Toxics) is a gas dispersion model, originally developed by the UK Atomic Energy Authority (UKAEA), and subsequently maintained by ESR Technology, with the support of the UK Health and Safety Executive (HSE). DRIFT is used within HSE to model atmospheric dispersion of toxic and flammable substances for the purpose of providing public safety advice on hazardous substance consent applications and land-use planning. Model evaluation of DRIFT has been undertaken for a variety of release scenarios [7, 8]. A mathematical description of DRIFT can be found in the report by Tickle and Carlisle [9].

In total, given all possible combinations of the sensitivity inputs, there are 99 runs for the long-duration release, and 396 runs for the instantaneous release.

Concentrations are measured along the centerline ($y = 0$) for a range of downwind distances ($x = 0$ to 10 km). To assess cloud lift-off for the instantaneous case, concentrations at each downwind coordinate x are recorded at 3 separate heights: $z = 1, 10, 100$ m.

The distance to AEGL-3 (10 mins exposure, 2700 ppm) are reported in the analysis as a measure of the clouds' extent. According to the National Research Council [10]: "AEGL-3 is the airborne concentration (expressed as ppm or mg/m^3) of a substance above which it is predicted that the general population, including susceptible individuals, could experience life-threatening adverse health effects or death".

Results: effect of liquid fraction on the instantaneous release

We first focus our attention on the instantaneous release. To determine if the cloud becomes buoyant and lifts-off from the ground, concentrations are reported at 1, 10, and 100 m. At each downstream location x, the height at which the maximum concentration is reached is recorded. We then assess the cloud behavior according to the following three characteristics:

1. The cloud is initially buoyant and remains buoyant. The height at which the maximum concentration is found is always greater than 1 m.
2. The cloud is initially buoyant, but further downwind the maximum concentration is recorded at the lowest height of 1 m.
3. The cloud travels along the ground, such that the maximum concentration is always recorded at 1 m.

By applying these criteria for each case and assigning a value on the behavior, we can then correlate this to the values given to the input parameters.

A Pearson correlation test is performed for each variable and reported in *Table 3*. The Pearson correlation coefficient (r) measures the strength and direction of a linear relationship between two variables, ranging from -1 (perfect negative correlation) to +1 (perfect positive correlation), with 0 indicating no linear relationship. The P-value assesses the statistical significance of this correlation; a low P-value (typically below 0.05) implies a statistically significant relationship. Conversely, a high P-value indicates a lack of statistical evidence for a significant correlation.

The strongest correlation between lift-off behavior and input parameter is observed with the liquid fraction. This suggests that liquid fraction is a dominant input, which greatly influences cloud buoyancy and therefore whether it lifts off from the ground.

Input variable	Correlation Coeff (r)	P-Value	Correlation
Liquid fraction	-0.90	0.00	Strong negative correlation.
Relative humidity	0.11	0.00	Weak positive correlation.
Temperature	-0.05	0.13	No statistical significance.

Table 3 Pearson correlation coefficient and P-value for testing correlation between input variables and cloud lift-off.

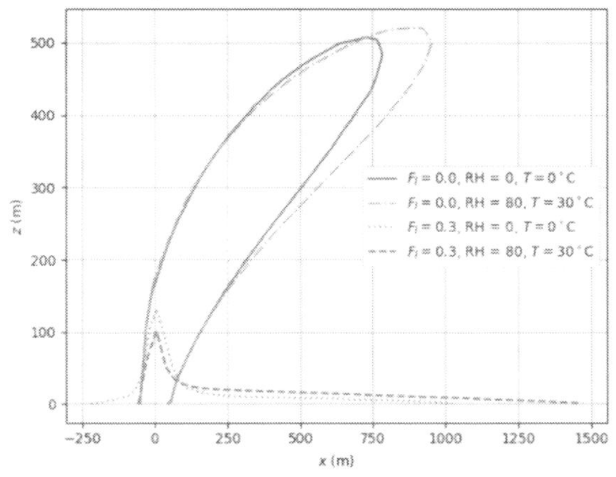

Figure 1 Contours of AEGL-3 (2700 ppm) for various liquid fractions $\boldsymbol{F_l}$*, temperatures, and relative humidities.*

Contours of AEGL-3 (2700 ppm) from a side-view of the cloud are presented in *Figure 1*. In these figures the wind is travelling left to right. These represent the maximum extent at which 2700 ppm is recorded, and are therefore not a snapshot of the cloud at a particular time. *Figure 1* illustrates the findings of *Table 1*- here the main difference between the contours of AEGL-3 (2700 ppm) stems from the value of the liquid fraction. For a gaseous release where $F_l = 0$, the cloud is initially buoyant and remains buoyant. Increasing the liquid fraction to 0.3 results in dense-gas dispersion behavior.

Going forward, we only consider the instantaneous release with a liquid fraction of 0.3. This is in order to compare concentration values with the long-duration release at a height of $z = 1$ (m).

Results: effect of weather conditions on dispersion

Figure 2 is a plot of concentration as a function of downwind distance at a height of $z = 1$ m. A select number of cases have been chosen to illustrate the difference between the two types of release. The instantaneous release produces concentrations near the source that are nearly two orders of magnitude lower than the long-duration leak. In order to understand this significant reduction between the long-duration and instantaneous release, an additional simulation is run with an averaging time of 0 s (effectively no time-averaging) for the case with temperature 30 °C and relative humidity 80%. It is demonstrated that an averaging time of 600 s reduces concentrations compared to the case where there is no averaging. This effect is particularly marked for the instantaneous release due to its resulting concentration time profile being relatively short duration.

For the instantaneous case, lower ambient temperatures lead to higher concentrations in the region $x < 10$ m. However, the opposite is true for $x > 10$ m, where higher ambient temperatures produce higher concentrations. A reasonable agreement is found between the current data (instantaneous, 30 °C, 80% RH) and the results of [2], who modeled a similar release.

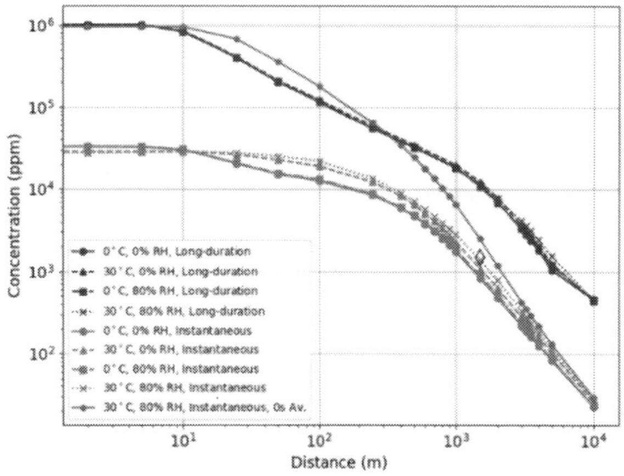

Figure 2 Concentration profiles at a height of 1 m. A black diamond marker represents the distance to 1500 ppm reported in [2]

Figure 3 and *Figure 4* present the distance to AEGL-3. The color bar limits are automatically pinned to the minimum and maximum of the data. As was identified in *Figure 2*, the long-duration release results in greater distances to AEGL-3, because a larger cloud is produced. Increasing the temperature from 0 to 30 °C and the relative humidity from 0 to 80% leads to an increase in the distance to AEGL-3 of: (i) 15% for the long-duration release, (ii) 35% for the instantaneous release. Although the concentrations predicted from an instantaneous release are lower, the ambient temperature and relative humidity have a greater effect on this type of release, in relative terms.

However, the contours also reveal how the parameters temperature and relative humidity affect the cloud extent. In the long-duration case, the distance to AEGL-3 increases as both temperature and relative humidity increases. The distribution is smooth, unlike the contour produced for the instantaneous case where there is a sharp transition in the temperature range 10 to 15 °C.

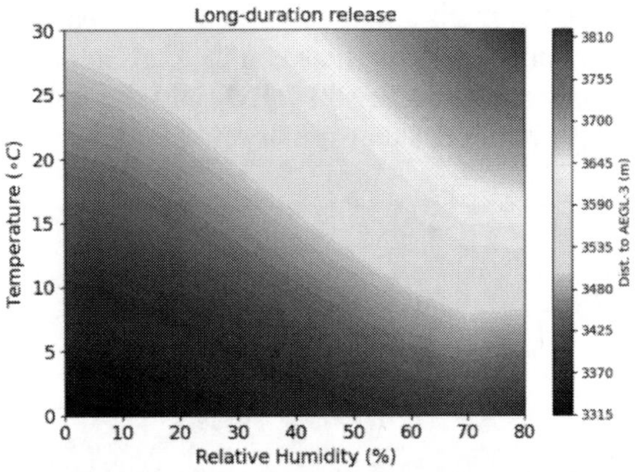

Figure 3 Contours of distance to AEGL-3 for a long-duration release.

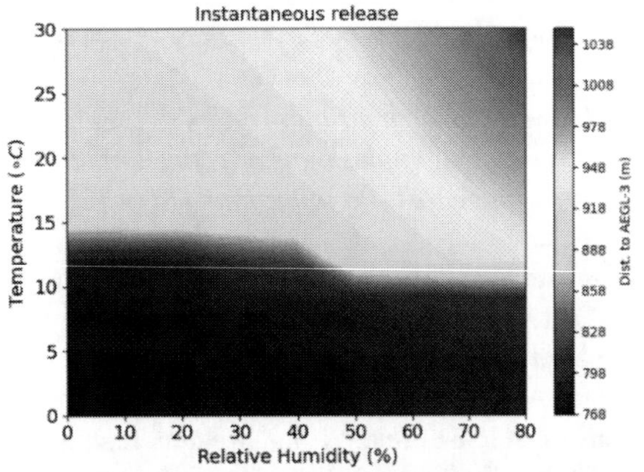

Figure 4 Contours of distance to AEGL-3 for an instantaneous release.

To further explore this, we extract the data at ambient temperatures of 12 °C, and plot the relative concentration as a function of distance and relative humidity in *Figure 5* and *Figure 6*. Concentrations are normalized by the 0% humidity case, which is why the relative concentration is equal to 1 at 0% RH.

In the long-duration case, increasing humidity has little effect for $x < 1$ km. However, a dip in the profiles of relative concentration at $x = 1$ km indicates that an increase in humidity leads to a slight reduction in concentration. Further downwind, the relative concentration increases to a maximum of approximately 1.2. Therefore, increasing the humidity from 0 to 80% results in a 20% increase in the concentration in the region $1 < x < 10$ km.

For the instantaneous release, an increase in the relative humidity has a more pronounced effect. In fact, there is a maximum increase in concentration of roughly 40%. There are two peaks observed in the profiles of relative concentration in *Figure 6* at approximately 100 m and 1500 m.

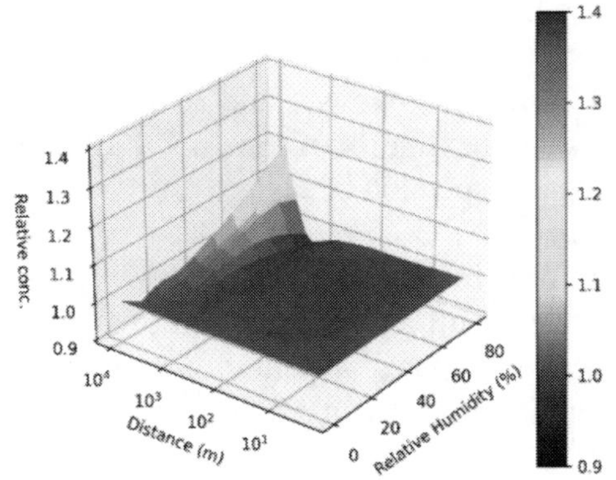

Figure 5 Relative concentration profiles as a function of distance and relative humidity for the long-duration release. Data are normalized by the concentration at 0% relative humidity.

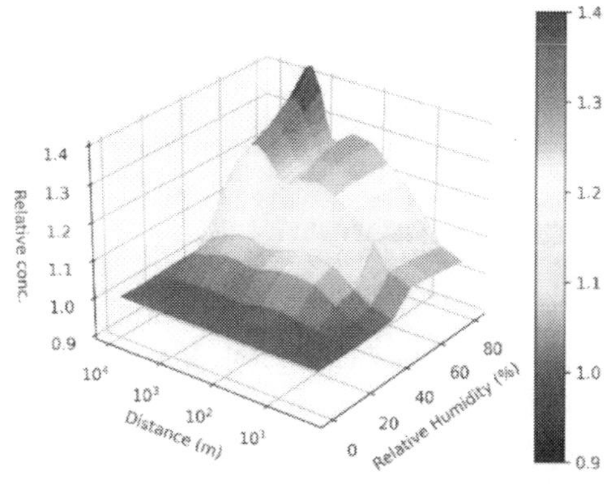

Figure 6 Relative concentration profiles as a function of distance and relative humidity for the instantaneous release. Data are normalized by the concentration at 0% relative humidity.

Conclusions

Atmospheric conditions, such as ambient temperature and relative humidity can affect dispersions of toxic chemicals, including ammonia which is hygroscopic and therefore has complex interactions with water vapor present in the atmosphere.

We have simulated ammonia dispersion for two types of release: (i) long-duration, typical of a leak from a hole in a vessel; (ii) instantaneous release, typical of a catastrophic release. These two cases are somewhat idealized representations of what can happen during loss of containment.

The integral model DRIFT has been utilized to predict ammonia dispersion and downwind concentrations for a range of temperatures and humidities. For the instantaneous release, the liquid fraction in the initial cloud was varied, and this was shown to have a strong impact on the cloud's behavior. A liquid fraction of 0.3 was found to result in a dense blanket cloud of ammonia for all inputs tested. Conversely, a liquid fraction of 0 produces a buoyant cloud, due to the lower density of ammonia compared to air.

The relationship between dispersion behavior and ambient conditions (temperature and humidity) was complex. The distance to AEGL-3 (2700 ppm) has been used as a marker to assess the cloud size. Generally speaking, for both long-duration and instantaneous releases, increasing the temperature and humidity resulted in a greater distance to AEGL-3. However, when considering a fixed temperature, the effect of varying humidity was dependent on distance from the source. Concentrations in the far field ($x > 1$ km) for the high humidity case at 15 °C were approximately 20% higher for the long-duration release, and 40% higher for the instantaneous release.

This study has shown that humidity and temperature have a modest effect on ammonia dispersion behavior. The effect is more pronounced for the case of an instantaneous release compared to a long-duration release. In addition, the effect is dependent on downwind distance.

Further work could explore the sensitivity of liquid fraction on the buoyancy and lift-off behavior of the cloud in the range 0 to 0.3.

References

1. French Ministry of the Environment 2006. "Explosion of an ammonia tank, March 24, 1992, Dakar, Senegal". DPPR/SEI/BARPI ARIA No. 3485. https://ammonia.co.nz/wp-content/uploads/2021/10/Senegal1992-1.pdf. (accessed 19 March 2024).
2. Dharmavaram, S. and Pattabathula, V., The Dakar Ammonia Accident: Analysis of the Worst Incident at an Anhydrous Ammonia User. https://www.uvu.edu/es/docs/paper4a-dakar_accident_final.pdf. (accessed 19 March 2024).
3. Rispens, J.R., 2020. Anhydrous Ammonia Chemical Release—Lake County, Illinois, April 2019. MMWR. Morbidity and Mortality Weekly Report, 69.

4. Hazardous Materials Investigation Report: HZIR-22/01 www.ntsb.gov/investigations/AccidentReports/Reports/HZIR2201.pdf. (accessed 07 March 2024).
5. Carissimo B., S.F. Jagger, N.C. Daish, A. Halford, S. Selmoer-Olsen, K. Riikonen, J.M. Perroux, J. Wurtz, J.G. Bartzis, N.J. Duijm, K. Ham, M. Schatzmann and R. Hall, 2001: The SMEDIS database and validation exercise. International Journal of Environment and Pollution. 16 (1-6), p614-629.
6. Coolpropgit web2py https://ibell.pythonanywhere.com/. (accessed 19 March 2024).
7. Coldrick, S. and D. Webber, 2017: Evaluation of the DRIFT gas dispersion model version 3.6.4. HSE Research Report RR1100. https://www.hse.gov.uk/research/rrhtm/rr1100.htm (accessed 07 March 2024).
8. Cruse, H., McGillivray, A., Chaplin, Z., Coldrick, S., Tickle, G., Tucker, H. and Macbeth, R., 2016. Gas dispersion modelling using the DRIFT 3 model to assess toxic and flammable chemical major hazards. In IChemE Hazards 26 Conference. https://www.icheme.org/media/11764/hazards-26-paper-27-gas-and-vapour-dispersion-modelling-using-the-drift-3-model-to-assess-the-safety-from-toxic-and-flammable-chemical-major-hazards.pdf (accessed 07 March 2024).
9. Tickle, G. and J. Carlisle, 2008: Extension of the dense gas dispersion model DRIFT to include buoyant lift-off and buoyant rise. HSE Research Report RR629. https://www.hse.gov.uk/research/rrhtm/rr629.htm. (accessed 07 March 2024).
10. National Research Council, Division on Earth, Life Studies, Board on Environmental Studies, Committee on Toxicology and Committee on Acute Exposure Guideline Levels, 2007. Acute Exposure Guideline Levels for Selected Airborne Chemicals: Volume 6. https://www.epa.gov/sites/default/files/2014-11/documents/ammonia_final_volume6_2007.pdf. (accessed 19 March 2024).

Acknowledgement

The contributions to this publication by HSE staff were funded by HSE. Its contents, including any opinions and/or conclusions expressed, are those of the authors alone and do not necessarily reflect HSE policy.

IMTP Packing Carry Over: Massive Damage in 2nd Packed Bed of Ammonia Plant CO_2 Stripper

The Kaltim 5 plant utilizes the KBR Purifier process to produce 2500 MTPD of Ammonia. The CO_2 removal unit decreases CO_2 content in process gas by using two stages amine absorption-stripping process, which utilizes columns with random packed beds. In 2020 this plant experienced massive damage in the 2nd IMTP packed bed of the CO_2 stripper when an emergency shutdown was activated, leading to fifteen days of plant downtime. This phenomenon teaches a lot of lessons for both operability and system reliability.

Muhammad Fahmi, Harianto Tondeang
PT. Pupuk Kalimantan Timur

Introduction

Ammonia Plant-5 (Kaltim-5 PT. Pupuk Kaliman Timur) uses the KBR-Purifier Ammonia process technology. A CO_2 removal unit is used to decrease the level of CO_2 in process gas form 17-18% to 0.01% through an aMDEA absorption-stripping process. The absorber and stripper are equipped with Intalox Metal Tower Packing (IMTP) random packing beds to facilitate vapor-liquid contact.The stripper column contains 140 m^3 of packing, installed in two section. In the stripper, CO2 vapor is stripped from rich aMDEA solution using a reboiler with heat provided by process gas and steam. The balance between liquid down-force and vapor up-force determine mechanical stability which impacts CO2 removal reliability.

Failure and Operational Symptoms

On November 11, 2020, the ammonia plant was operating normally at 102% of design production rate, around 2560 MTPD. The CO_2 removal unit operated at 105% of design circulation rate which consist of 720 m^3/hour (72000 Liter/hour) lean solution and 3840 m^3/hour semi lean solution, with sufficient concentration of activated MDEA around 39.8%.

At 4.30 p.m., the ammonia plant experienced an emergency shutdown. There was significant pressure decrease in feed gas supply. The feed gas supply pressure suddenly decreased from 38 kg/cm^{2}g to 22 kg/cm^{2}g and activated the interlock system (IS) of FALL-1042, low-low flow of feed gas. An Uninterruptable Power Supply (UPS) power failure in motorized solenoid shut-off control valve for feed gas distribution in the upstream battery limit of ammonia plant was the main cause.

There were some operational parameters and events changes in CO2 removal unit when the emergency shutdown occurred. These were:

1. There was a fluctuating differential pressure followed by an increase in the differential pressure indications of PDI-1028 and PDI-1029, pressure drop indicator for 1st and 2nd bed of CO_2 Stripper. This happened a couple of minutes after emergency shutdown.
2. Solution level in CO_2 Stripper significantly decreased from 55% to 0% for a few minutes and then suddenly increased to more than 90% level.
3. Lean solution flow suddenly increased from 720 m^3 /hour (72000 Liter/hour) to 800 m^3/hour.

4. CO_2 Stripper bottom temperature seemed to be slightly decreased from 119.5°C (392.65 K) to 113°C.
5. Stripper column pressure decreased from 0.7 kg/cm^{2}g to 0.2 kg/cm^{2}g.
6. The CO_2 compressor had shutdown after 8 minutes.

The process system stabilization procedures were applied to secure catalysts and rotating equipment, such as compressors and turbines. A few hours later, the feed gas supply returned to normal supply pressure. The ammonia plant was then restarted. The start-up process was normal and smooth until the plant reached 60% natural gas feed rate, and heat-up of aMDEA solution and circulation of lean solution from CO_2 stripper to CO_2 absorber through lean solution pump began. In this moment, several operational obstacles came up, such as:

1. Cavitation of lean solution pump. The discharge pressure and flow were fluctuating and a vacuum condition was observed when a pump casing vent was opened.
2. A couple of minutes before those cavitation events of lean solution pump, pressure drop indications of all CO_2 stripper packed bed were fluctuating followed by an increased differential pressure
3. Operators verified the phenomenon in point 1 by trying to switch mechanical three way filter by the lean solution pumps, but this effort was useless. The three-way valve switcher jammed. It seemed to be stuck by a hard object.

The CO_2 removal unit was deactivated/shutdown and all solution was drained to the slop tank for further inspection. The mechanical team opened and checked mechanical filter STR-1026 in outlet piping of CO_2 stripper column.

Shocking Findings: Massive Damage

The first shocking finding was the mechanical filter STR-1026. It was full of IMTP packings from the stripper packed bed.

Figure 1. First finding. CO_2 Stripper IMTP packing filled the mechanical filter STR-1026

After confirming those first findings of what seemed to be massive structural damage/failure in CO_2 stripper packed bed, management decided to shutdown the ammonia plant for further inspection. The technical inspector conducted a comprehensive inspection of all CO_2 removal systems, especially the stripper column. The inspection found the 2nd bed structure was totally damaged. Additional findings of further inspections were:

1. IMTP CO_2 stripper's packing carried over to these locations (Figure 2):
   - Mechanical filter STR-1026.
   - Outlet piping of CO_2 stripper.
   - Inlet liquid piping of CO_2 stripper reboiler.
   - Shell side of CO_2 stripper reboiler.
   - Outlet vapor-liquid piping of CO_2 stripper reboiler.

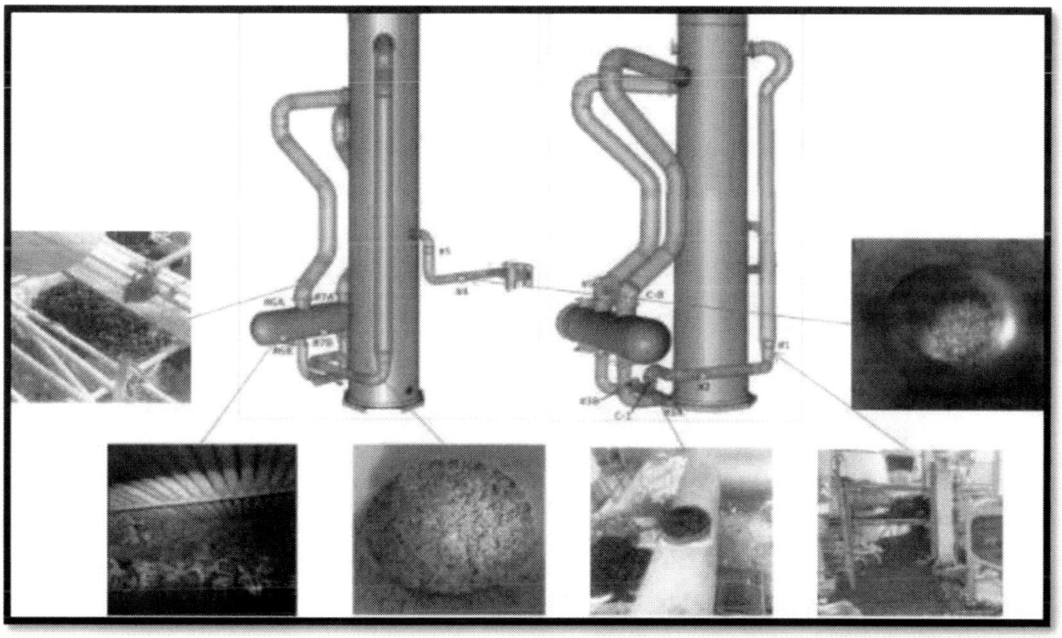

Figure 2. Layout of packing carry over

2. Support grid of 2nd packed bed collapsed into reboiler draw off pan.
3. Reboiler draw off pan plate was damaged with a heavy dent.
4. Dislocation and deformation of 2nd packed bed liquid distributor.
5. Dislocation and deformation of liquid distributor centre beam support.
6. Dislocation and deformation of beam stopper and bed limiter.
7. Damage to draw off pan.
8. Dislocation of some liquid distributor nozzles for both 1st bed and 2nd bed.
9. Looseness in 1st bed limiter.

Figure 3. Support grid and bed limiter collapsed into reboiler draw off pan

Figure 4. Deformation of support grid

10. Damaged, dented, corroded and eroded to the carbon steel Intalox Metal Tower Packing (IMTP) as shown in Figure 5:

Figure 5. Carbon steel IMTP ran suffer corrosion

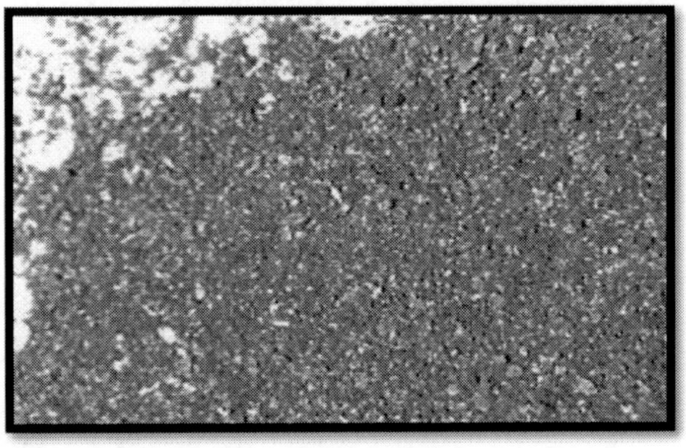

Figure 6. Corroded debris dirt

Root Cause Analysis (RCA) and Process Evaluation

1. For these massive damages, the root cause analysis and process evaluation concluded several root causes, which are:Abundance of debris dirt/carbon corrosion product in 2nd packed bed. The corrosion products made the bed void fraction to be smaller and caused a resistancy for liquid to flow. At the end resulting in liquid flooding in packed bed may occurred.
2. The fluctuating pressure in CO_2 removal system when the emergency shutdown occurred. The aggravating factor is the existence of heat load in the stripper reboiler. Up-force through excessive vapor from bottom can not pass through packed bed due to point 1. At the point when the up-force is getting too large and ravaged the packed bed structure resulting in massive damage.
3. The other aggravated factor is the decrease of stripper pressure due to CO_2 compressor suction. The CO_2 compressor created a "pull effect" on at this condition resulting in excessive "push force" (boiled up vapor) to destroy the bed structure.
4. The support grid and packed bed structure which isa floating design can not withstand this fluctuating pressure.

Main Repair Program

The restoration / repair program took around fifteen days. Based on these massive damages to both internal column structure, piping and reboiler, there were several main repairments, which were:

1. New fabricated bed limiter was 2 mm thicker than the old one.
2. Support grid was enhanced by additional clamp to fix packed bed and prevent floating during operation.
3. Replacement of all Intalox Metal Tower Packing or carbon steel IMTP packing with stainless steel IMTP packing.
4. Replacement of the broken draw off pan floor.
5. Recondition of all deformed parts such as beam support, support grid, liquid distribution and beam support.
6. Recondition all temporary unloading/access windows.

Figure 7. Support grid reinstallation and additional clamp in bed support grid

Figure 8. Replacement of stainless steel IMTP packing

Recommendations

The Reliability and RCA team also suggested technical and process recommendations, which are:

1. Replacement of carbon steel IMTP to stainless steel IMTP. **(done)**
2. Apply additional clamp support for 2nd bed support grid to strengthen and fix bed position during normal operation. **(done)**
3. Replacement of Uninterruptable Power Supply (UPS) with industrial grade UPS. **(done)**
4. Rearrange logic instrumentation for feed gas supply shut-off control valve from fail close to fail open mode. **(done)**
5. Re-HAZOP CO_2 removal unit with following recommendations:
   - Implement alarm of CO_2 compressor shutdown in CO_2 removal DCS of ammonia plant **(done)**
   - Revise emergency shutdown work instruction, to activate/open stripper reboiler bypass line. The purpose is to minimize heat load in stripper reboiler. **(done)**
   - Apply alarm annunciator for an increase in packed bed pressure drop indications. **(done)**

Operating Experience and Improvement in Operational Practices in 32 years-old SS-316 Lined Urea Reactors

This paper describes the operating experience of 32-year-old SS-lined Urea Reactors installed in Fatimafert Limited, Pakistan. Two Urea reactors designed by Urea Technologies Incorporated (UTIC) were installed in 1991 during the revamp of the ammonia–Urea complex. The reactors are multi-layered vessels with SS-316 lined inner shell fabricated by Nooter, Austria. Corrosion is always a big challenge due to the corrosive nature of fluids in this service and severe operating conditions (Pressure = 3200 psig (22,063 kPa) and Temperature = 383°F (195°C)) coupled with older metallurgy. The reactors with a design life of 20 years were operated with close monitoring and day-to-day adjustment of all operating parameters and passivation air to protect the liner and internals from acute thickness loss and to avoid catastrophic release of hazardous substances like Ammonium Carbamate, Ammonia and/or Urea. Over the period, the control regimes were also reviewed and benchmarked with other industries and the standard operating procedures (SOPs) were modified to achieve a better life. Reactors holding time after the plant shut down, as recommended by Process Licensor, was also reviewed and benchmarked with other plants and optimized to reduce liner corrosion. All the control measures and vital monitoring reduced the reactors' corrosion potential by 35%. Stamicarbon also carried out the remaining life assessment. The details of the Life assessment report and measures taken are also presented in the paper.

Muhammad Usman Liaquat
Fatimafert Limited, Lahore, Pakistan

Adeel Ahmad
Fatimafert Limited, Lahore, Pakistan

Introduction

Fatimafert Limited, an Ammonia and Urea manufacturing complex situated near Lahore, Pakistan, was commissioned in 1971 (previously Dawood Hercules Chemicals). Mitsui Toatsu originally designed its Urea plant having capacity of 1100 MTPD. The plant's total capacity was increased to 1350 MTPD in 1989 and it was redesigned and upgraded by Urea Technologies Incorporated (UTIC). Two new Urea reactors, fabricated by Nooter, Austria, were installed in this revamp and have been in service since then. The plant recently achieved 1630 MT per day of urea production after another inhouse debottlenecking and essential equipment replacement activity.

Urea Manufacturing

Urea is the first organic compound formed artificially by Wohler in 1828. This started the new era of manufacturing several other organic compounds, as well as research into newer and more efficient ways to manufacture Urea. This resulted in the utilization of the reaction where ammonia and carbon dioxide were used to form ammonium carbamate at high temperatures and pressures. This ammonium carbamate was dehydrated to give Urea and water. The process was first described by Bassarow in 1870 [1] and gained wide commercial success with the Haber-Bosch process of ammonia manufacturing, where carbon dioxide was produced in high quantities [1].

Urea today is commercially manufactured using Bassarow's process. Many Urea manufacturing technologies have been used commercially. The process employed at Fatimafert Limited is "a total heat recycle process".

Fatimafert Urea Plant:

One of the main features of this process is its energy conservation in two ways:

1. The introduction of medium-pressure carbon dioxide reduces power load.
2. Using the heat of reaction of ammonia and carbon dioxide as heating media.

The urea plant comprises two parallel urea Synthesis lines: Urea lines A & B (CO_2 Compressors, HP Ammonia and HP Carbamate pumps, and Urea Reactors). The 2 lines join right after reactors and decomposition, recovery and Finishing sections are typical.

Both of the reactions (ammonium carbamate and urea formation) take place in a reactor with a special configuration to accommodate these two reactions, which is explained further in the next section.

Reactors Configuration

Urea reactors installed at the Fatimafert Limited plant facility were designed and fabricated by Nooter Corporation, Austria. The internals of the reactor are designed to incorporate both intermediate and final reactions of the urea manufacturing process. Figure 1 shows a schematic of the reactor. There are 28 tubes of a high-grade austenitic stainless-steel material called SANDVIK 2RE69. Raw materials enter in the tube side from the top, where the first exothermic reaction of ammonium carbamate formation takes place, and this heat from the tubes is then utilized for the second chemical reaction of urea formation inside the shell. Tubes are further divided into two sets of 14 tubes each, increasing the contact area for reaction as well as heat transfer.

The outer shell of the reactor is made up of carbon steel and needs a protective layer to protect it from corrosive conditions. Another high-grade austenitic stainless-steel material, SA-240-316L SS, protects the inner side of the reactor. There are two layers of this material installed in the reactor; one layer has a thickness of 0.512 in (13 mm), and another layer beneath it is 0.24 in (6 mm). There are 11 baffle plates inside the shell to increase contact area and residence time, and finally, ~33% urea product solution leaves from the reactor top.

Tube bundles are supported by baffle plates, and the whole assembly of 28 tubes and 11 baffle plates, along with the top distributor and bottom collector, act as a single unit inside the reactor shell.

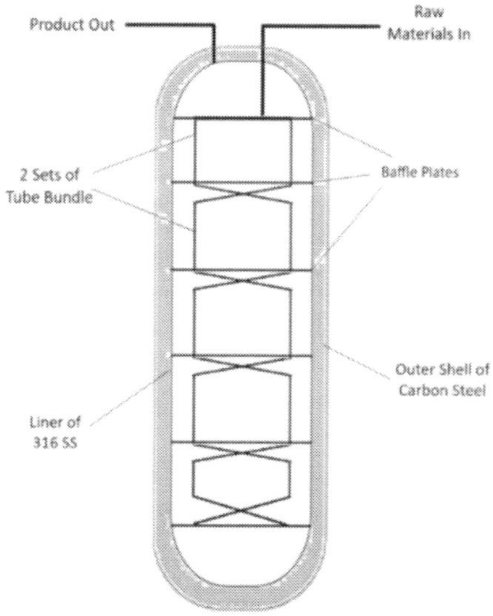

Figure 1 Schematic Layout of Urea Reactor at Fatimafert Plant Facility

Following are the operating parameters for the urea reactors at the Fatimafert plant facility:

Pressure:	~3200 PSIG (22,063 kPa)
Temperature:	~383 ^{0}F(195 ^{O}C)
Ammonia-CO_2 N/C ratio:	4.5

Corrosion and Urea Manufacturing

As explained in previous sections, ammonium carbamate is produced as an intermediate product during urea manufacturing. This ammonium carbamate is highly corrosive, especially at high temperature and pressure conditions. This is why urea plants require unique materials for vessels and piping [2]. Such special stainless-steel materials are collectively known as Urea-grade stainless steel.

Urea reactors still have high corrosion rates and, if not managed properly, can significantly decrease the service lives of these vessels. There have been some disastrous incidents worldwide due to corrosion in urea reactors. For instance, one occurred in a fertilizer manufacturing facility in the Pingyin county of Shandong province of China on the 21st of March 2005, which resulted in 4 deaths [3].

This paper will focus on the strategies employed to control the corrosion rates of the two reactors and their results.

Urea Reactors Monitoring for Corrosion Protection

Urea reactors require extra vigilance and monitoring of the corrosion rates of their internal liners. Careful acquisition of dependable data on corrosion is needed. There are two ways in which this data is collected at the Fatimafert plant facility. One is by conducting lab analyses for the presence of iron contents in the outlet streams of both reactors individually, and the second is the thickness monitoring and inspection of the liners during turnarounds.

Lab Analysis

Turnarounds usually occur every two or three years depending upon the plant conditions; thus, relying only on this data is not the best choice due to infrequent data updates. To further support corrosion monitoring, compositional analysis of the outlet stream is carried out to check the amount of iron.. The Fatimafert plant has also developed a regime for monitoring iron contents once a month based on a lab analysis since the commissioning of the reactors. The in-house lab is available in the plant facility to analyze the required samples as per the quality plan prepared in light of the specifications given by the vendor. This quality plan states that iron contents in the outlet stream of urea reactors must not exceed the limit of 0.7 ppm.

Historical Analysis of Corrosion Data

New urea reactors were installed in 1989 as part of the plant revamp. Initially, high corrosion of the urea reactors was observed due to inadequate monitoring and control regime. This is shown in

Figure 2. The two trends shown in Figure 2 represent the iron contents as detected in the reactor outlet streams from 1997 to 2014.

The green trendlines show the overall trend of the iron contents detected in the outlet stream of each reactor. Both trendlines display an increasing trend. This necessitated immediate corrective actions and improvement in monitoring regimes to counter this high corrosion observed.

The average iron content measured before 2014was 0.59 ppm for Reactor A and 0.66 ppm for Reactor B. These were both within the control limit of 0.7 ppm but considered high considering the corrosion rates obtained at different turnarounds explained in next section.

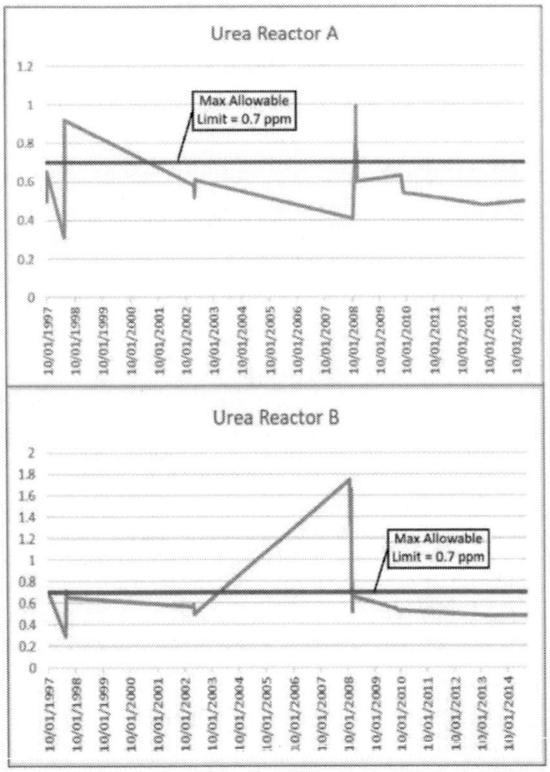

Figure 2 Trend showing Iron Contents Data in Urea Reactors Outlet Streams from 1997-2014

Inspection Reports

Inspection of both the urea reactors' internals and liner was carried out every two to three years during annual turnarounds.

To inspect the liner, the top head of the reactors and the tube bundle had to be removed. This is an extensive and risky exercise. Major risks include safety issues while carrying out this job and removing the tube bundle because any imbalance can damage the liner material. This is the reason inspection of the liner is not carried out more often.

However, table 1 shows the corrosion rates of Reactor A and Table 2 for Reactor B.

Years	Corrosion Rate of Reactor A	
	mm/yr	in/yr
1989-2006	0.1	0.00393
2006-2009	0.25	0.00984
2009-2010	0.2	0.00787
2010-2019	0.2	0.00787
2006-2023	0.09	0.00354

Table 1 Corrosion Rate of Reactor A

Years	Corrosion Rate of Reactor B	
	mm/yr	in/yr
1989-2006	0.14	0.00551
2003-2004	0.3	0.01181
2004-2006	0.12	0.00472
2006-2017	0.16	0.00630
2020-2023	0.04	0.00157

Table 2 Corrosion Rate of Reactor B

Root Cause Analysis of High Corrosion Rates

The plant team carried out a detailed RCA to identify the root and contributory causes of this corrosion in the urea reactors and remedial actions to be taken to reduce it considering the thickness of the liner and to increase remaining lives of both reactors. The maximum allowable iron contents at the reactors' outlet was 0.7 ppm, per the process licensor.

Identification of the Corrosion Factors

The following were the main factors identified during the RCA that resulted in higher corrosion rates:

1. Lesser N/C (<4.5/1 design)
2. Lean Passivation air monitoring control regime
3. High Reactors holding time in case of shutdown with reference to other fertilizer plants, as revealed in a benchmarking study

Control Strategies

To control these three factors identified, the following strategies were devised:

1. Process controls
2. Administrative controls

The following will explain how the three contributing factors were controlled using the above mentioned strategies.

Maintaining Ammonia to Carbon Dioxide Ratio of 4.5 to 1

As per the design, urea reactors of Fatimafert plant N/C is to be maintained above 4.5/1 against the stoichiometric requirement of 2/1 to get maximum yield. Moreover, this excess ammonia also acts as a protective layer in the ammonium carbamate and urea-rich, highly toxic and corrosive environment. Process control strategies were implemented to ensure that excess ammonia in the reactor was met. In order to maintain the N/C ratio inside the urea reactors, two flowmeters are installed on each line for ammonia flow into reactors. Flowrates of CO_2 are also monitored via flowmeters installed at the reactor feed line; the normal operating HP and MP CO2 flows are 665,000 SCFH (19,918 Kgmole / hr) and 290,000 SCFH (8,337 Kgmol/hr) respectively. There are no specialized N/C meters installed, and thus, specialized techniques are used to monitor this ratio. One scheme is by monitoring the reactor top and letdown temperature difference, which should be in the range of 120-130 ^{0}F. This temperature difference occurs due to the flashing of excess ammonia when the pressure after the reactor is reduced from 3200 PSIG (22,063 kPa) to 320 PSIG (2206 kPa). Log sheets, first-hour checklists, and shift logbooks monitor this differential temperature. Another parameter for monitoring the N/C ratio is the concentration of ammonium carbamate solution, which is recycled into the reactors. This concentration is monitored via an online concentration meter, lab analysis once per shift, and local analysis by the area operator every two hours.

Outlet Temperature of Urea Reactors

Another process parameter that requires close monitoring is the temperature of the urea reactor, as the corrosion rate increases exponentially with an increase in temperature [4] Ensuring both the operating temperature was maintained at 383 ^{0}F (195 °C), and a heating rate of 80 F^0 (44 °C) per hour during startups, as recommended by the vendor, was followed.

Monitoring these two parameters became easier when the conventional analog-based operating system was replaced in 2006-2009 in stages with the Distributed Control System (DCS).

Further introduction of log sheets and startup/shutdown checklists ensured a finer control of these process parameters, decreasing the reactors' corrosion factor.

Passivation Air Monitoring and Control

Urea vessels are fed with air to form a passivation layer in the form of oxides of iron, chromium, nickel and molybdenum [5]. This layer acts as a barrier between corrosive Urea and ammonium carbamate solution and the liner of the urea vessels, including Urea reactors. In the Fatimafert plant facility, this passivation air to the urea reactors is fed in the following three different ways:

1. Via CO_2 compressors
2. Via HP Ammonia pumps
3. Via HP Ammonium Carbamate pumps

The air flow to the urea reactors was ensured by developing a checklist to monitor proper airflow to the equipment mentioned above. Another important check to ensure adequate airflow is lab analysis carried out at the purges of the high-pressure circuit; the amount of Oxygen available was maintained at a minimum 4% by volume of the constituent of the gases in the purges, which are primarily hydrogen and nitrogen coming along with CO_2 from ammonia plant, ensuring excess passivation air. This development of the checklist and its monitoring regime proved very beneficial in reducing corrosion rate. It must be noted that, as per the UTIC design, O2 must be 0.14% of the total CO2 flow towards the Urea reactors.

Reactors holding time in case of shutdown

Another important corrosion protection factor was reactor holding time after plant shut down. During emergency or short-term shutdowns, the pressure inside the urea reactors is maintained so that it remains filled with the urea and carbamate solution and is not required to be pressurized again. This saves time for pressurization and saves raw materials. However, it carries a serious drawback of corrosion of the reactor liner because of the corrosive material inside the reactor and the lack of a passivation layer during this holding period. As per the Process licensor of the Urea plant, the reactor can be held up for 7 days with high Ammonia contents (short-term shut down) while it is to be drained after 7 days, but management of the Fatimafert plant facility decided to keep this holding time to be 3 days instead. However, the detailed analysis of the reactor content revealed high iron concentrations (0.8-1.8 ppm) that indicated high corrosion potential during such episodes. Therefore, a benchmark was carried out with other fertilizer plants, and it was highlighted that most plants drain the reactors after 24 hours. Consequently, it was decided to reduce this reactor holding time from 72 to 24 hours.

The shutdown SOP was also modified to reduce the depressurizing time of the reactors from 72 hours to 48 hours.

These changes helped to ensure reduced corrosion rates of the reactors and ultimately increased lives. Results are further presented in the next section.

Results and Discussions

The results of different actions taken to control the corrosion factors of reactors proved to be very promising and are explained in the following sections.

Lab Analysis Data

While implementing the actions and ensuring proper monitoring over the years, the lab analysis of the data was also continuously monitored. Figure 3 shows the data obtained of the iron contents in the reactor outlet stream over the years and shows a significant decrease. The net result of these actions was a decrease in the iron at the outlet of the Urea reactors from 0.7 ppm to <0.4 ppm (35% decrease in the corrosion). Table 3 shows the iron contents of Urea reactors for the year 2018 to 2023 period (the plant remained shut down most of the time in 2015 to 2017 due to gas supply interruption from Natural gas suppliers).

Some spikes in the trend are the conditions of plant startup or shutdown, during which plant parameters sometimes fluctuate, but a point of attention is that the iron contents remained within the limits of 0.7 ppm as compared with the previous years, where iron contents went up to 1 ppm during these upset/unsteady state operations.

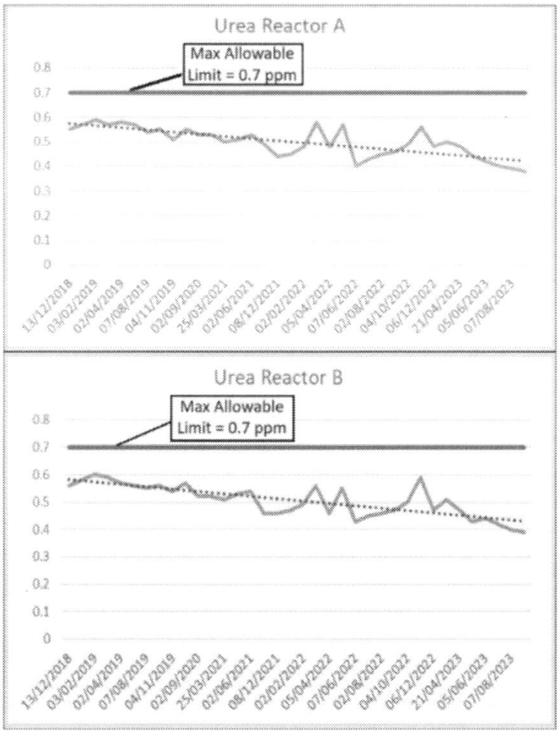

Figure 3 Iron Contents Analysis in Urea Reactors Outlet Streams from 2018-2023

Inspection Report of Turnaround 2023

The thickness of the liner observed during the last inspection carried out during the turnaround of the year 2023 is given in Table 3. It is important to note that the design thickness of the liner was 13 mm, while the lowest thickness observed at one segment of the liner, which is at the top of the reactor, is 7.25 mm. There is still another layer of 6 mm present beneath this liner.

Reactor	Minimum thickness (mm)	Maximum thickness (mm)
A	7.78	11.4
B	7.25	11.7

Table 3 Minimum and Maximum Thickness of the Liner Recorded During Turnaround 2023

Study Carried out by M/s Stamicarbon

After the 2023 turnaround, the inspection data was shared with M/s Stamicarbon Technologies, a leader in the urea industry, to carry out a life assessment of the two reactors. M/s Stamicarbon not only allowed the usage of the vessels for 5 more years but recommended inspecting after 3 years and then deciding about liner replacement for the top part of the reactor. This study is a testament of the actions carried out from 2019 to 2023 in controlling the corrosion factors and ultimately resulting in increasing the remaining life of the reactors that are already more than 30 years old.

Conclusion

Urea reactors installed in the Fatimafert plant facility, designed and fabricated by the M/s Nooter Corporation, are based on a multilayer structure. The innermost layer, also known as liner, is comprised of SA-240-316L SS material. Over the years, its liner with its original thickness 13 mm, has reduced to 7.25 mm due to corrosion. Different steps were taken to control the corrosion of the reactors from 2019 to 2023.

Three main factors were identified which can help reduce the corrosion rate if controlled properly:

1. Maintain an ammonia-to-carbon dioxide ratio of 4.5 to 1 as per the design of M/s UTIC
2. Ensure the flow of passivation air,
3. Reduce the maximum allowed reactor holding time in case of short-term or emergency shutdown.

The first two factors were controlled by developing administrative controls for monitoring the ammonia and air flows to the reactors. The third was controlled by reducing reactor holding time from 72 to 24 hours.

These actions reduced the reactors' corrosion rates by more than 35%, as evidenced by the data obtained from turnaround reports and lab analysis reports. This reduction in corrosion rates has increased the remaining lives of both reactors by an estimated 5 years, as supported by M/s Stamicarbon as well.

Acknowledgements

We are grateful to all the Fatima Group colleagues, mainly General Manager Manufacturing Mr. Asghar Naveed, for his continuous support and guidance while carrying out this study. We would also like to thank our Chief Executive, Mr. Fawad A. Mukhtar, for providing us with the support and opportunity to present this paper.

Literature Cited

[1] C. WARREN, "METHODS FOR THE MANUFACTURE OF UREA," in *Urea As a Protein Supplement*, Pergamon, Elsevier Ltd., 1967, pp. 33-71.

[2] A. Galakhova, F. Kadisch, G. Mori, S. Heyder, H. Wieser, B. Sartory and S. Burger, "Corrosion of Stainless Steel by Urea at High Temperature," *Corrosion and Materials Degradation,* pp. 461-473, 2021.

[3] A. L. Y. Z. X. Y. Y. L. Z. C. Weiqiang Wang, "The explosion reason analysis of urea reactor of Pingyin," *Engineering Failure Analysis,* pp. 972-986, 2009.

[4] K. Matsumoto and T. Shinohara, "Electrochemical Study of Stainless Steels in High-Temperature Urea Solutions," *Corrosion The Journal of Science & Engineering,* pp. 387-393, 1984.

[5] W. A. H. Yan Fie Sun, "A Study of Surface Film Formed on 316L Stainless Steel in Industrial Urea Media," *Key Engineering Materials,* pp. 2485-2492, January 1988.

Enhancing Reliability and Performance in Ammonia Plant Operations: Managing WHB Leaks and Catalyst Preservation

On June 16th, 2023, the ammonia plant started up after previously experiencing a Waste Heat Boiler (101-C) leakage. The tube bundle was already repaired, and no leakage was detected prior to start-up being commenced. However, during the startup, high pressure drop and high CO slip in the High Temperature Shift Converter (104-D1) were observed, namely 15.21 kg/cm² (261 psi) and 5%, respectively. The design pressure drop for that reactor is only 0.5 kg/cm² (7 psi) with a CO slip 3.65% so the plant inevitably had to be shut down. The inspection was carried out to find that the catalyst was being damaged due to being submerged in boiler feed water from the previous leakage. It took nine days to replace the High Temperature Shift Converter (104-D1) catalyst with a new charge. Afterward, the startup was executed and shortly the indication of Waste Heat Boiler (101-C) leakage was again spotted by a sudden pressure decrease of high-pressure steam and a sudden temperature decline downstream of the Waste Heat Boiler (101-C). Learning from previous events, the operator quickly responded to open the drains upstream and downstream of the High Temperature Shift Converter (104-D1) to prevent the catalyst submersion by boiler feed water leakage and subsequently plant was shut down again. The inspection results of Waste Heat Boiler (101-C) tube bundle B showed short-term overheating, which might have been caused by a quick temperature increase that went unobserved during startup. Inspection, repair, and replacement has been carried out. The ammonia plant is now producing well with no more leaks in the Waste Heat Boiler (101-C), and the performance of the High Temperature Shift Converter (104-D1) could be reverted to normal condition with a pressure drop of 0.23 kg/cm² (3.3 psi) and CO slip of 2.04%.

Raynal Rahman, Agung Swandaru, Muh. Makki Maulana, Gelar Panji Gemilar, Alfin Ferdiawan, Zuhroni Ali Fikri, Andi Lusdiono
Petrokimia Gresik, Indonesia

Introduction

EStablished in 1972, Petrokimia Gresik (PG) has emerged as the largest fertilizer company in Indonesia. With a staggering total production capacity of 8.9 million metric tons annually, it encompasses 5 million metric ton per year of fertilizer products (Urea, NPK, SP36, Kalium Sulfate (ZK), Ammonium Sulfate (ZA)) alongside 3.9 million tons per year of non-fertilizer products (Ammonia (NH_3), Sulfuric Acid (H_2SO_4), Phosphoric Acid (H_3PO_4), AlF_3, Liquid CO_2, Dry Ice, Purified gypsum, surfactant). Petrokimia Gresik spanning more than 450 hectares in the Gresik Regency of East Java, Indonesia.

Petrokimia Gresik are operating 2 Ammonia Plants. The first 1350 MTPD Ammonia Plant commissioned in 1994, Ammonia 1A, was licensed by MW Kellogg Company now known as KBR. The latter 1725MTPD plant, commissioned in July 2018 Ammonia 1B, was also licensed by KBR and includes the additional Purifier section c.

On June 16th, 2023, the ammonia 1A plant restarted following repairs to a previously detected Waste Heat Boiler (101-C) leakage. However, during startup, abnormal conditions including high pressure drop and CO slip in the High Temperature Shift Converter (104-D1) were observed, leading to a shutdown.

This paper will describe PG experiences and challenges related to managed Waste Heat Boiler (101-C) leaks and catalyst preservation to restore High Temperature Shift Converter (104-D1) catalyst performance.

Process Description

Ammonia 1A is a MW Kellogg licensed ammonia plant. As a typical ammonia plant, several reactions are needed to convert natural gas, air, and steam to form Ammonia. Each reaction has different required conditions and thus the raw material needs to be conditioned in specific operating conditions. For the process gas, compressors are utilized to control the pressure for the sake of reaction optimization. Most compressors in the Ammonia Plant use a steam turbine to rotate the compressor shaft and produce the desired pressure. The steam in the Ammonia plant is either produced outside the Ammonia Plant using a boiler in the utility plant or generated inside the Ammonia Plant by utilizing waste heat from the process gas.

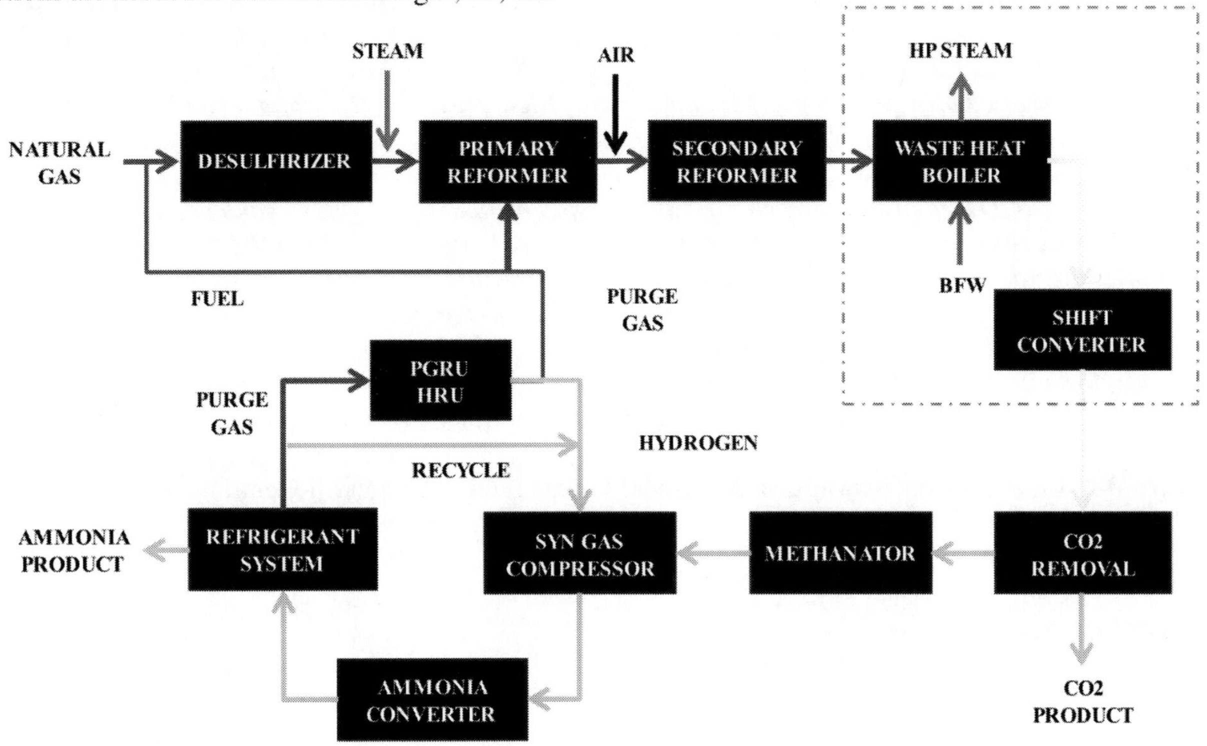

Figure 1. Ammonia Plant 1A Block Diagram

The process of ammonia production can be categorized into three primary stages: preparation of raw synthesis gas, purification of raw synthesis gas, and the synthesis of ammonia.

In the initial stage of raw synthesis gas preparation, natural gas is introduced into a desulfurizer to eliminate sulfur content, as depicted in Figure 1. This is followed by a steam reforming reaction in both the primary and secondary reformers, which results in the production of synthesis gas with a reduced methane content.

The synthesis gas output from the Secondary Reformer (103-D), which has a temperature of approximately 1,005°C (1,841°F), is utilized to generate steam in a shell and tube Waste Heat Boiler (101-C). The synthesis gas traverses the shell side under a pressure of about 36 kg/cm^2 (512 psi), while boiler feed water is fed through the tube side under a pressure of 123 kg/cm^2 (1,749 psi) in order to evaporate it for steam production.

In the subsequent stage, the CO containing gas produced from steam reforming undergoes a conversion process in the shift converter, which consists of a High Temperature Shift Converter (104-D1) and a Low Temperature Shift Converter (104-D2), to transform the CO into CO_2 via the water gas shift reaction. The design CO slips on High Temperature Shift Converter (104-D1) and Low Temperature Shift Converter (104-D2) are 3.65% and 0.4%, respectively.

Chronology of Events

On June 16th, 2023, a significant event occurred in the operation of the ammonia plant. The plant had previously been experiencing a Waste Heat Boiler (101-C) tube bundle A leakage, a serious issue that required immediate attention. This leakage had the potential to disrupt the plant's operations and impact its efficiency.

In response to this issue, tube bundle A was repaired and replaced with spare tube bundle B. Repairs were carried out for 11 days to replace tube bundle B with tube bundle A, which has been declared clean after mechanical cleaning. This repair process was meticulous and thorough, ensuring that no leakage was detected prior to the commencement of the start-up. This was a critical step in ensuring the smooth operation of the plant.

However, during the startup, unexpected issues arose. A high pressure drop and high CO slip in the High Temperature Shift Converter (104-D1) were observed. These were namely 15.21 kg/cm^2 (261 psi) and 5% respectively, values that were significantly higher than the norm.

The design pressure drop for that reactor is only 0.5 kg/cm^2 (7 psi) with a CO slip of 3.65%. These deviations from the design parameters indicated a problem with the operation of the plant. As a result, the plant was inevitably shut down. This shutdown was a necessary step to prevent further issues and to allow for the investigation and resolution of the problem.

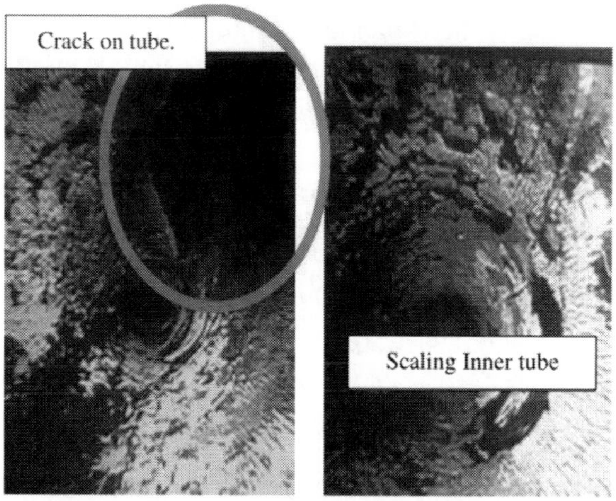

Figure 2. Waste Heat Boiler Leakage (101-C) for Bundle A

An inspection was carried out and revealed that the catalyst pallet with 6x6mm size turned into slurry. This was attributed to the operating catalyst being submerged in boiler feed water from the previous leakage as depicted in Figure 2 and 3, led to thermal shock of the catalyst. This was a significant finding, as it pointed to the root cause of the problems experienced during the startup.

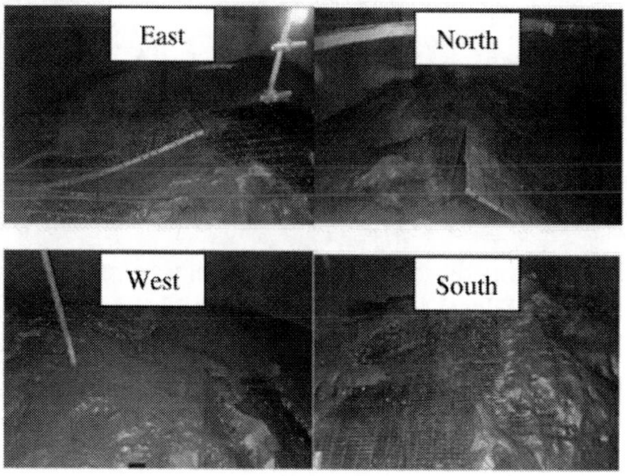

Figure 3. Internal Condition in High Temperature Shift Converter (104-D1)

The resolution of this issue was not a quick process. It took nine days to replace the high-temperature shift converter catalyst with a new charge. On June 27th, after the replacement of the catalyst, the startup was executed again. However, shortly after the startup, the indication of Waste Heat Boiler (101-C) tube bundle B leakage was again spotted by a sudden pressure decrease of high-pressure steam from 118 kg/cm^2 (1,678 psi) to 61 kg/cm^2 (868 psi) and a sudden temperature decline on the downstream of the Waste Heat Boiler (101-C) as depicted in Figure 4.

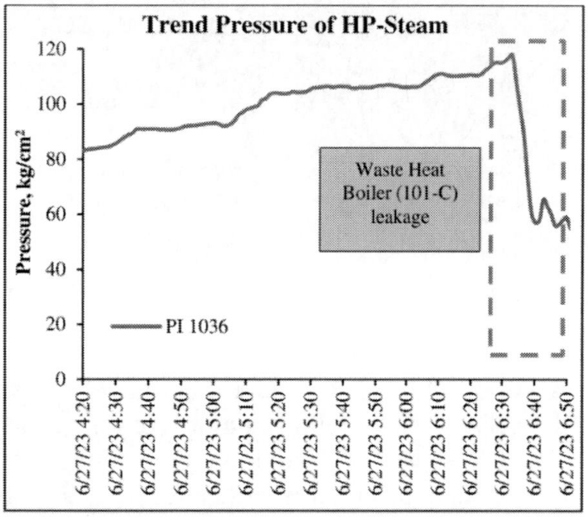

Figure 4. Trend Pressure of HP-Steam.

Learning from the previous events, the operators quickly responded. They opened the drains upstream and downstream of the high temperature shift converter. This action was taken to prevent catalyst submersion by boiler feed water leakage. Despite these efforts, the plant had to be shut down again.

Root Cause Analysis

Based on the result of the inspection carried out on the Waste Heat Boiler (101-C) tube bundle, it was found that damage was in the form of cracks on one of the tubes with a "fish mouth" shape and bulging on the pipe as shown Figure 5. One of the possible cause of this fish mouth shape crack is Short-term overheating because it is characterized by several distinct features. These includes swelling of the tube, ruptures resembling a fish-mouth, and thin-lipped ruptures. Also such overheating failures are commonly observed during the startup phase of a boiler.

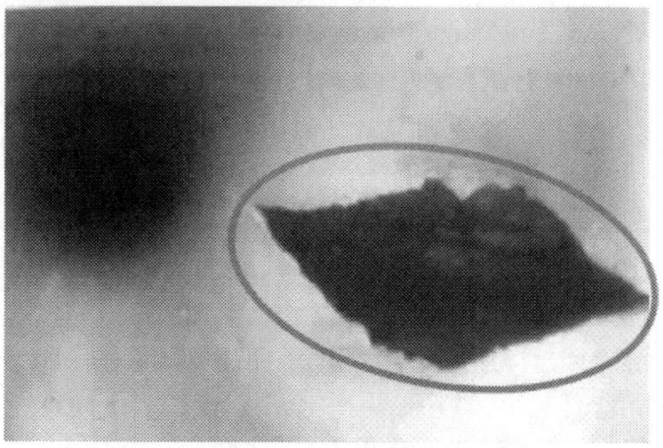

Figure 5. Crack Finding on One of Tubes "Fish Mouth"

There were some possible causes of the overheat which have been identified by a root cause analysis (RCA) carried out internally by PG:

- Process gas temperature is too high.
- Not enough Boiler Feed Water (BFW) flow.
- Scaling on inner tube.
- Tube Blockage.
- The rate of increase in the temperature of the Waste Heat Boiler (101-C) inlet gas at startup is too high.
- Impingement plate/Faulty distributor.

Following a concise root cause analysis of the issue at hand, it has been determined that the most probable root cause is an excessively rapid increase in the temperature of the Waste Heat Boiler (101-C) inlet gas during startup. This rate of increase surpasses the threshold limit, leading to potential mechanical integrity failure.

The failure is triggered when the metal tube's temperature experiences an extreme rise. This rise can be attributed to two main factors: a lack of adequate cooling by steam and insufficient water flow. The steam forms a film layer on the tube wall, which serves as a cooling mechanism. In the absence of this layer or in the event of inadequate water flow, the tube's temperature can rise drastically. In addition, the condition of tube bundle B was stored with inadequate preservation during 2018–2023.

Referring to Figure 6, it is evident that during startup, the rate of gas temperature increase inlet the Waste Heat Boiler (101-C) exceeds the limit of 85°C (185°F) per hour. Furthermore, the Distributed Control System (DCS) indicator does not provide a temperature rate indication. This lack of critical information makes it challenging to control the temperature rise at the Waste Heat Boiler (101-C) inlet effectively. Future work should focus on improving these indicators to prevent such failures and ensure optimal system performance.

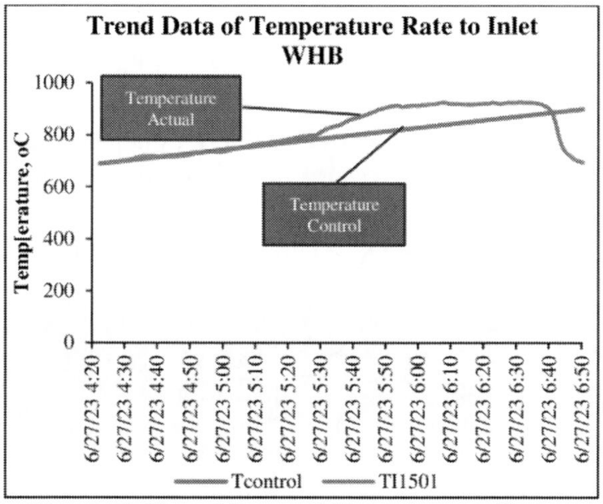

Figure 6. Trend Data of Temperature Rate to Inlet Waste Heat Boiler (101-C).

Maintenance Plan and Action

Below the details of the maintenance plan are explained, starting from plant equipment protection procedure during shutdown and maintenance action after the plant had shutdown.

1. Plant Equipment Protection Procedure.

Safety measures are initiated when indications of Waste Heat Boiler (101-C) leaks are detected. These events are typically characterized by a decrease of pressure in high-pressure vapors and an increase in boiler feed water (BFW) flow.

a. Remove Hazardous Gas (Feed Gas and Air)

The first step is to prevent hazardous (feed gas and air) from entering the High Temperature Shift Converter (104-D1).

b. Draining the High Temperature Shift Converter (104-D1)

After Remove Hazardous Gas then The High Temperature Shift Converter (104-D1) Converter inlet and outlet were drained for catalyst protection.

c. Assessing Catalyst Damage

If there is a rapid decrease in temperature in the catalyst bed and a significant amount of condensate emerges from the High Temperature Shift Converter (104-D1) outlet and inlet drain, the likelihood of catalyst damage increases, potentially necessitating its replacement.

d. Visual Inpsection of the Catalyst

A visual inspection of the catalyst is performed to determine whether it is still functional or needs to be replaced.

e. Adherence to SOPs and Temperature Control

During the startup process, strict adherence to existing Standard Operating Procedures (SOPs) is crucial, especially concerning the increase in Waste Heat Boiler (101-C) inlet temperature. The temperature of the high-temperature shift inlet is carefully maintained at a rate of 10-15°C (50-59°F) per hour to avoid excessive water evaporation if water is trapped in the catalyst. This comprehensive safety procedure ensures the efficient and safe operation of the system.

2. Maintenance Plan and Action

"Maintenance Plan and Action" was implemented during the shutdown on June 27 and startup from July 12-14, 2023, with the following details:

a. Draining the High Temperature Shift Converter (104-D1)

The High Temperature Shift Converter (104-D1) Converter inlet and outlet were drained for catalyst protection. As seen in Figure 7, water was discharged, indicating a Waste Heat Boiler (101-C) leak into the High-Temperature Shift Converter (104-D1). The drain was continuously operated to prevent the High-Temperature Shift Converter (104-D1) from being submerged in water.

Figure 7. Internal Condition in High Temperature Shift (104-D1)

b. Installing the Temperature Rate Monitor

A temperature rate monitor was installed in the Distributed Control System (DCS) as shown in Figure 8. The installation was carried out at the Secondary Reformer (103-D) and Waste Heat Boiler (101-C) inlet to facilitate monitoring of the temperature rise rate during startup.

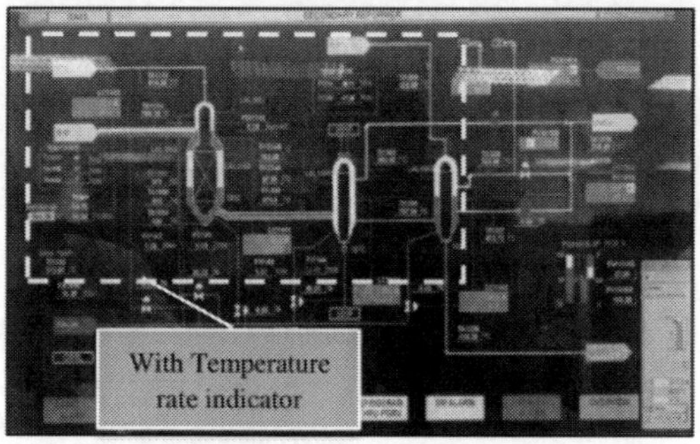

Figure 8. Converter Temperature rate monitor on display DCS

c. Visual Inspection of the Catalyst

A visual check was performed on the catalyst's condition to determine whether it could still be used. This check was carried out by opening the High Temperature Shift Converter (104-D1), manhole under nitrogen conditions. The inspection results showed that the catalyst appeared to be still in good condition (see Figure 9).

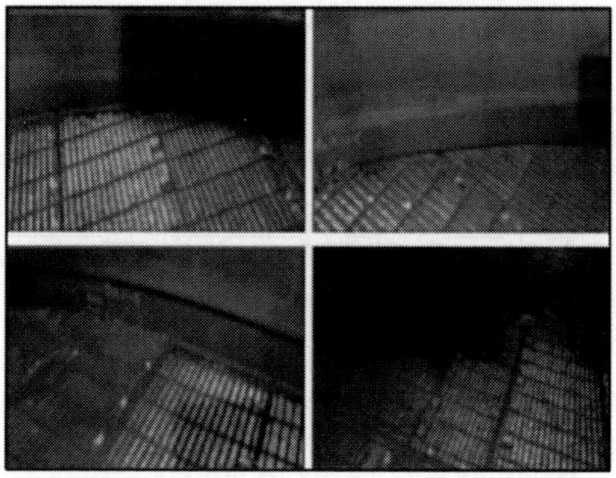

Figure 9. Condition in vessel High Temperature Shift Converter (104-D1)

d. Startup Process

The startup process from July 12-14, 2023, went quite well, where the waste heat boiler (100-C) inlet temperature heat up rate was effectively controlled to be below 85°C (185°F) per hour.

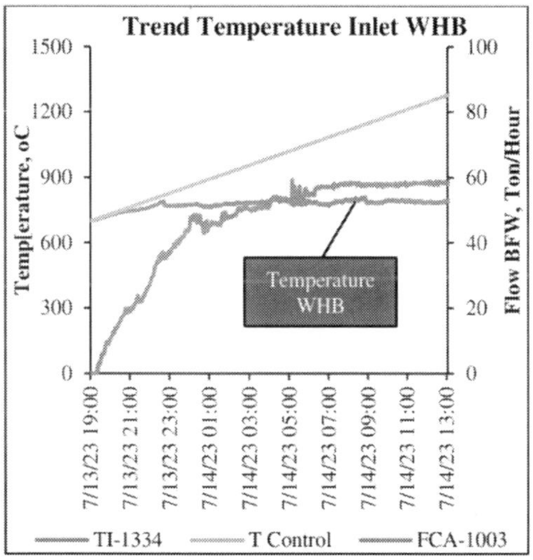

Figure 10. Temperature rate inlet Waste Heat Boiler (101-C).

e. **High Temperature Shift Converter (104-D1) Inlet Temperature Control**

The High Temperature Shift Converter (104-D1) inlet temperature was maintained at a rate of 10-15°C (50-59°F) per hour to avoid excessive water evaporation if there was water trapped in the catalyst.

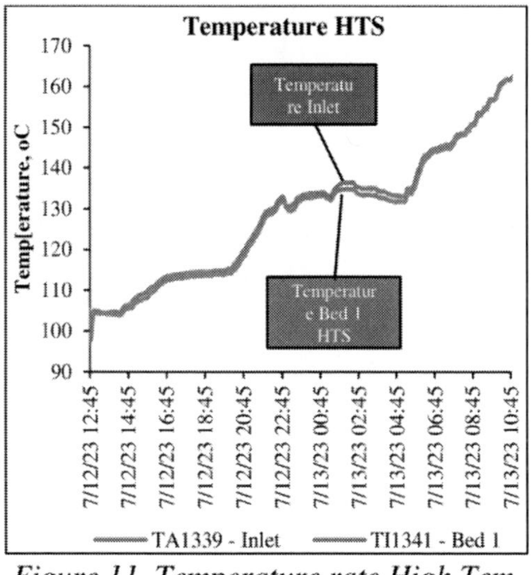

Figure 11. Temperature rate High Temperature Shift Converter (104-D1).

Result of Maintenance

The performance of the High-Temperature Shift Converter (104-D1) converter has been satisfactory after the start-up. For the period of July to October 2023, the High-Temperature Shift Converter (104-D1) operated with a pressure drop of 0.23 kg/cm² (7 psi). This pressure drop is within acceptable limits, ensuring that the converter can function effectively without causing operational issues.

Additionally, the carbon monoxide (CO) slip was recorded at 2.04%. CO slip, the percentage of carbon monoxide in the converter's outlet stream compared to the inlet stream, is a crucial parameter in assessing the performance of the High-Temperature Shift Converter (104-D1). A CO slip of 2.04% indicates that most of the carbon monoxide was successfully converted, contributing to the efficient production of ammonia.

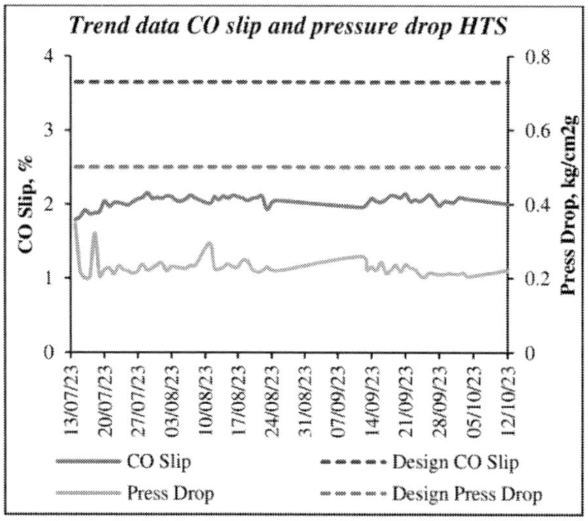

Figure 12. Trend data CO slip and pressure drop High Temperature Shift (104-D1).

Conclusion

In 2023, the ammonia plant faced a series of challenges, including a Waste Heat Boiler (101-C) leakage that caused catalyst damage, leading to a high pressure drop and CO slip in the High Temperature Shift Converter (104-D1). However, upon startup, indications of another Waste Heat

Boiler (101-C) leakage were detected, causing another shutdown. Further inspections showed short-term overheating, due to a rapid, unobserved temperature increase during startup. During the shutdown, the High-Temperature Shift Converter (104-D1) was drained, a Temperature Rate Monitor was installed, and a visual inspection of the catalyst was conducted. During the startup, an improved start up procedure and High-Temperature Shift Converter (104-D1) Inlet Temperature Control were carried out. These steps successfully prevented a recurrence of the Waste Heat Boiler (101-C) leakage and maintained the performance of the High-Temperature Shift Converter (104-D1) catalyst, allowing the ammonia plant to operate normally.

Replacement of Superheater Module in Primary Reformer Convection Section

Yara India experienced extensive damage to the intermediate tube sheets of a convection section superheater module, which posed a high risk to plant reliability. The replacement was an arduous task, as it involved a large volume of alloy steel welding, working at height, space constraints, and project completion within an ambitious period of four weeks during a turnaround.

This paper describes Yara's approach to meet this challenge of safe and timely execution, by evaluating multiple alternatives, preparing a procurement strategy to overcome the uncertainty of material pricing (due to Russia-Ukraine Conflict) and executing this meticulous plan, without compromising safety and work quality.

Saurabh Mathur, Amit Singh Chauhan
YARA Fertilizers India Pvt Ltd, Babrala

Introduction

Yara India observed damage on the intermediate tube sheet of a superheater module in the convection section of an ammonia plant primary reformer, which resulted in tube sagging of several tubes and posed a high risk to the sustenance of uninterrupted plant operation. Due to the large amount of other work near the waste heat section, the task of replacing a superheater module was challenging in many ways. This paper offers insights into the precise approach adopted and executed within the allocated plant turnaround duration in 2023.

Background

Yara India operates an Ammonia-Urea Complex, which was commissioned in 1994 and was originally designed to produce 1520 MTPD of Ammonia and 2620 MTPD of Urea. The process licensors are Haldor Topsoe A/S, Denmark, for the Ammonia Plant and Snamprogetti S.p.A, Italy, for the Urea Plant. In 2008, a de-bottlenecking project was undertaken to enhance Ammonia and Urea production capacity to 2000 MTPD and 3500 MTPD respectively. During the de-bottlenecking project, a new fired heater (H-202) was installed upstream of the primary reformer to partially off-load the primary reformer, which otherwise would have been overloaded by the revamped conditions.

The ammonia plant primary reformer (H-201) is divided into two sections: A radiant section and convection section. The radiant section is comprised of 300 numbers of vertical tubes, installed in two radiant chambers. Heat is supplied to the catalyst filled tubes with the help of 432 side-mounted burners, arranged in six rows on each furnace wall. The convection section recovers heat from the flue gas leaving the radiant section, and consists of series of heat exchange coils to preheat various process streams. No major changes or upgrades have been made to the convection section since its inception.

The configuration of the reformer waste heat recovery section is shown in Figure 1 and Figure 2.

Each module in the convection section consists of series of tubes (some bare and some finned, depending on process conditions) and each bundle of tubes is supported by four tube sheets (tube supports); two on the ends (End Tube Sheets, or ETS) and two in the middle (Intermediate Tube Sheets, or ITS). The material of construction of each module of tubes and ETS/ITS within waste heat recovery section (WHRS) are different, as shown in Table 1.

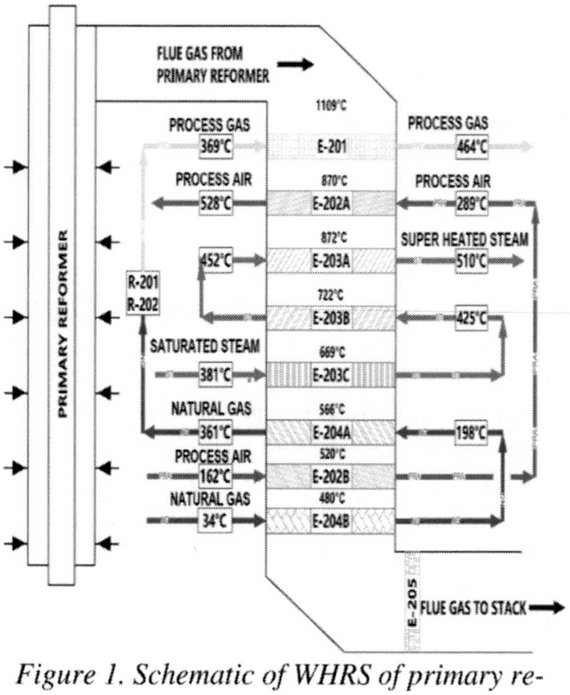

Figure 1. Schematic of WHRS of primary reformer

The Problem

The WHRS heat exchangers have been in operation since 1994 and based on the experiences and learning from other Indian nitrogenous fertilizer plants, some extent of degradation is expected in the hot flue gas path components, especially in steam superheater coils (E-203 A/B/C).

Figure 2. Photograph of WHR section (only east side front available for work)

Equipment	Fluid	Tube material	ITS material
E 201	Process Gas	SA312 TP321H and SA335 P22	PG25/35 Nb
E202A	Process Air		
E203A	Steam	SA312 TP347H	ASTM A351 HK40
E203B	Steam	SA335 P22	
E203C	Steam		ASTM A447 Gr-II
E204A	Natural Gas	SA 335 P1	
E202B	Process Air		
E204B	Natural Gas		
E205	Combustion Air	SA 53 Gr B	Carbon Steel

Table 1. Coil process and metallurgy details

A thorough inspection of all the convection section coils was conducted in 2012, and since then the observation on coils and ITS degradation were consistently noted in scheduled plant turnarounds, which occur every two years. In 2018, a Fitness For Service assessment (FFS) was performed for the superheater coils and the likelihood of creep failure in coils E-203A and E-203C was reported.

While all coils and their ITS in the Waste Heat Recovery System (WHRS) have experienced some degree of damage, the severity was notably higher in the steam superheater coils. This is particularly significant considering their location in the middle of the hot flue gas path. Inspection during the 2021 turnaround revealed damage to E-203C, as shown in Figure 3.

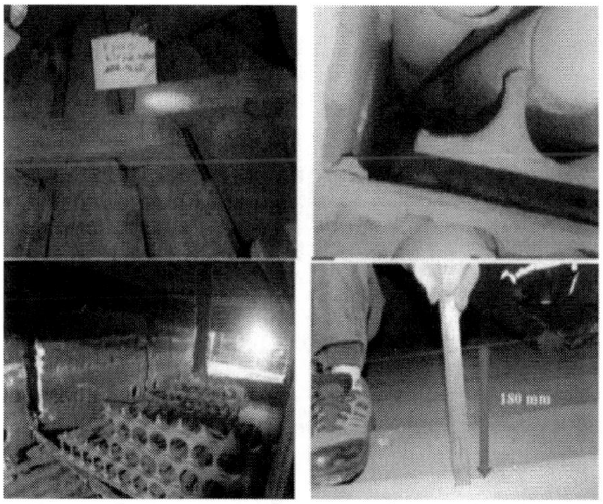

Figure 3. Damage of intermediate tube sheets and sagging of tubes.

The inspection findings included:

- Tubes sagging (approx. 180 mm (7.1 in) from adjacent tube)
- Both ITS were found broken from top and bottom
- Thermography from outside showed higher wall temperature, indicating damaged wall refractory. (Figure 4)

This discovery led the plant team to begin planning to replace superheater coil E-203C during the next turnaround, planned for 2023.

Damage casual factors

The major damage was observed on the ITS, which caused sagging of tubes. The dimension of the ITS are 3.21 m x 1.15 m x 0.12 m (10.5 ft x 3.8 ft x 0.4 ft) and the tubes pass through a series of holes in the ITS block. During the 2021 inspection, two pieces of the E-203C ITS were found lying on the top of module below. The crack/fracture appeared to be brittle in nature and with no plastic deformation.

The team worked with internal and external

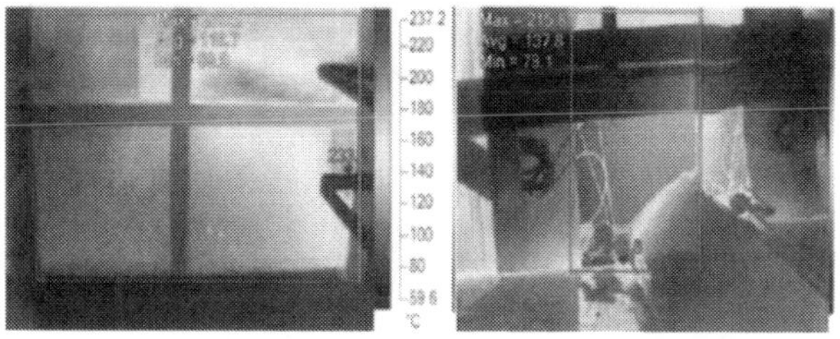

Figure 4. Thermography report of E 203C The shell plate (after refractory) temperature near/at tube sheet is on higher side, 170-233°C (338-451°F) compared to the recommended limit of less than 150°C (302°F)

specialists and evaluated possible causes:

Error in tube sheet design: Other than E-203A and E-203C ITS, the other coils' ITS were in acceptable condition. The middle super heater coil's (E203B) ITS was also intact and this possibility is ruled out.

Wrong selection of metallurgy: The internal tube sheet of E-203C was made of ASTM A447 Gr II, a cast heat-resistant ferrous alloy containing 25% Cr and 12% Ni. According to Table 10 of API 560 standard [1], this metallurgy is designed for a maximum temperature of 980°C (1796°F), whereas the operating temperature of the flue gas is 592°C (1098°F). Both the thickness of the ITS and visual inspection reports confirmed that the correct metallurgy was in place.

Quality issue with internal tube sheet: Sampling during operation wasn't feasible, but the team conducted a thorough review of old quality documents. Fortunately, no deviations were identified from these documents, providing assurance regarding the integrity of the fabrication process.

Process disruption: During discussions with the process licensor, the possibility of sudden load occurrences on the superheater could not be dismissed, with dual-phase flow potentially contributing to damage to the tube sheets. It was noted that during plant startup, improper draining of condensate or water could lead to pressure surges from water hammering, resulting in damage to the ITS and tubes.

Before the start of the biennial plant turnaround, a thorough physical inspection was conducted in accordance with piping isometric drawings. It was observed that the drain point or drip leg from E-203C, the lowest coil, was absent, raising concerns about potential water hammering during plant startup.

The Approach

References of similar damage in primary reformer waste heat sections have been documented in other Indian fertilizer plants. Some of them replaced complete modules with long execution times or made ineffective repairs by installing temporary supports beneath the tubes.

Yara management viewed this project as posing a risk to safe and reliable operation and was committed to addressing the issue with a permanent solution. Consequently, in 2021, a project team was assembled with the specific responsibility for the repair of E-203C and tasked with identifying the best available solution. The team outlined an exhaustive approach to ensure deployment during the 2023 turnaround. The outline included these steps:

1. **Establishment of project objective:** Team reviewed the Inspection Team's past observation reports and validated the same during plant's short shut down in 2021. Eventually, after compiling and analyzing all the facts, the project objective "*Restoring the safety and reliability of the Waste Heat Recovery section through the complete replacement of E-203C and strengthening of the E-203A intermediate Tube sheet*" was established.

2. **Performing top level risk assessment and mitigation plan:** As a first step, the Team prepared and presented the top-level risk assessment and acceptance/mitigation plan with clear delegation of project risk ownership.

Project Risk No. 1: Selection of supplier
Acceptance/Mitigation plan:

- Only process licensor approved bidders to be considered.

Project Risk No. 2: Quality: Raw material sourcing to site execution
Acceptance/Mitigation plan:

- Raw material sourcing from process licensor's approved European suppliers only.
- EN 10204 Type 3.2 certification for tubes, fittings, and ITS/ETS.
- Quality test witness by competent third-party inspectors.

Project Risk No. 3: Project timeline: Ordering to implementation
Acceptance/Mitigation plan:

- Sought support from Yara central project team for material expediting, whenever required.
- Detailed execution plan with service provider agency and challenging each critical milestone timelines.

Project Risk No. 4: Project cost and technical scope
Acceptance/Mitigation plan:

- Identification of internal and external stakeholders and recording their observations.
- Bucketing of inputs and linked with proposed solutions for validation and/or approval with site senior team.
- Finalization of proposed solution by management of change.
- No scope change without approvals.
- Review of project plan versus actual expenses during fortnightly review meeting with turnaround steering committee.

Project Risk No. 5: Safety risk during site execution activities
Acceptance/Mitigation plan:

- Detailed job safe analysis with all stakeholders under supervision of Fire and Safety coordinator to ensure no deviation from Yara HESQ policy.

- Deputation of dedicated fire watch and safety stewards on every shift.

Project Risk No. 6: Legal compliances, like, Indian Boiler Regulation Act (IBR), Factory Act, etc.
Acceptance/Mitigation plan:
- Complete job to be approved by Indian Boiler Regulations and no deviation is acceptable.
- No deviation against Indian Factory Rules for deployment of manpower during project cycle.

Project Risk No. 7: Project synchronization with plant's operational requirements
Acceptance/Mitigation plan:
- No change in heat transfer surface area and revalidation with process licensor.
- Capturing operational problems, if any, and addressing them with newer installation.

Project Risk No. 8: Learning from past and recommendation for future
Acceptance/Mitigation plan:
- Capturing past learning: Discussion with Yara central team, Indian fertilizer plants and experienced service providers.
- Depicting the lesson learn from this experience and submit to technical library and long-term asset management team for future references.

3. ***Scope finalization through evaluation of alternatives:*** Possible concerns and observations were collected with stakeholders and finalized the best available solution in systematic manner. (Table 2) Following stakeholder feedback, alternatives were assessed, leading to the decision on the optimal solution (Table 3).

4. ***Impact on HESQ and establishment of detailed execution plan:*** Involvement of Plant's Fire and Safety Team during micro-planning of site execution and separate Safe Job Analysis, with key site execution team members planned.

The Deployment

1. **Safety-during site execution activities:**

To adhere to Yara safety policy and procedures, the safety aspects were incorporated in the project activities from planning stage, e.g.

a) *Heavy lifting and shifting:* There were two options evaluated – the first was to lift each module one by one from the top, while the second option was to carry out work from the side. Both options had pros and cons and after evaluation, work from the side was deemed more feasible. This strategy minimized the heavy lifting and shifting risk.
b) *Work at height:* A temporary platform was designed using existing structural columns and scaffolding material. This provided more stability and working area.
c) *Working in confined space:*
   - Job sequencing was planned in a way that no overhead parallel activities were carried out inside the WHRS.
   - Before going inside the WHRS, the end plates were cut and removed for effective air circulation.
d) *Slips and trips:* All electrical cables routed through overhead path and the old/removed material was immediately moved to the scrap yard.
e) *Use of grinding machines for pipe cutting and weld edge preparation:* A beveling machine was used in place of handheld grinding machines. This removed the exposure to the high-speed grinding/cutting wheel, sparks, and vibration hazards.
f) *Exposure to weld light and falling of welding spark:* 100% coverage of welding area by fire blanket and adherence of PPE usage were checked through surprise visits and daily toolbox talks. Site support services team of Babrala were assigned responsibility of capturing safety observations for all major project activities on daily basis.

S.No.	Concerns / Observations collected from Stakeholders	Production	Maintenance	Inspection	Subject matter expert	Implemented solutions
1	Creep failure of E-203C (FFS)			√	√	E203C: Complete replacement with Tube, Header Pipe, Fitting, ETS (no metallurgy change) and ITS (upgraded metallurgy)
	Significant damage of E-203A/C ITS	√	√			
	Damage of ITS is on increasing side E-203C	√		√		
	E-203C: ITS is fully disconnected and tube sagging is significant			√		
2	The shell plate temperature near/at ITS on higher side ***(Figure 4)***		√			New refractory lining done on walls & ETS
3	Possibility of improper draining of condensate during Plant start up	√				Providing a drain with drip leg at bottom most section of E 203C Inlet
4	E203A: ITS damage is visible with no tube sagging			√		E203A: Design and incorporation of new ITS supporting beam of similar metallurgy & localized refractory repairing
5	Un-used Naphtha option steam header resulting in more condensate formation	√				Old naphtha steam line removed
6	Temporary repairing in case of E203A/C sudden failure may lead to sub-standard work				√	Preventive action of replacing the E-203C taken
7	Any modification/improvement to be aligned with IBR requirement		√		√	IBR certification & approval taken from authorities
8	Descaling of tubes, if any, from E-203A/B/C tubes to improve heat transfer	√				Chemical Cleaning carried out on Superheater Section- E203A&B and Natural Gas Section: E204A&B

Table 2. Capturing and addressing stakeholder concerns.

Alternative	Details	Decision
Complete replacement of WHRS from E-201 to E-204 (WHRS sections removal from top to bottom)	Improve reliability with minor energy saving, but this option is not commercially viable in cost-benefit analysis. Also, each section of WHRS is bolted as well as welded, and there are chances of further damage to all modules, as it would be opened for the first time in 29 years.	Rejected
E-203 C ITS replacement only	Reliability may improve to some extent, but it can't be implemented under tube's sagging condition. Further, the tube sagging issue will not be addressed with this option.	Rejected
Temporary supports (Make-shift arrangement) in E-203C	Plant reliability will not be improved	Rejected
E-203 C complete module replacement (severely damaged)	Improve reliability, due to complete replacement of observed sections and turnaround timeline is best suitable for the replacement of the same	Accepted
Strengthening of E-203A intermediate tube sheet (comparatively lesser damage)	Presently no tube sagging is observed, so strengthening of intermediate tube sheet will restore the present condition and give sufficient time to implement during other plant turnaround	Accepted

Table 3. Alternatives evaluation

g) *Lifting and shifting of material:* Ensured competent rigging foreman and third-party load testing of all rigging tools/equipment.

h) *Contact of radioactive radiation:* There were 296 weld joints which required radiography, as it was a requirement of the IBR. As there were multiple projects occurring in the same vicinity, such as Primary Reformer burner and refractory replacement, Superheater-1 (E-207) tube repairs, etc., and as these projects were critical in nature, all radiography of E-203 C weld joints were performed in early morning hours, so that other jobs would not be impacted. In addition to this, 3% of the weld joint of pipe headers having thickness of more than 36 mm (1.4 in), were checked with PAUT (Phased Array Ultrasonic Testing) and TOFD (Time of Flight Diffraction), due to limitation in radiography. During radiography testing, the affected radius was declared a no work zone to ensure safety.

i) *Hazard during steam blowing:* Steam blowing is associated with very high noise leading to utter discomfort to workers. A new method of using silencers was adopted to drastically reduce the noise level to a maximum of 90 dB during steam blowing (black arrows show steam blowing path whereas red arrows show normal steam production path). (Figures 5 and 6)

j) *Electrical Safety Hazards:* Sub-contractor electricians underwent written tests and interviews by Yara's Electrical team. Only upon satisfactory performance, they were qualified to work on electrical system.

Figure 5. Steam blowing

2. **Technical Scope:**

After deliberation with all stakeholders, the following scope was implemented:

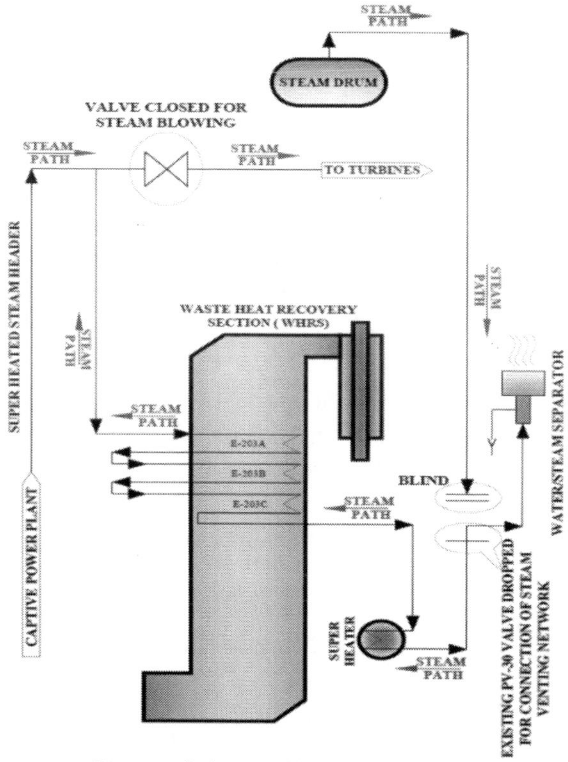

Figure 6. Steam blowing scheme.

- Replacement of E-203C with ITS material up-grade (from ASTM A447GrII to ASTM A351 HK40).
- De-scaling of E-203A/B and E-204A/B (additional scope) coils from inside.
- Incorporation of new supports to E-203A intermediate tube sheet (ITS damage visible with no/minor tube sagging) after proper design.
- Providing drip leg for proper draining of condensate from E-203C.

3. Quality- raw material sourcing to final delivery:

Process licensor approved sub supplier selection made mandatory in contract, and the following quality requirements (other than general chemical composition and mechanical properties) were finalized at the pre-order phase to get standard quality of material:

- ITS designing by finite element analysis.
- Approval of all drawings by Yara India.

- Only seamless pipes / tubes and fittings were selected.
- J Factor = 120 max.
- Impact test at -29°C confirming to minimum Impact Energy of 27 J,
- Chemical restriction P + Sn < 0.018% Max.
- Ultrasonic testing (longitudinally) of tubes after hydrotesting and magnetic particle testing of all fittings.
- Radiography testing of critical and repair section (if any), of ITS/ETS castings.
- The EN 10204 Type 3.2 certification for tubes/pipes, fittings and ITS/ETS.
- Third party witness of physical properties of refractory material.
- Approval of refractory casting procedure and witness the implementation by Yara third Party Inspector.

4. **Timelines- Ordering to implementation:**

a) Material delivery

The inquiry document was issued in November 2021. After finalizing the technical phase, commercial discussions commenced in late February 2022. However, amidst the Russia-Ukraine conflict in February 2022, global metal trading halted, leading to delays in negotiations. Talks resumed in late March 2022, and the contract was finalized in April 2022.

Following contract finalization, timely deliveries from European sub-suppliers were jeopardized by the energy crisis resulting from sanctions on Russia. Yara requested the subcontractor to dispatch their Italian counterpart to expedite progress. Additionally, Yara's inspection team in Europe visited the sub-supplier shop to ensure progress aligned with the plan. Meanwhile, in India, the Yara team committed to reviewing and approving documents within 2-3 days.

The detailed discussion and clarity in quality requirements during the pre-order technical phase minimized scope changes or additions. This ensured the readiness of all materials by the first week of December 2022. However, due to the Christmas holidays, transportation time for tubes increased, and they were finally received at the site on February 23, 2023, although the plant shutdown was planned from March 3, 2023.

b) Site execution

The microplanning of site execution started from October 2022 with the service provider and it was certain to all parties that any deviation in planned activities in terms of safety and quality would extend the execution timelines. Thusly, the following norms were developed:

For Safety:

- Replaced grinding activities by Cold Cutting machine for cutting and beveling.
- Fulltime safety supervisor deputed on shift basis.
- Use of certified crane for lifting and shifting made mandatory, i.e., no manual lifting/shifting.
- Timely removal of scrap from project site.
- Zero tolerance with Yara safety work permit program.

For Quality:

- 100% witness of welding procedure establishment and welder qualification test by Yara quality team.
- Deputed independent CSWIP 3.1 certified welding inspectors on shift basis for 100% witnessing from material identification to Final welding steps against finalized welding procedure.
- 100% Positive Material Identification (PMI) of weld joints.
- 100% review of radiography films (after stress relieving).
- Close monitoring of the baking temperature of welding electrodes.
- Hydrotesting of complete loop.

Other than above, following challenges are identified in planning phase and taken care in microplanning:

i. Limited distance between adjacent tubes

During the planning stage, a SWOT analysis (strengths, weaknesses, opportunities, and threats) identified space constraints as a major risk for site execution activities, posing serious threats to quality and potentially leading to rework or schedule delays. To mitigate this risk, a demonstration specimen was prepared at the subcontractor's shop, simulating work conditions similar to those encountered on-site. All welders underwent testing and training on this demonstration specimen to enhance their skills for on-site constraints, as depicted in Figures 7 and 8.

This demonstration specimen played a crucial role in finalizing the requirements for left-hand and right-hand welders, selecting different sizes of welding electrode holders, and adopting zig-zag welding techniques with nearby joint's heat treatment.

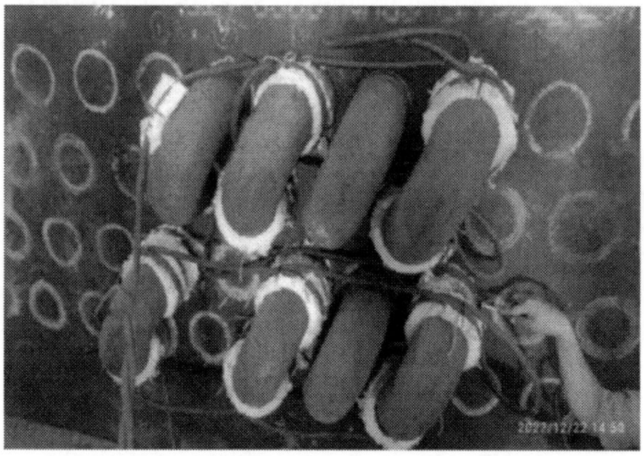

Figure 7. Demonstration specimen

ii. 296 Nos of High-pressure welding joints of 5 Inch and Above:

The individual length of the tube was approximately 10.3 m (33.8 ft), and insertion was only feasible from the east side (Steam Header end). Given this constraint, inserting the tubes in a hairpin (U) pattern instead of a hockey stick (J) pattern was preferred. This decision reduced the efforts required for welding as well as tube removal and insertion during turnaround by approximately 40%, as outlined in Table 4.

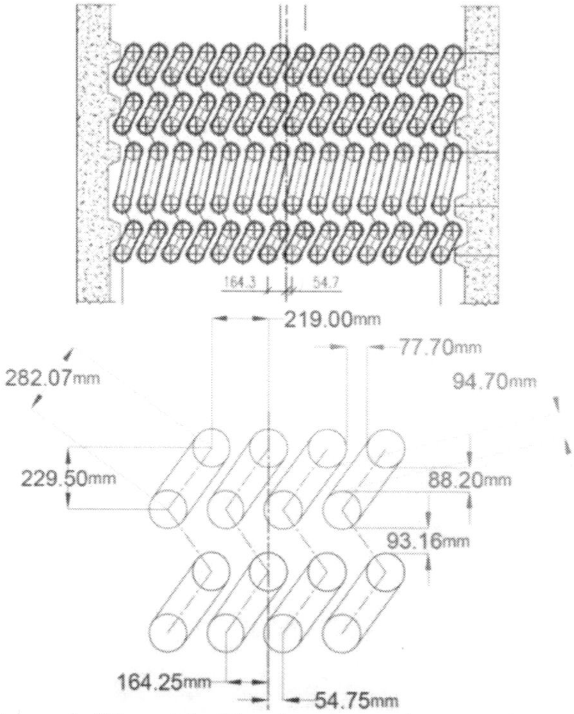

Figure 8. West side (Non header end) view of tube's configuration Gap

Size (In Inch)	Metallurgy	Thickness (mm)	Number of weld		
			Total	During Pre-Turnaround	During Turnaround
5	SA 213 T22 / SA 335 P22	12.5	280	140	140
16		36.52	5		5
16		52	1		1
18		39.67	4	2	2
18		58	2		2
10	API5LGr B	7.8	2		2
10	SA335P1	12.7	2		2
		Total Nos	**296**	**142**	**154**

Table 4. Weld joint details

iii. Avoid electrical load tripping through:

- Using rectifier welding machines consuming 12A/phase, instead of traditional ones at 18A/phase.
- Continuously monitor phase current at incoming panel to balance load across phases (R, Y, B). Crosscheck relay settings for maximum allowable imbalance.
- Portable devices equipped with dedicated 30mA RCBOs (Residual current operated circuit breaker) to prevent total supply failure.

- Schedule maximum stress relief cycles at night during radiography testing with no welding or cutting at the site.

iv. ***Others:***

- The tube pulling and reinsertion was carried out using a VFD (Variable Frequency Drive) operated electrical winch machine.
- The dry-out of refractory was planned through natural air circulation and controlled staging of primary reformer burners during startup.
- The duration of steam blowing was reduced by adopting hydro-lazing (to remove mill scales) of all tubes prior to erection.
- Project office shifted near to site, improving effective control and real time decision making.

Figure 9. Non header end (West side) tube to tube welding with 180° bends

The Outcome

With meticulous planning and precise adherence, the E-203C upgrade project was successfully completed, showcasing remarkable achievements:

Safety: The project maintained Yara theme of "***Safe by Choice***" approach, executing activities round-the-clock without a single safety incident or near miss across approximately 33,000 man-hours, supported by 115 toolbox talks.

Cost: Despite scope additions, the project achieved an impressive 94% budget utilization. Effective negotiation with subcontractors preserved scope and quality, while stringent inspection ensured minimal exigency expenses.

Milestones: Through efficient planning and teamwork, zero defects and incidents were recorded. Notable accomplishments include:

- Completing 78% of pre-turnaround welding within a week, overcoming a 20-day delay in tube receipt (February 23 to March 3).
- Recovering from a half-day delay during old tube dismantling by completing 140 tube joints welding between March 16-30, 2023 through demonstration specimen work out.
- Agile planning and resource optimization mitigated last-minute changes in steam blowing location, setting new standards of excellence in project execution.

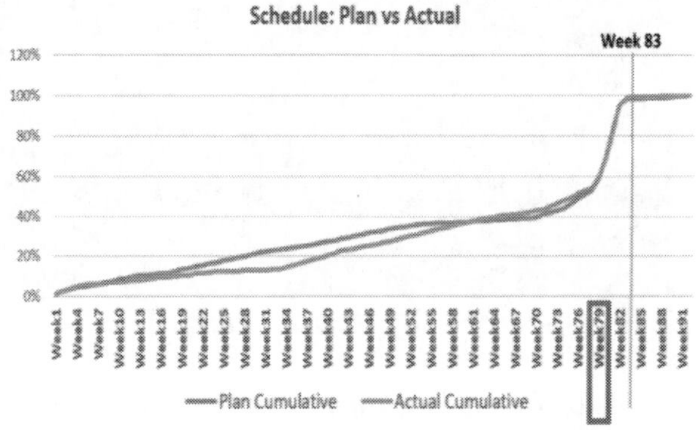

*Figure 10. S curve: Planning vs Adherence**

Note: 1). Contract finalization for material supply delayed by 16 weeks; 2). 90% of pre turnaround prefabrication done in **week 78 as tubes reached site by start of this week; 3). Plant shut down started from **week 79** (March 3, 2023) and E-203C job handed over after modification by April 3, 2023 (**week 83**).*

Conclusion

Yara India's unwavering commitment to asset integrity and long-term asset management has yielded 29 years of consistent production performance. Identification of potential issues with super-heater tubes and ITS mitigated risks of unplanned outages, which could have had dire consequences. Facing the challenge head-on, Yara and its subcontractor efficiently replaced the module within the allocated turnaround time, executing the project safely and with exceptional quality. Each activity was meticulously planned, with demonstrations for critical tasks to ensure timely completion. Despite numerous constraints, including intricate alloy steel piping welds, successful project completion transformed the approach to plant reliability, shifting away from temporary solutions towards sustainable improvements in the convection zone.

References

1. API 560: Fired Heaters for General Refinery Service, Feb 2016
2. ASME Section V: Non destructive Examination, 2013
3. ASME Section IX: Welding, Brazing, and Fusing Qualifications, 2013
4. The Indian Boiler Regulation-1950

Ammonia as a Fuel: Evaluating safety aspects, addressing key gaps, and indicating the way forward

The focus of this paper is on evaluating the state-of-the-art safety guidelines, and the proposed safety barriers from various safety classification societies for ammonia-fueled ships. A critical analysis is carried out to examine ammonia hazards such as toxicity, chemical corrosion, and fire/explosion. Valuable insights have been gained by detailed reviewing of ammonia incidents as in ship transportation cargo and where ammonia is used as a refrigeration coolant. In addition, this research provides a holistic analysis of ammonia-fueled combustion engine emission management technologies such as Ammonia Slip Catalyst (ASC) and Selective Catalytic Reduction (SCR), where novel catalysts and ammonia itself are utilized to reduce N_2O and NO_x emissions. This work exemplifies how conducting leak and dispersion studies by using the Process Hazard Analysis Software (Phast) Tool yield information in terms of ammonia hazard distances.

V. Alipour Tabrizy, S. H. Nilsen, T. Ryberg, J. Belgaroui, and B. O. Neeraas
Equinor ASA, Norway

Introduction

The pressure and regulatory urgency to reduce greenhouse gas emissions from industry have increased in the last few decades. The path to decarbonization of shipping industry includes alternative fuel technologies using carbon free, carbon-neutral, and low carbon fuels. Ammonia ranks as one of the best alternatives due to its commercial scalability, convenient storage and transportation and its favourable volumetric energy density. However, the safety hazards of utilizing ammonia as a fuel are fundamentally different from conventional marine fuels or recently adopted low flash point fuels such as Liquified Natural Gas (LNG), or Bio/E-methanol. Ammonia has a lower fire and explosion potential than other fuels. The main safety risk is its toxicity connected to respiratory failure, and it also poses health hazards such as cryogenic and chemical burns. Ammonia is also labelled as a very toxic substance to aquatic and edaphic environment with long lasting effect. The maritime industry has no experience for the use of ammonia as a fuel, on board, and international safety regulations are not yet established. As of now, the international code of safety for ships using gases or other low-flashpoint fuels (known as IGF code) includes only LNG, Liquefied Petroleum Gas (LPG), methanol and ethanol fuels. This study reveals that bunkering process, fuel preparation and the engine rooms are the most probable sources for accidental releases of ammonia, due to the number of leak points. The environmental and possibly climate consequences connected to N_2O, and NO_x emissions are dominated from combustion engines, only. The tolerance for ammonia releases, proposed safety distances and the safety barriers for different nodes vary among safety classification societies. The paper summarizes insights for the operators, safety professionals, and policymakers in the future development of safety in design and operation connected to utilization of ammonia as a ship fuel. The work sheds light on the knowledge gaps and the need for further research within the area.

Health and toxicity risk

Ammonia is an acutely toxic gas, that can cause breathing difficulty, chest pain, bronchospasms and at higher concentrations, also pulmonary oedemas, and respiratory failure, leading to fatal outcome at high concentrations [1]. Ammonia in atmospheric conditions is a gas and lighter than air (specific gravity is 0.6). However, if it released from pressurized liquefied conditions, it will form a two-phase cloud, consisting of small liquid droplets. In the entrainment zone, the turbulent jet resulting from a release induces the entrainment of ambient air in the jet. The energy supplied by this air, which is warmer than the jet, is initially used to vaporize the liquid droplets present in the jet. This vaporization cools the jet which then behaves like a cold heavy two-phase cloud in the atmosphere. The temperature in such a cloud can be significantly lower than the boiling point temperature of ammonia, theoretically down to -72 °C (- 97.6 °F). This is explained theoretically and measured in experiments ([2], [3], [4 and [5]). The heavy gas effect is well documented by experimental data for pressurized liquefied ammonia and leads to significant hazard distances. The cloud from a release of refrigerated liquified ammonia will be a gas, less heavy, and even buoyant [2]. Smaller hazard distances are therefore expected for refrigerated liquefied ammonia. Very little experimental data exist though from such releases. Ammonia is also readily soluble in water. Thus, in case of leakage, ammonia can readily be absorbed by a considerable amount of moisture, causing the resultant mixture to become a heavy cloud and prone to stay along the ground. The dissolved ammonia in water (NH_4^+) is dangerous for fauna and flora in proximity if the concentration exceeds the lethal concentration [6]. The high solubility of ammonia in water will influence the dispersion both above and below the sea surface. Experimental data are needed to close the knowledge gaps related to this effect and is of special importance for utilizing ammonia as ship fuel.

Flammability and explosion risk

Ammonia flammability interval is between 15-28 vol. % mixture with air, compared to flammability limits for other fuels which are lower [7] (e.g., flammability limit of LNG is between 4-15 vol.% and for Marine Gas Oil (MGO), it is between 0.4-8 vol. %, the lower end flammability limit of hydrogen is also around 4 vol. %). This implies that the volume of flammable gas clouds caused by ammonia releases will be relatively smaller compared to other flammable liquefied gases. The low-end flammability limit of ammonia (LFL) concentration also is much higher than the concentration levels connected to the risk of toxic injuries. The low laminar burning velocity (0.07 m/s or 0.23 ft/s) of ammonia [7] (compared to laminar burning velocity of other fuels such as LPG (0.43 m/s or 1.41 ft/s), LNG (0.37 m/s or 1.21 ft/s), methanol (0.36 m/s or 1.18 ft/s), and hydrogen (3.51 m/s or 11.52 ft/s)) lead to a difficulty in combustion of ammonia [7]. The measured ignition energies for ammonia have been reported from 8 mJ to 680 mJ (7.582×10^{-6}-0.00064 BTU) in the literature [8, 9, 10], compared to 0.28 mJ (2.65×10^{-7} BTU) for LNG, 0.14 mJ (1.32×10^{-7} BTU) for methanol, and 0.011 mJ (1.042×10^{-8} BTU) for hydrogen [7]. This can be seen as a safety advantage when it comes to fire and explosion risk, implying that accidental ammonia fires or explosions in open unconfined areas are less likely to occur as compared to other conventional fuels. However, in confined areas the fire and explosion risk of ammonia must be considered, especially in presence of other combustible materials, as seen in the 1983 Borden ice-cream factory accident in Houston, TX, USA [11].

Corrosion

Ammonia, unlike conventional maritime fuels, is corrosive and can cause chemical corrosion. Stress corrosion cracking (SCC) is possible when susceptible materials, corrosive conditions and stress are present at the same time. Details related to stress corrosion cracking are explained elsewhere [12, 13, 14]. Important take aways are:

- Liquid ammonia becomes corrosive when oxygen concentration is more than 0.05 ppm.
- Water level of up to 0.2% (2000 ppm) is used to avoid SCC in liquid phase. The higher the oxygen content, the higher the water content is necessary to prevent NH_3 SCC.

• Both the probability for SCC and the severity of the cracking is much less for -33°C and atmospheric pressure storage tanks than in pressurized and ambient temperature storage tanks [12, 13].
• The susceptibility to stress corrosion cracking increases with increasing yield strength of the steel. Det Norske Veritas (DNV) specifies that actual yield strength shall not exceed 440 N/mm^2) [13].
• Nickel steel containing more than 5% nickel and carbon-manganese steel, not complying with the requirements of above, are particularly susceptible to NH_3 SCC and shall not be used in containment and piping systems for the carriage of NH_3.
• Material such as copper, zinc, silver, gold, mercury, and thallium must not be used [12, 14].
• Post weld heat treatment reduces the probability of cracking.

Ammonia as a fuel

Combustion of ammonia will eliminate the emission of other pollutants such as carcinogenic soot (diesel exhaust), SO_x (Sulphur oxides), CO and particulate matter (PM) which may come from conventional hydrocarbon-based fuels. Although, ammonia has a higher volumetric energy density (12.7 MJ/L or 340 BTU/cf) than liquid hydrogen (8.5 MJ/L or 227 BTU/cf) [7]. The net storage volume required for ammonia as a fuel is 3 times more than for conventional fuels for delivering a similar amount of energy [1].

Although ammonia is a carbon neutral fuel and relatively easy to store and transport, it does have certain limitations that make its use as a standalone marine fuel solution challenging. These limitations include a low calorific value, low cetane number, a high ignition energy requirement and high latent heat of vaporization, with a low flame speed. When the pure ammonia is ideally combusted completely, only H_2O and N_2 are formed. However,, under real conditions, the reaction kinetics of ammonia combustion tend to form nitric oxide (NO_x) which has high health risk hazard as well as nitrous oxide (N_2O), which is a strong greenhouse gas, with a global warming potential (GWP) that is about 300 times higher than that of carbon dioxide (CO_2) over a 100-year time horizon [15]. This indicates that such releases must be limited to a minimum.

Ammonia has been tested successfully as a fuel in two-stroke, four-stroke spark-ignited and compression ignition engines with a pilot fuel injection of diesel or other high-cetane fuels (around 5 % energy share) to initiate the combustion by several marine engine manufactures such as MAN, Wärtsilä, and WinGD, recently [16]. In addition to internal combustion engines, ammonia has been tested directly as a fuel in high temperature Solid Oxide fuel cells (SOFC) or cracked into H_2 and then used in hydrogen Proton-exchange membrane (PEM) fuel cell [17].

Ammonia incidents at sea

Ammonia is a common ship cargo with about 18-20 million tons of transported ammonia at sea (10 % of global production) every year [7]). It is the third largest cargo product among marine liquefied bulk chemicals (after LNG and LPG). International safety regulations for sea transportation of ammonia were developed and fully described by the International Maritime Organization (IMO) under International Code of the Construction and Equipment of Ships Carrying Liquefied Gases in Bulk (IGC) code from 1986 [7, 18]. Few incidents are reported/known for ammonia cargo transport at sea and shore/ship interface, indicating the robustness of the safety regulations for sea transportation of ammonia [7,18]. Most of the accidents occur on fishing vessels where ammonia is used as refrigerant in refrigeration system [7,18]. Some incidents were reported for container ships/LPG tankers, where there were ammonia leakages as cargo [7,18]. Lack of training for crew, inadequate preventive maintenance procedures, improper operating procedures, component failure, mechanical wear and stress corrosion cracking are among the main causes [7]. Due to the toxicity of ammonia, in most cases there were fatalities associated with its release while the impact on the environment pollution was negligible [7].

Safety barriers - NH_3 as marine fuel

The process for supplying ammonia as a marine fuel to the engine is from bunkering station→ bunkering manifold → fuel tank → connection of the fuel tank with the valves and fuel supply pipes (tank connection space) → fuel supply equipment such as the compressor, pumps, and heat exchanger (fuel preparation room) → fuel valve unit → gas safe machinery space and engine [15]. In this section, the functionality of each node is explained, and the main safety barriers proposed by major safety classification societies (Lloyd's Register (LR), American Bureau of Shipping (ABS), Bureau Veritas (BV), Det Norske Veritas (DNV AS), Korean Register (KR), Nippon Kaiji Kyokai (NK), and Registro Italiano Navale (RINA)) are reviewed.

A **bunkering station** on a ship is a location where fuel (bunkers) can be loaded onto or discharged from the vessel's tanks. That includes both facilities on-board the vessels and onshore facilities. Proposed main safety barriers for bunkering systems are as follows [20-28]:

1) Ammonia leakage detection system
2) Automatic and manual emergency shut down (ESD) system between bunkering supply station and receiving ship tank. Activation time of ESD is suggested to be within 5 sec.
3) Water spray system above the bunkering manifold to reduce toxic vapors
4) Spill tray below the bunkering manifold
5) The bunkering hose is proposed to be equipped with dry-disconnect couplings and break-away devices that will prevent overstressing hoses and manifold in case of a drift-off scenario. Leakage from the hoses or pull-away incidents is one of the most common likely causes of the loss of containment for ammonia loading/unloading.
6) Personal Protective Equipment (PPE)
7) Emergency showers and eyewashes
8) Electrical grounding
9) Bunkering control station located in a safe remote location
10) Purging and inerting of pipes after bunkering
11) Mechanical spray shielding
12) Bunding system for onshore ammonia storage tanks

Ammonia can be stored either in a cryogenic system or in a semi pressurized/ pressurized system. The proposed main safety barriers for the **storage tank** are as follows [20-28]:

1) Primary and secondary containments as barriers, depending on the type of tank
2) Pressure relief valves (PRV)
3) Emergency shutdown system
4) Gas/leak detection system
5) Filling degree limitations
6) Reliquefication of ammonia vapor
7) Thermal oxidation of ammonia vapor
8) Energy consumption by the ship (engine, gas turbines, boilers, etc.)
9) Water curtain for ammonia vapor releases
10) Drip trays and spray shield
11) Cofferdam to separate ammonia hold space from the machinery space
12) Compatible materials for Ammonia SCC
13) Bulkheads to compartmentalize the tank into several compartments to reduce slugging

The filling limit shall be linked to the set point pressure of the pressure relief valves (PRV). Figure 1 demonstrates the loading limit curves for PRV set point pressure values of 2 barg, 4 barg, and 10 barg, in addition to the ammonia saturation curve. The figure shows that, as an example, for loading temperature of -33 °C (and vapour pressure of 0 barg), for PRV= 10 barg, the loading limit is 86%, while for PRV=4 barg, the loading limit is 91%, and for PRV= 2 barg, the loading limit is 93.5%. i.e., by reducing the setpoint pressure of PRV, the loading limit could be increased for the ship tank. For loading temperature of 0 °C (and vapour pressure of 3.3 barg), for PRV= 10 barg, the loading limit is 91.8%, while for PRV=4 barg, the loading limit is 97%. Pressure relief valve of 2 barg, obviously could not be used for this condition as it is lower than the ammonia vapour pressure inside the tank.

Fuel tank connection space (TCS) is a designated area in a ship where the fuel tanks are connected to the ship's piping system and thereafter fluid supply system. TCS contains all fittings, flanges, and tank valves, which are required to be enclosed in a gas tight space, unless the tank connections are on an open deck. The goal of the **Fuel Preparation Room (FPR)** and **Fuel Supply System (FSS)** is to provide ammonia fuel to the engine within a specific range of temperature, pressure, and purity. This is achieved by means of Low-Pressure (LP), High-Pressure (HP) pumps, flow meter, filters, ammonia catch tank, and heat exchanger and filters. The proposed safety barriers for TCS and FPR are as follows [20-28]:

1) Emergency shutdown systems
2) Leak detection systems
3) Water curtain/water screen
4) NH_3 vapor processing systems, such as scrubber, combustion units, diffusion tanks and dilution systems
5) Leak detection with automatic closing of isolation valves
6) Ventilation system with a minimum of 30 air changes per hour of FPR
7) Gas tight boundaries and compartments
8) Bolted hatch access to TCS with additional lobby with a conventional gastight door in front of the entry hatch
9) Double wall piping and duct systems
10) Fire safety mitigation systems
11) Drip trays and water spray shield
12) Compatible materials to avoid low temperature embrittlement and ammonia SCC
13) Bilge and drainage systems

The **machinery space** on a ship is a designated area on a vessel where the machinery, main ship engine and all other equipment necessary for the ship's propulsion and operation are located. There is an increased risk of explosions in machinery space and other enclosed areas, and barriers to prevent and reduce the explosion risk may need to be reviewed in order to present further escalation on the vessel. The proposed safety barriers for machinery space are as follows [20-28]:
1) Emergency shutdown systems
2) Gas detection systems
3) Water curtain/water screen
4) Ammonia vapor processing systems
5) Leak detection with automatic closing of isolation valves in case confirmed leak
6) Ventilation system with a minimum of 30 air changes of machinery space per hour
7) Gas tight boundaries and compartments
8) Double wall and duct systems
9) Fire prevention and mitigation systems
10) Drip trays and water spray shield
11) Compatible material to avoid ammonia SCC
12) Bilge and drainage systems
13) Gas Consumption Unit (GCU)/Boiler
14) SCR and ACS
15) Exhaust Gas Recirculation (EGR)

Ammonia fuel cell engines

The safety barriers listed for internal combustion engines are applicable also to fuel cells, except for the last node, which involves the use of fuel cells instead of combustion engines in gas safe machinery spaces. The proposed specific safety barriers for ammonia **fuel cell engines** are as follows [29].:

1) Hydrogen leak detection: Hydrogen gas detectors are installed within the fuel cell system and in the surrounding area, to detect hydrogen gas leaks. If hydrogen is detected, an automatic shutdown of the system is carried out to eliminate the risk of ignition and explosion.
2) Hydrogen venting systems: PEM fuel cell systems onboard ships is proposed to be designed with hydrogen venting systems that can remove hydrogen in a safe manner in case of leakage or high pressure in the system. The venting system ensures that the hydrogen gas is released in a controlled manner to avoid any risk of explosion or fire.

3) Non-Sparking (ex-proof/ex-classified) components
4) Pressure-relief devices are proposed to prevent overpressure and protect the fuel cell stack.
5) Cooling and heating system: PEM and SOFC fuel cells require an effective cooling and heating system to maintain safe operating temperatures of the fuel cell stack. Proper thermal management of the fuel cell stack is essential to prevent thermal runaway reactions.
6) Fire suppression systems are proposed systems in PEM and SOFC fuel cell applications.
7) Double-wall insulation
8) Emergency shutdown system in case of accidental release of NH_3 and H_2
9) Catalyst poisoning prevention and catalyst regeneration program
10) Isolation and containment: Isolating and containing hazardous materials reduces the likelihood of contamination from spills or leaks. This containment can be achieved using leakproof containers, closed systems, and double containment.

The IGF code's regulations and guidelines are broadly used by classification societies for establishing safety guidelines for ammonia as a fuel. However, there are differences in the suggested requirements for different nodes between the various class societies, resulting in somewhat different proposals for safety barriers and preventative protocols. Table 1 shows ammonia concentration thresholds for alarms and recommended safety action by four classification societies for different nodes. The recommendations for alarm limits and response, proposed solutions for ammonia releases vary among safety classification societies. For instance, for the fuel preparation room, ABS requires alarm activation at a maximum detection level of 25 ppm of ammonia release, followed by the activation of a water screen guard and fuel supply shutdown at detection of 150 ppm. In contrast, RINA suggest ammonia detection levels up to 50 ppm. DNV and NK suggest detection levels of 25 ppm, followed by fuel supply shutdown. Additionally, NK recommends increased ventilation at 300 ppm, whereas DNV suggests ventilation combined with fuel shutdown at detection of 350 ppm. There is also a variance between the international standards and classification societies' recommendations related to the distance of the Pressure Relief Valve (PRV) from the air intake of accommodation (service space), concerning ammonia-powered ships. This distance for LNG-powered ships is prescribed at least 10 m according to IGF code; however, for ammonia fueled ships, KR mandates this distance as a minimum of 15 m, while ABS requires a minimum of the ship's width or 25 m, whichever is less.

Ammonia fuel emissions management

Ammonia combustion and utilization of ammonia fuel in fuel cells presents three specific emissions risks: Ammonia slip, NO_x, and N_2O.
The mitigation technologies include [15]:

1) Reliquification and boil off gas (BOG) management system
2) Engine for combustion of excess NH_3 vapour from fuel tank
3) Gas Consumption Unit (GCU), boiler, water catcher and chemical adsorber for NH_3 emission risk management from fuel supply system
4) Engine tuning for ammonia, N_2O emission management from combustion engine
5) SCR for NO_x and N_2O emissions management and ACS for ammonia slip emission management. In SCR, NH_3, as reducing agent, is used in the exhaust gases. During this process, NO_x present in the exhaust reacts with NH_3 over a catalyst, as the following reactions:

$$4NO + 4NH_3 + O_2 \rightarrow 4N_2 + 6H_2O$$
$$2NO_2 + 4NH_3 + O_2 \rightarrow 3N_2 + 6H_2O$$
$$N_2O + 2NH_3 \rightarrow N_2 + 2H_2O$$

ASC is used to remove the excess amounts of unreacted ammonia from the exhaust gas stream in SCR, according to the following reactions:

$4NH_3 + 5O_2 \rightarrow 4NO + 6H_2O$
$2NO + 2NH_3 + O_2 \rightarrow 2N_2 + 3H_2O$
The most used catalysts are based on noble metals such as platinum (Pt), palladium (Pd), or rhodium (Rh), or non-noble metals such as copper (Cu), iron (Fe), cobalt (Co), or manganese (Mn)

6) Exhaust Gas Recirculation (EGR) for ammonia and NOx emissions management

Among these technologies, reliquification, SCR, EGR are matured technologies for NH_3 and N_2O emission management, whereas other technologies such as Plasma, GCU, water catcher, engine tuning are under development [15].

Leak and dispersion analysis

Leak and dispersion calculations have been carried out for release scenarios that should be evaluated for an ammonia bunkering facility, utilizing the Phast code. Phast (v.8.7) is a software tool consequence modelling package, developed by DNV. Phast calculates the progress of a potential release of gas or liquid from the initial release to the far-field dispersion including the modelling of rainout and subsequent vaporization. Phast is not a CFD (computational fluid dynamics) model and cannot predict the effect of local obstructions and complex terrain directly. However, models for basic dispersion kinetics and thermodynamics are included and verification of results against experimental data have been done for several chemicals, including ammonia experimental release data. Phast is validated by ref. [30] and [31] for releases of pressurized liquefied NH_3. Lately small-scale experiments including atmospheric refrigerated and pressurized NH_3 at boiling point temperature have been carried out [2].
The leak scenarios calculated by Phast are based on the following assumptions:

• Two horizontal cylindrical storage tanks, each with a volume of 1000 m^3, diameter 6 m and length of 35.5 m are considered with refrigerated ammonia at atmospheric pressure, and T~ -33.5 °C. Only one tank is assumed to leak or fail at a time.
• The bund is large enough to contain the volume of both tanks, and there is space for inspection, assumed 2 m between the tanks and 3 m on the outside of the tanks
• Bund area assumed as 41.5 x 20 m = 830 m^2, bund wall height of 2.4 m.
• Outlet valves (ESD) are located at the very bottom of the tanks. Liquid head in the tank assumed 5.5 m
• Hose/loading arm diameter is assumed as 100 mm with length of 25 m.
• The flow monitoring/detection system detects a leak relatively quickly, within 30 sec. for a small leak, and 15 sec. for large leak- (hose/valve rupture).
• Pump pressure is 12 barg, with max pumping flowrate about 250 m^3/hr.
• Release height for all scenarios assumed 1.5 m, in the horizontal direction, and in line with the wind direction. All releases are assumed to be on land (not on water), since the effect of water on dispersion is uncertain.
• Atmospheric temperature is assumed to be 10 °C. Three weather conditions are considered: F1.5 (stable weather with moderate clouds and wind speed of 1.5 m/s), F3 (stable weather with moderate clouds and wind speed of 3 m/s), and D5 (neutral weather and higher wind of 5 m/s).
• The 0.1% fatality is based on the following probit function for ammonia $Pr = -17,16 + 0,99 \times \ln(C^{2,02} \times t)$, [32] where C is concentration in ppm and t is exposure time in minutes.

Description of leak scenarios:
1) Tank rupture (all ammonia content released within fixed duration of 10 minutes, assumed inside bund).
2) 100 mm leak from the tank bottom, all tank content for one tank leaks out inside the bund.
3) Hose rupture outside bund, without pumping.
4) Hose rupture outside bund during pumping.
5) 20 mm leak from hose outside bund during pumping
6) 4 mm leak from hose during pumping outside bund

The Phast results are summarized in Table 2. Three main observations can be highlighted as follows:

• Tank rupture is the worst-case scenario in terms of leakage flow rate, toxic lethality footprint and maximum distance to IDLH (Immediately Dangerous for Life and health, as defined by NIOSH – the National Institute for Occupational Safety and Health), followed by the hose rupture scenario outside bund during pumping operation, leakage scenario from the tank bottom, the house rupture scenario outside bund without pumping operation and finally the hose leakages scenarios.

• The bund reduces the hazard distances (distance to IDLH concentration and toxic lethality footprint) for the largest releases since the pool diameter and thereby evaporation rate is significantly reduced.

• The hose rupture (scenario 3 and 4) gives significantly longer hazard distances for the pumping (12 barg) compared to the same release without pumping. This is caused by the higher release rate, the larger pool, and the heavy gas behavior for the 12-bar scenario. The evaporation rate also enhances as wind speed increases.

In general, higher wind speed leads to lower hazard distances for heavy gases, but for some scenarios, the gas cloud lifts from the ground (buoyant behavior) in the stable atmosphere and low wind, and the D5 weather condition causes longer hazard distance (scenario 2).

The distance to 0.1% lethality is significantly shorter than the distance to the IDLH value (300 ppm). This indicates that the IDLH concentration for NH_3 is rather based on the confusion caused by the irritation of the eyes and respiratory tract, rather than being an acute lethal concentration.

Conclusions and the way forward

The concept and design of ammonia as a fuel, in terms of its safety risks, are still not matured as international regulations for ammonia as a marine fuel are not yet finalized. This paper summarizes proposed safety barriers/guidelines from various safety classification societies for ammonia fueled ships. The bunkering process, fuel preparation and the engine rooms are the most probable nodes for accidental releases of ammonia. N_2O and NO_x emissions could be caused by the combustion engines, only. All these three risks can be reduced to acceptable limits through developed SCR and ACS technologies. The recommendations for alarm limits and response, proposed solutions for ammonia releases vary among safety classification societies. To address these inconsistencies, effective collaboration, knowledge, and experiences sharing are necessary, among different stakeholders. Leak and gas dispersion analysis via Phast shows significant hazard distances. The bunding/ammonia liquid collection systems is an effective safety barrier to limit the pool surface, and thereby ammonia evaporation rate and the toxic lethality footprint. These results indicate that the safety barriers for ammonia as a ship fuel are important and that education/competence building among ship crew, bunkering personnel and the public is needed.

Acknowledgement

Equinor AS is appreciated for financial support and granting permission to publish this paper.

References

[1] Mengyao, Y., Clara N. K. L., and Ming L. Ammonia as a Marine fuel– Bunkering, Safety and Release Simulations (2022). Nanyang Technological University.

[2] Dharmavaram, S., Carroll, M. J., Lutostansky, E. M., McCormack, D., Chester, A., & Allason, D. (2023). Red Squirrel Tests: Air Products' ammonia field experiments. Process Safety Progress.

[3] Bouet, R. (2005). Ammonia Large-scale atmospheric dispersion tests., INERIS – Accident Risks Division.

[4] Bouet, R., Duplantier, S., and Salvi o. (2005). Ammonia large scale atmospheric dispersion experiments in industrial configurations. Journal of Loss

Prevention in the Process Industries 18(4-6): p. 512-519.
[5] Blanken, J., Behavior of Ammonia in the Event of a Spillage (1980). American Institute of Chemical Engineers.
[6] Tsang, S. C. E., Ayvali, T., and Van Vrijaldenhoven, T. (2021). The Position of Ammonia in Decarbonizing Maritime Industry: An Overview and Perspectives: Part II: Costs, safety and environmental performance and the future prospects for ammonia in shipping. Johnson Matthey technology review, 65(2).
[7] Jang, H. et al. (2023). Regulatory gap analysis for risk assessment of ammonia-fuelled ships. Ocean Engineering, 287, 115751.
[8] Wisløff I. and Nervold E. H. (2021) Report for Pilot "Ammonia as Fuel". Green Shipping Program.
[9] Harris, G. F. P., and MacDermott, P. E. (1977). Flammability and explosibility of ammonia. In Icheme symposium series (Vol. 49, pp. 29-36).
[10] B. Lewis and G.v. Elbe: "Combustion, Flames and Explosion of Gases", London, Academic Press, 1987
[11] McRae, M. H. (1987). Anhydrous ammonia explosion in an ice cream plant. Plant/Operations Progress, 6(1), 17-19.
[12] Lunde I. and Nyborg R. (1989) Prevention of stress corrosion cracking of carbon steels in ammonia, Safety in Ammonia Plants and Related Facilities Symposium, AIChE, Paper No. Aiche-30-007.
[13] DNV Rules for Classification (2020) - Ships, Part 5 Ship types, chapter 7 Liquified gas tankers.
[14] IGC Code (as amended in 2014)- International Code for the Construction and Equipment of Ships Carrying Liquefied Gases in Bulk.
[15] Mærsk McKinney Møller Center for Zero Carbon Shipping (MMMCZCS) (2023) Managing emissions from ammonia-fuelled vessels.
[16] Handbook of 44th Motorship Propulsion & Future Fuel Conference November (2023). Hamburg, Germany.
[17] https://almacleanpower.com/news/partners-alma-clean-power-odfjell-and-dnv-to-start-testing-of-power-system-with-solid-oxide-fuel-cells-sofc-for-deep-sea-shipping
[18] Cheliotis, et al. (2021). Review on the safe use of ammonia fuel cells in the maritime industry. Energies, 14(911), 3023.
[19] Ammonia as a marine fuel safety handbook (2023), green shipping program, DNV and sjøfartsdirektorat.
[20] Lloyd's Register 'Rules and Regulations for the Classification of Ships using Gases or other Low-flashpoint Fuels' (LR, 2021)
[21] American Bureau of Shipping - 'Requirements for Ammonia Fuelled Vessels' (ABS, 2021a)
[22] American Bureau of Shipping- 'Guide for Ammonia Fuelled Vessels' (ABS, 2021b)
[23] American Bureau of Shipping- 'Rules for Building and Classing Marine Vessels (MVR) - Part 5C, Specific Vessel Types (Chapters 7–18)' (ABS, 2022)
[24] Bureau Veritas - 'Ammonia-Fuelled Ships' (BV, 2022)
[25] Det Norske Veritas AS (2021) - 'Part 6 additional class notations - Chapter 2 Propulsion, power generation and auxiliary systems' (DNV AS, 2021)
[26] Korean Register - 'Guidelines for Ships Using Ammonia as Fuels' (KR, 2021)
[27] Nippon Kaiji Kyoka – 'Guidelines for Ships Using Alternative Fuels' (Class NK, 2021)
[28] Registro Italiano Navale – 'Amendments to the 'Rules for the Classification of Ships' (RINA, 2021)
[29] DNV, G. (2017). Study on the use of fuel cells in shipping. Retrieved from European Maritime Safety Agency (EMSA): http://www. emsa. europa.eu/news-a-press-centre/external-news/item/2921-emsa-study-on-the-use-offuel-cells-in-shipping. html.
[30] Witlox, H. W., Fernandez, M., Harper, M., Oke, A., Stene, J., and Xu, Y. (2017) Verification and validation of consequence models for accidental releases of toxic or flammable chemicals to the atmosphere. In Mary K O'Connor Process Safety Symposium. Proceedings.
[31] Butchvarova, E. (Ed.). (2004) Harmonization within atmospheric dispersion modelling for regulatory purposes. Interscience Enterprises.
[32] Voie Ø.A. and Aaneby J. (2020) Vurdering av nye og gamle probitfunksjoner for innånding av akutt giftige stoffer, FFI rapport, 2020-10-10 (Title translation Assessment of new and old probit functions for inhalation of acute toxic substances).

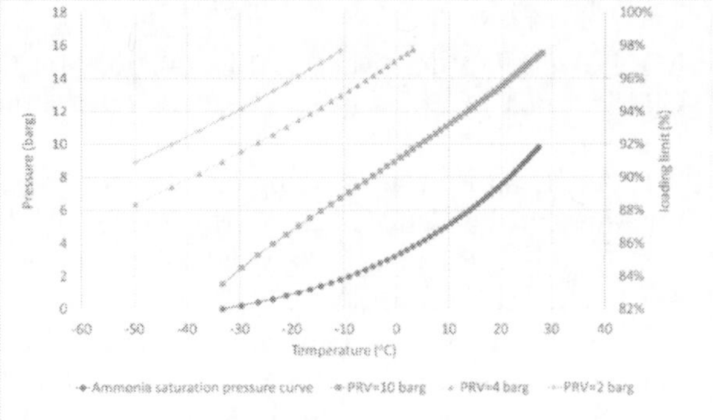

Figure 1. Loading limit curves for PRV set point pressure of 2 barg, 4 barg, and 10 barg

Table 1. Ammonia thresholds for alarming according to four different marine safety classification societies.

	ABS	BV	DNV	NK
Fuel storage hold space	25ppm⇒alarm	NO requirements stated	150ppm⇒alarm	NO requirements stated
Fuel Preparation Room	5ppm⇒alarm 150ppm⇒ alarm/water screen/ increased ventilation/ shutdown of fuel supply	30ppm⇒ alarm	350ppm⇒alarm shutdown of fuel supply/increased ventilation	25ppm⇒alarm 300ppm⇒ alarm/shutdown of fuel supply/ increased ventilation
Tank Connection Space	25ppm⇒alarm 150ppm⇒ alarm water screen/ increased ventilation/ shutdown of fuel supply	30ppm⇒ alarm	350ppm ⇒alarm shutdown of fuel supply/increased ventilation	25ppm⇒alarm 300ppm⇒ alarm/ shutdown of fuel supply/increased ventilation
Pipe ducting or secondary enclosure	150ppm⇒ alarm 300ppm⇒ alarm shutdown of fuel supply	30ppm⇒ alarm	350ppm⇒alarm/ shutdown of fuel supply/purge in the machinery space 350ppm ⇒alarm/ shutdown of fuel supply/purge	25ppm⇒alarm 300ppm⇒ alarm/shutdown of fuel supply/increased ventilation
Vent mast	300ppm⇒ alarm	NO requirements stated	Alarm for high value	NO requirements stated
Exhaust	10ppm⇒alarm 50ppm⇒alarm shutdown of fuel supply	NO requirements stated	NO requirements stated	NO requirements stated
Service space	25ppm⇒alarm 50ppm⇒alarm Close the inlet and opening	NO requirements stated	NO requirements stated	NO requirements stated
Bunker station	150ppm⇒ alarm	30ppm⇒alarm/water spray	350ppm⇒alarm	NO requirements stated

Table 2. Leak and dispersion analysis by Phast

Scenario	1			2			3		
	Tank rupture in bund /without bund, all content released within 10 mins			**100 mm leak from the tank bottom in bund /without bund**			**Hose rupture outside bund, no pumping**		
Mass flowrate (kg/s)	1136.4			37.7			23.2		
Leak duration (s)	600			3600			15		
Velocity(m/s)	11.7			11.7			4.3		
Liquid fraction	1			1			1		
Droplet size (μm)	2502			2502			10000		
Weather category	**F1.5**	**F3**	**D5**	**F1.5**	**F3**	**D5**	**F1.5**	**F3**	**D5**
Rainout (kg/s)	1105	1104	1102	35.3	35.2	35	22	22.5	22.4
Min liq. temperature in cloud (°C)	-41.6	-41.8	-42	-52	-53	-54	-42	-43	-44
Pool max evap. rate (kg/s)	3/ 442	5/ 490	7.4/ 524	3.9/ 27	5.2/ 28.5	6.2/ 29.5	1.3	0.9	1.16
Pool diameter (m)	32.5/ 429	32.5/ 416	32.5/ 405	32.5/ 170	32.5/ 132	32.5/ 125	11	11.1	11
Pool duration (s)	3600	3600	3600	3600	3600	3600	3600	3600	3600
Distance to 50 % lethality (m)	188/ 802	167/ 592	88/ 508	60/ 60	72/ 72	104/ 181	1.5	1.5	1.5
Toxic lethality footprint distance to 0.1 % lethality (m)	774/ 3706	717/ 2304	203/ 1070	170/ 224	172/ 280	222/ 467	90	60	21
Max footprint Distance to IDLH/ 300 ppm (m)	2918/ 15650	2667/ 12730	1079/ 5760	489/ 486	437/ 2960	631/ 1705	668	660	322

Scenario	4			5			6		
	Hose rupture outside bund during pumping. pressure 12 barg			**20 mm leak from hose outside bund during pumping**			**4 mm leak from hose during pumping outside bund**		
Mass flowrate (kg/s)	132.8			8.6			0.35		
Leak duration (s)	15			30			30		
Velocity(m/s)	25.5			66.9			66.9		
Liquid fraction	0.99			1			1		
Droplet size (μm)	284.48			76.8			76.8		
Weather category	F1.5	F3	D5	F1.5	F3	D5	F1.5	F3	F1.5
Rainout (kg/s)	119	118.7	117.9	2.7	2.3	0.42	No	No	No
Min liq. temperature in cloud (°C)	-57.7	-58.1	-58.9	-70.1	-70.3	-71.1	No	No	No
Pool max evap. rate (kg/s)	2.8	3.5	4.3	0.11	0.12	0.03	-73	-73.4	-74
Pool diameter (m)	25.5	25.3	25	5.4	5	2	NA	NA	NA
Pool duration (s)	3600	3600	3600	3600	3600	2400	NA	NA	NA
Distance to 50 % lethality (m)	40	30	14	18	18	17	5.2	5	5.5
Toxic lethality footprint distance to 0.1 % lethality (m)	224	170	146	71	66	59	9.0	9.3	10.7
Max footprint Distance to IDLH/ 300 ppm (m)	1285	1325	755	512	546	647	162	197	168

Improvement of Mitigation Barriers in Fire Case: A Study Case from the Ammonia Converter

Ammonia Converter is the main equipment of ammonia synthesis which has been operating for 21 years without experience of catalyst change. The Ammonia converter leaked from its bolt and nut connection on the top cover in 2015 and created a continuous leak of flammable syngas during normal operation. A hazard identification study was carried out and concluded the condition as "High Risk". Tightening the connection during shutdown window did not solve the problem. Following the leakage, the fire occurred five times from some of the leakage points. Some mitigation barriers are in place to manage released gas safely.

Zaki S. Rahman, Gilang H. Pratomo, Dimas D. Ananda, Rahmatullah Q. Gobel, & Baskara A. Nugraha

PT. Kaltim Parna Industri, Bontang, East Kalimantan - Indonesia

Introduction

PT. Kaltim Parna Industri (KPI) based in Bontang, East Kalimantan – Indonesia, is an ammonia producer using Topsoe license for the process with the capacity of 1,500 MTPD and built by Mitsubishi Heavy Industries (MHI). The ammonia converters (R-0501 and R-0502) are radial-type converters with the gas flowing through the catalyst bed in a radial direction. R-0501 is an S-200 converter which contains two catalyst beds and an interbed heat exchanger (see Figure 1). R-0502 is an S-50 converter consisting of one catalyst bed. The S-50 converter is installed downstream of an S-200 converter with a steam superheater between the two converters.

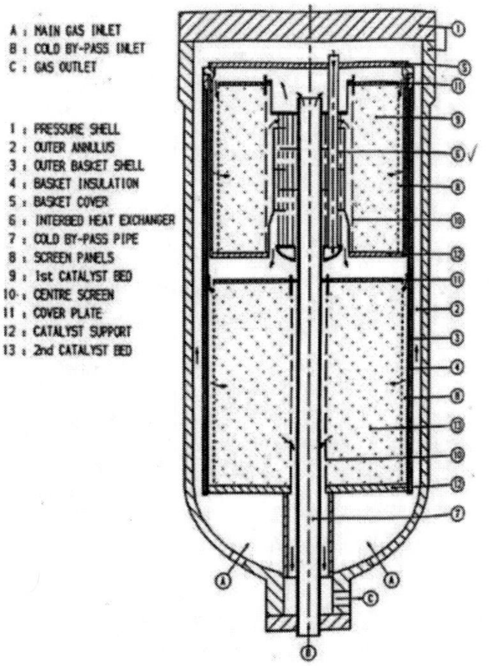

Figure 1. S-200 Ammonia Synthesis Converter [1]

The ammonia synthesis loop has been designed for a maximum pressure of 2204 psig (155 kg/cm^2g). The normal operating pressure will be 1191 psig (140 kg/cm^2g) in the 1st ammonia converter and 1963 psig (138 kg/cm^2g) in the 2nd ammonia converter, depending on load and catalyst activity. Normal operating temperatures will be in the range of 680-900^{0}F (360-482^{0}C) for the first bed and 697-822^{0}F (369-439^{0}C) for the second bed in R-0501. For R-0502, the range will be 375-481^{0}C.

History

The ammonia converter (R-0501) has been operating for over 20 years since this company operated commercially in 2002 without any catalyst changed. In December 2015, the first fire was found in the top cover of ammonia converter towards the primary reformer. Fire cases at the top cover ammonia converter can be shown in the table below.

Year	Source
December 2015	Thermowell 4 leakage
February 2018	Bolt No.8-13 and No.19-29
July 2019	Bolt No. 23-25
April 2021	Bolt No. 26

Table 1. 1st Ammonia Converter Fire Cases History

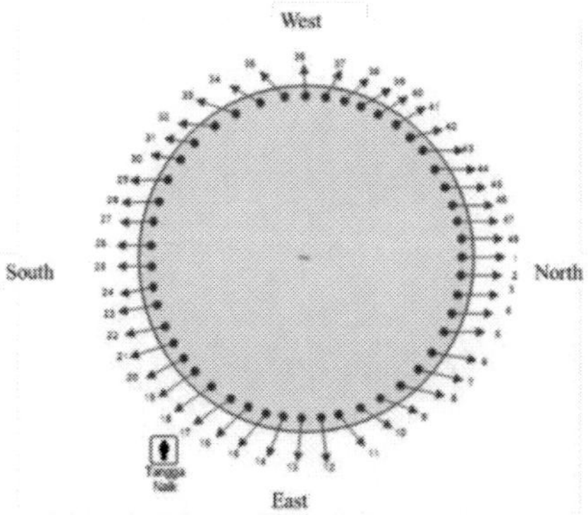

Figure 2. Sketch of Top Cover Ammonia Converter

The gas composition of flammable gases inlet 1st ammonia converter and physical properties of these gases can be shown in the table below.

Gas	Gas Composition (%Vol)	Lower Flammability Limit (%)	Upper Flammability Limit (%)	Auto-Ignitic (oF)
H2	55,0	4,0	74,2	1076

Table 2. Physical Properties of Flammable Gas of Inlet 1st Ammonia Converter

Since syngas contains flammable gases such as hydrogen and methane, the daily gas check is carried out to monitor and ensure the concentration of gas release does not exceed its flammability limit. Ammonia was also monitored, but the gas detector used can only detect maximum at 200 ppm or only 1,3% of its Low Flammability Limit so that the flammability of ammonia can't be identified.

Case Study

In December 2018, after the fire happened, the gas check was carried out and found the release gas around the leakage point exceeded the lower flammability limit of hydrogen (>4%) at 0 cm and 5 cm from the leakage point. The gas check result and leakage mapping can be shown in Table 3 and Figure 3.

Bolt	0cm			5 cm		
	H_2 (ppm)	NH_3 (ppm)	%LEL CH_4	H_2 (ppm)	NH_3 (ppm)	%LEL CH_4
Bolt 8	10000	200	70	0	0	0
Bolt 9	>40000	>200	100	>40000	>200	100
Bolt 10	>40000	>200	100	>40000	>200	100
Bolt 11	>40000	>200	100	>40000	>200	100
Bolt 12	>40000	>200	100	>40000	>200	100
Bolt 13	>40000	>200	100	0	0	0
Bolt 19	>40000	>200	100	>40000	>200	100
Bolt 20	>40000	>200	100	>40000	>200	100
Bolt 21	>40000	>200	100	>40000	>200	100
Bolt 22	>40000	>200	100	>40000	>200	100
Bolt 23	>40000	>200	100	>40000	>200	100
Bolt 24	>40000	>200	100	>40000	>200	100
Bolt 25	>40000	>200	100	>40000	>200	100
Bolt 26	>40000	>200	100	>40000	>200	100
Bolt 27	13000	200	70	0	0	0
[illegible]	7000	200	[illegible]	0	0	0

LEL : Lower Explosive Limit

Table 3. Gas Check Result on Bolt 8 to 29 After Fire Case in 2018

In July 2019, the concentration of hydrogen at the fire point (bolt 23 to 25) was above its flammability limit. The gas was detected around 5 cm from the bolt.

In April 2021, the fire happened for the third time. The fire was found around bolt 23 to 26 and caused severe damage to the elements of temperature indicator (TICA-0514).

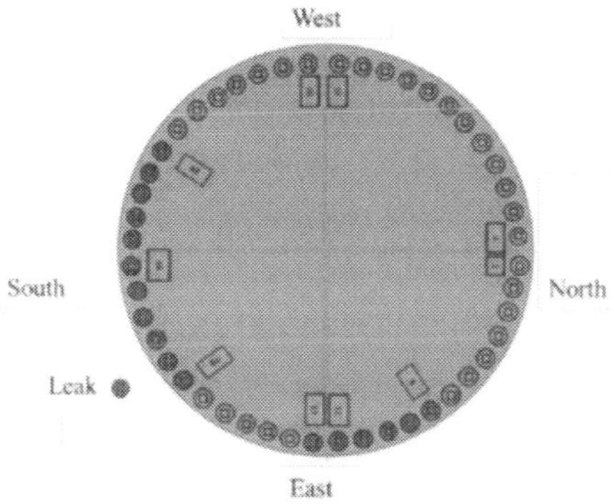

Figure 3. Leakage Mapping After Fire Case in 2018

Date	Bolt 24			Bolt 25			Pressu Synloo (psig
	H2 (ppm)	NH3 (ppm)	CH4 (%LFL)	H2 (ppm)	NH3 (ppm)	CH4 (%LFL)	

Table 4. Gas Check Result on Bolt 24 and 25 After Fire Case in July 2019

The indication became apparent when the alarm on DCS alerted the operator, informing them that the operating mode of controller TICA-0514 changed from AUTO to MANUAL. Field Operator also found the droplets of aluminum from the top cover 1st ammonia converter that indicated the fire catching the cover of insulation below the grating and caused the aluminum sheet to melt.

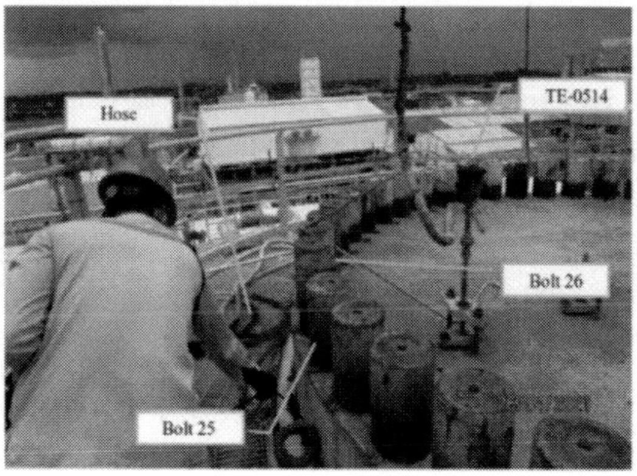

Figure 4. Field Check on Top 1st Ammonia Converter After Fire in April 2021

Risk

The risk analysis was done qualitatively through Hazard Identification (HAZID) Study. The worst credible scenario identified from this case was continuous release of syn-gas leading to fire and explosion with potential for fire escalation resulting in personnel fatality and asset loss, if ignited. This scenario was categorized as High Risk or Consequence in the Company Risk Matrix.

The scenario was further investigated in the Quantitative Risk Assessment (QRA) and was identified as one of the highest risk outcomes from the study. The study used a continuous leak of flammable gas from the ammonia converter top cover through a 10 mm hole size. The consequences analysis of this scenario shows that the jet fire will generate length of flame of only 4.75 meters and the radiation heat will no higher than 4 kW/m^2 which the level of lethality less than 3% of population. The impacted radius from the flash fire is about 2.8 meters from the leakage point.

The risk raised from this scenario is the flammable gas is released continuously and if a delayed flash fire happened is about 4.50 E-3/year and if followed by explosion, the risk is 3.00 E-3/year.

The scenario contributes higher Individual Risk per Annum (IRPA) for the ammonia field operator which works around the leakage point daily. The IRPA for the ammonia field operator is 7.01E-03/year, higher than the acceptable level for personnel at 1.00 E-03/year.[2] The overall risk contour from the Quantitative Risk Assessment study can be seen in Figure 5.

Both of those studies provided the recommendation to repair the leak on 1st ammonia converter to lower the risk. However, limited options for repairing the connection on top cover ammonia converter became an issue since the repair required removal of the top cover, unloading of the catalyst and replacement once contacted with air.

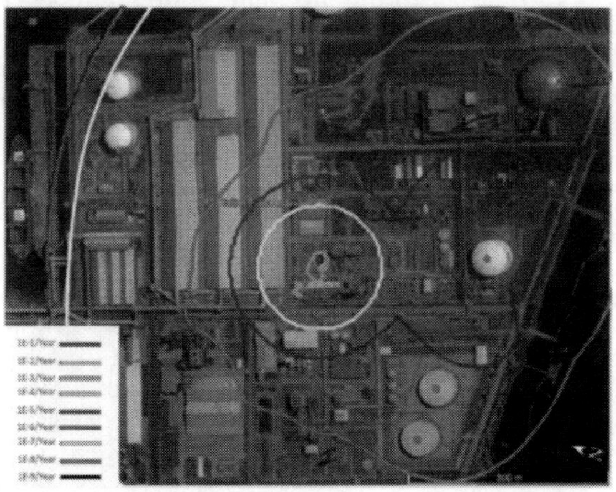

Figure 5. Overall Risk Contour of PT Kaltim Parna Industri [3]

The identification of ignition sources of the fire is excluded in this paper.

Post Release Mitigation

Preventive action was taken to eliminate or reduce the leakage. During the shutdown window in 2018, the physical check and cold bolt tightening at bolt was carried out. However, the result was insignificant, the flammable gases were still found at the bolts. The gas check result before and after bolt tightening are shown in Table 5.

Since the preventive action could not solve the leakage, post-release mitigation was required to reduce the probability of releasing gas that could ignite fire and minimize the impact of worst-case scenario. Center of Chemical Process Safety (CCPS) defined a countermeasures technique when the released material had formed a vapor cloud.[4] The technique includes containment or suppression, which involves limiting the amount of material that is released into the air and countermeasures, which are applied once the released material has formed a pool and/or a vapor cloud.

No	Monitoring Location	ppm H_2		ppm NH_3		%LEL CH_4	
		Before CB (*)	After CB (**)	Before CB(*)	After CB (**)	Before CB (*)	Aft CI (**
1	Bolt 1	0	20000	0	200	0	10
2	Bolt 2	0	18000	0	200	0	10
3	Bolt 3	0	3600	0	200	0	25
4	Bolt 8	10000	0	200	0	70	0
5	Bolt 9	40000	0	200	0	100	0
6	Bolt 10	40000	0	200	0	100	0
7	Bolt 11	40000	0	200	0	100	0
8	Bolt 12	40000	0	200	0	100	0
9	Bolt 13	40000	0	200	0	100	0
10	Bolt 19	40000	0	200	0	100	0
11	Bolt 20	40000	0	200	0	100	0
12	Bolt 21	40000	0	200	0	100	0
13	Bolt 22	40000	0	200	0	100	0
14	Bolt 23	40000	0	200	0	100	0
15	Bolt 24	40000	0	200	0	100	0
16	Bolt 25	40000	0	200	0	100	0
17	Bolt 26	40000	0	200	0	100	0
18	Bolt 27	13000	0	200	0	70	0
19	Bolt 28	7000	0	200	0	50	0
20	Bolt 29	2000	0	50	0	5	0

Table 5. Gas Check on Top Cover 1st Ammonia Converter Before and After Cold Tightening

From those techniques, there are several additional mitigation barriers and actions determined to safely manage this condition.

Syn-loop Pressure Reduction

The main purpose of reducing the syn-loop pressure is to reduce the leakage rate. The mass flow rate of gas charged through a hole or leakage point is determined by several factors. CCPS has defined the equation to estimate the mass flow rate in Figure 6. [5] From the equation in Figure 6, it can be known that the mass flow rate (m) is directly related to upstream or internal pressure (P_1)

$$\dot{m} = C_D A P_1 \sqrt{\frac{2 g_c M}{R_g T_1} \frac{k}{k-1} \left[\left(\frac{P_2}{P_1}\right)^{2/k} - \left(\frac{P_2}{P_1}\right)^{(k+1)/k} \right]}$$

where

- $\dot{m}$ is mass flow rate of gas through the hole (mass/time)
- C_D is the discharge coefficient (dimensionless)
- A is the area of the hole (length^2)
- P_1 is the pressure upstream of the hole (force/area)
- g_c is the gravitational constant (force/mass-acceleration)
- M is the molecular weight of the gas (mass/mole)
- k is the heat capacity ratio, C_p/C_v (unitless)
- R_g is the ideal gas constant (pressure-volume/mole-deg)

Figure 6. Gas Mass Flow Rate Through A Hole Equation [5]

The mass flow rate will decrease when the pressure of the syn-loop is decreased. Since the hole size of leakage is unknown, with this concept, reducing the syn-loop pressure can reduce the gas release. The mass flowrate can be reduced up to 7% when depressurizing the syn-loop from 1992 psig (140 kg/cm^2.g) to 1850 psig (130 kg/cm^2.g).

When the fire occurs, the operator will decrease the pressure of the syn-loop to 1850 psig by opening pressure control valve (PV-0213) to release some syn-loop feed through a cold vent stack. However, during a fire case in 2021, the operator needed to reduce pressure syn-loop from 1905 psig (133,9 kg/cm^2.g) to 1895 psig (133,9 kg/cm^2.g) to help extinguish the fire.

Since the mass rate of gas release reduces, the concentration of flammable gas exits the leakage point can be reduced to below its flammability limit after mixed with the steam dilution.

Steam Dilution System

As the gas check result found the concentration at leakage point above flammability limit, dilution system using non-flammable substance is required. The dilution system is intended to reduce the released flammable syn-gas concentration to below the lower explosive limit before contact with the air, so the fire can be prevented. Several non-flammable gases such as nitrogen, carbon dioxide, and steam can be used to dilute the leakage gases. [6]

After a fire case in December 2015, an ejector was installed around leakage point and several non-flammable gases were used as dilution system.

Fluid	Pressure (psig)	H2 Concentration (% vol	
		Inlet Ejector	Outlet Ejecto
Nitrogen Low Pressure Combined With Medium Low Pressure Steam	58.8	6.73	1.78
Nitrogen High Pressure	117.6	6.24	2.34

*With dry basis

Table 6. Non-Flammable Gas Dilution

In 2018, an additional steam ring was provided to dilute the flammable gases through Bolt 7 to 30. The used steam was low pressure steam (3 kgf/cm^2). This steam ring successfully reduced concentration of flammable gases to below their flammability limit as shown in table below.

However, after the fire case in 2019, steam was switched to medium pressure steam and replacing the hose that used to deliver the low pressure steam to permanent piping.

Sampling Date	Bolt 23			Bolt 24			Bolt 25		
	5 cm			5 cm			5 cm		
	H_2 (ppm)	NH_3 (ppm)	%LEL CH_4	H_2 (ppm)	NH_3 (ppm)	%LEL CH_4	H_2 (ppm)	NH_3 (ppm)	%LEL CH_4
4-Mar-18	8000	50	4	6000	40	2	6000	40	2
5-Mar-18	8000	50	4	6000	40	2	6000	40	2
6-Mar-18	8000	50	4	5000	38	2	5000	37	2
7-Mar-18	8000	50	4	6000	40	2	6000	40	2
8-Mar-18	8000	50	4	6000	40	3	6000	40	2
9-Mar-18	8000	50	4	6000	40	3	6000	40	2
10-Mar-18	7990	50	2	6000	38	2	5950	40	2
11-Mar-18	8000	50	2	6000	38	2	6000	40	2
12-Mar-18	8000	48	2	6000	38	2	6000	40	2
13-Mar-18	8000	48	2	6000	40	2	6000	40	2
14-Mar-18	8000	50	2	6000	[illegible]	2	6000	40	2

Table 7. Flammable Gas Concentration After Diluted with Low Pressure Steam

Since the leakages are not found only in one spot, a cover that was made by aluminum and isolation was provided around lip seal of 1st ammonia converter to relocate the release gas. Two specific holes in the cover were made as the exit way for the diluted gas. The cover was connected to a permanent steam ring to dilute the release gas in the cover.

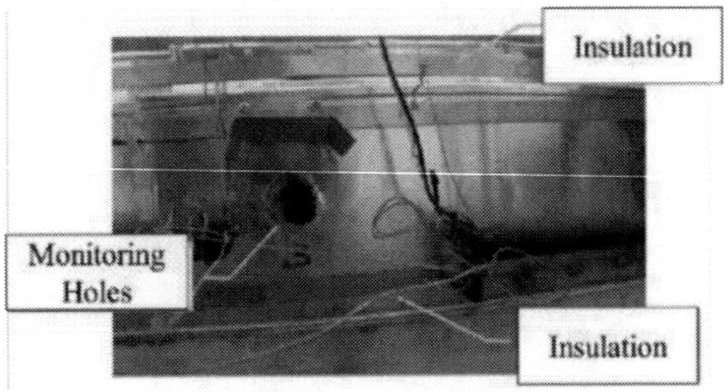

Figure 7. Cover Lip Seal 1st Ammonia Converter with Cover (above) and Steam Box (below)

The steam box was fabricated to replace the cover to enhance the integrity of the steam dilution system since the cover was made by aluminum sheet that has melting point around 1220^{0}F (660^{0}C) and exceeds the flame temperature of hydrogen that up to 4712^{0}F (2600^{0}C).

After the fire case in 2021, the monitoring area on the top cover of 1st ammonia converter was divided into 4 points as follows.

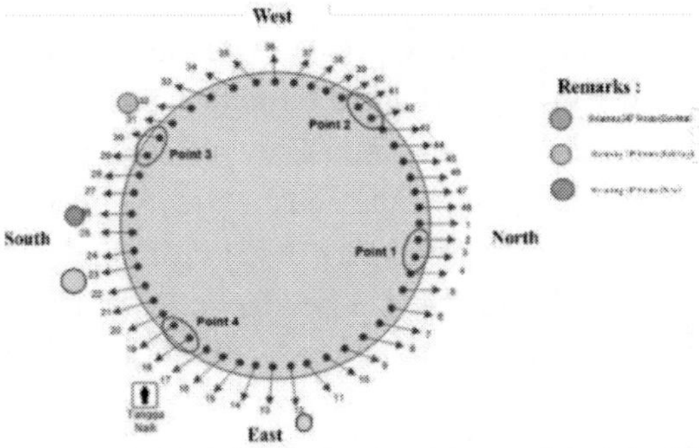

Figure 8. Flammable Gas Monitoring Point After Fire Case In 2021

To pull out the diluted gas in the cover, an ejector is also installed at the outlet of hole. A low medium pressure steam is used as the motive fluid. This facility is also intended to control fires and help the operator and the emergency response team to focus on specific locations to extinguish the fire.

Figure 9. Ejector From Steam Box Outlet

During the fire, the amount of steam dilution can be added into the steam box to help extinguish the fire.

Temporary Fire Monitor and Sprinkler

One of countermeasures technique is plume dilution. The aim of this technique is to minimize the impact area of hazardous vapor release. The dilution can be done by using air and steam curtains, water sprays and curtains, etc. Water sprays are simple, reliable, and inexpensive and can be either fixed or mobile. [4]

The water sprays and curtains also have a role as a fire suppression required to limit the fuel or oxygen supply or reduce the flame zone's temperature. The water will extinguish the flame by these mechanisms by heat extraction (cooling), oxygen displacement by water vapor, and blocking of radiant heat. [7]

The top cover of the ammonia converter is about 100 feet from the ground. This condition gives several challenges to the emergency response team and operations team to extinguish the fire as below.

a) KPI did not have a dedicated fire water pump and relied on fire water supply from another party. The distance between the fire water pump and KPI plant is about 1 km. On a standby basis, there is a jockey pump run to maintain the fire water supply pressure at 73 psig (5 barg). Based on fire water test, it needs 8-10 bar pressure for the fire water at KPI plant to achieve that elevation from a fire monitor at the ground level. The supplier shall switch from the jockey pump to the main fire pump to achieve that pressure and this operation takes more than 5 minutes.
b) The team needs more than five minutes to climb the monkey ladders to get to the location. The maximum time for the firefighting team to arrive at the fire location is 5 minutes and 20 seconds. [8]

Figure 10. Access Ladder to Top Cover 1st Ammonia Converter

Facing these challenges, the temporary fire monitor and sprinkler facility were installed as water sprays and curtains during the fire on the top of the ammonia converter. Fire monitor nozzles and hose lines can be used for vapor mitigation.[4]

The temporary fire monitor consists of two fire nozzles that are installed toward bolt 20-23 (east side) and bolt 41-42 (west side) where the specific holes from the cover are provided. The nozzles are connected to the nearest fire hydrant and valves are provided at the hydrant connection. The operator or emergency team should only open those valves to activate the nozzles. The nozzles are in spray position. Since the first fire occurred in 2015, the water curtain has already been installed. The illustration of water curtain activation can hbe seen in **Figure 11.**

Figure 11. Water Curtains Illustration on Top Cover 1st Ammonia Converter

Since the water curtain using fire hose is activated manually, a water sprinkler is provided as a water suppression that can be activated automatically in accordance with its thermal sensitivity. The specification of the water sprinkler is as follows.

Type	**Pendent**	
Size	Inch	1/2
Max Ceiling Temperature	F (C)	225 (107)
Thermal Sensitivity	F (C)	286 (141)
Finishing		Chrome

Table 8. Specification of Water Sprinkler

Thermal CCTV and Flame Detector

Since the ammonia plant is not equipped with a flame and fire detector, a real time monitoring is required to detect the fire on the top cover of ammonia converter as fast as possible. From 2015 to 2019, the fire was only detected if the operator went to routine check every two hours and might see fire occurred on the top cover of ammonia converter. Since the gas composition mostly contains hydrogen, the flame is invisible to the naked eye during daylight.

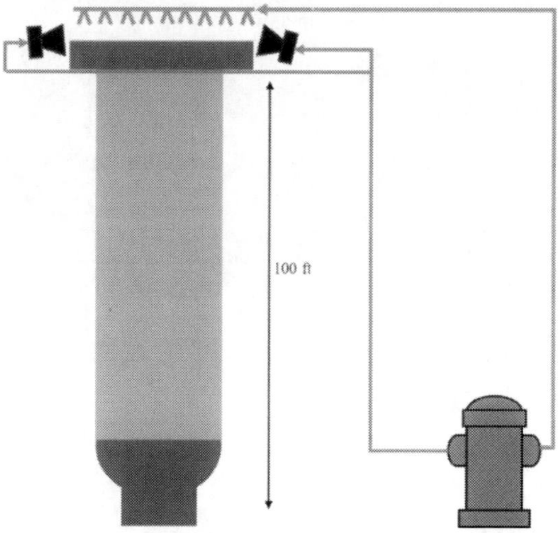

Figure 12. The Illustration of Water Sprinkler and Temporary Fire Monitor on The Top Cover Ammonia Converter

A thermal CCTV is installed to monitor the leaking converter and may found a fire on the top cover ammonia. The thermal CCTV shows temperature around the top cover. When the fire occurs, the temperature will increase until the set point at 572^{0}F (300^{0}C) and the alarm will alert the operator in the Control room. The thermal CCTV will show a red mark on the fire point.

Figure 13. Thermal CCTV and Normal CCTV With Top Cover 1st Ammonia Converter View

Figure 14. Flame Detector on The Top Cover 1st Ammonia Converter

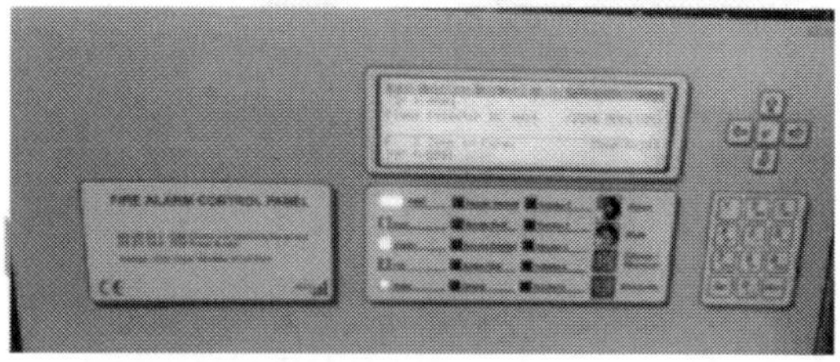

Figure 15. Existing Fire Alarm Panel When Detect Fire on The Top Cover of 1st Ammonia Converter

The specifications of the flame detector and design for flame detector installation are below.

Material	Stainless Steel
Detects	Hydrogen and Hydrocarbon Fires
Detection Range	Fuel Hydrogen – max. 131 ft (40 m)
Minimum Fire Dimension	0,75 High, 0,25 wide plume fire
Coverage area	58400 ft^2 (5426 m^2)
Standard Response Time	Typically, <2 secat (131 ft)

Table 9. Specifications of Flame Detector

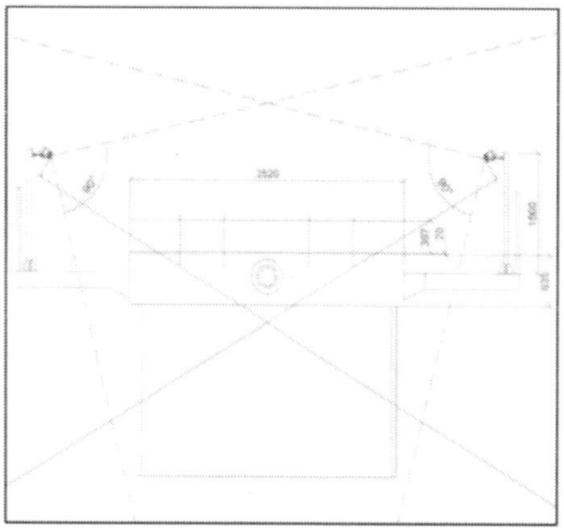

Figure 16. Flame Detector Installation Design on The Top Cover 1st Ammonia Converter

Evaluation

The post release mitigation is evaluated to know how effective the implementation.

a) Company experiences show that reducing syn-loop pressure during fire could help to reduce duration of fire. However, plant shutdown is still required if the fire escalates.
b) The steam dilution system can effectively dilute the flammable gases to below their lower flammable limit. The cover with two specific exit holes also gives opportunity help on fire detection and firefighting strategy.
c) The temporary fire monitor can be activated immediately by the operator and emergency response personnel and the fire water can be supplied in less than five minutes. The water sprinkler is also successfully activated and being an indication of fire on the top cover 1st ammonia converter.
d) The thermal CCTV and flame detector can improve the response time of operator in case of fire at the top of ammonia converter. Before the equipment installation, the fire could only be detected if any field operator sees the fire during field surveillance every two hours or if he found the

water sprinkler was active. When the CCTV thermal and flame detector alert the operator in control room then the mitigation tasks will be carried out immediately, starting with reducing the synloop pressure and activating the water curtains from the ground level.

Conclusions

The post release mitigation is successfully implemented to improve the mitigation barriers in case of fire on the top cover of ammonia converter. Notes should be taken that the mitigation barrier is active after the event of fire. A fire can still occur since the root causes of fire can't be identified.

References

[1] Operation Manual of 1500MTPD Ammonia Plant for KPI Project. 2000. Mitsubishi Heavy Industries, LTD.

[2] Reducing Risks Protecting People. 2001. HSE Executive UK.

[3] PT Kaltim Parna Industri Quantitative Risk Assessment (QRA) Report. 2023. PT Keselamatan Proses Indonesia

[4] Guidelines for Post Release Technology In The Chemical Process Industry. 1997. CCPS of AICHE.

[5] Guidelines for Consequence Analysis of Chemical Releases. 1995. CCPS of AICHE.

[6] Biennal Report, "Hydrogen safety barrier and measures". 2006. Hysafe.

[7] Alageel, K.S. 1999. Mitigation of Compartment Jet Fire Using Water Sprays.

[8] NFPA 1710. 2020. Standard for the Organization and Deployment of Fire Suppression Operations, Emergency Medical Operations, and Special Operations to the Public by Career Fire Departments.

AUTHOR INDEX

AIChE
120 Wall Street, FL 23
New York, NY 10005-4020

ISBN 979-8-3313-0264-1

CURRAN ASSOCIATES INC.
proceedings.com